Make: Sensors

Tero Karvinen, Kimmo Karvinen, and
Ville Valtokari

Make: Sensors

by Tero Karvinen, Kimmo Karvinen, and Ville Valtokari

Printed in Canada.

Published by Maker Media, Inc., 1005 Gravenstein Highway North, Sebastopol, CA 95472.

Maker Media books may be purchased for educational, business, or sales promotional use. Online editions are also available for most titles (*http://my.safaribooksonline.com*). For more information, contact O'Reilly Media's corporate/institutional sales department: 800-998-9938 or *corporate@oreilly.com*.

Editor: Brian Jepson
Production Editor: Kristen Brown
Proofreader: Julie Van Keuren
Indexer: Judith McConville
Cover Designers: Juliann Brown and Brian Jepson
Interior Designer: Nellie McKesson
Illustrators: Tero Karvinen, Kimmo Karvinen, and Ville Valtokari
Photographer: Kimmo Karvinen
Cover Photo: Kimmo Karvinen

May 2014: First Edition

Revision History for the First Edition:

2014-05-05: First release

See *http://oreilly.com/catalog/errata.csp?isbn=9781449368104* for release details.

ISBN: 978-1-449-36810-4

[TI]

Table of Contents

Preface

Welcome to *Make: Sensors*. Soon you'll be making gadgets that can sense it all—from dangerous gases to acceleration. In this book, you'll use sensors to measure the physical world, represent the result as a numeric value, and take some action based on that value.

For example, a sensor could measure heat, pressure, light, or acceleration and report a value such as 22 C, 1015 millibars, light is detected, or 2.3 g acceleration (in the case of light, notice that we represented it as a Boolean or yes/no value rather than a numeric quantity; you'll see examples of this later).

A microcontroller board is the brain of the robot, system, or gadget you're building. You'll write your own software to run on the microcontroller. In this book, you'll work with two very popular boards: Arduino and Raspberry Pi. Either of these makes it easy to write software code to work with electronics.

It's About Your Ideas

If your interest in electronics started with a desire to quickly learn some basics and then design your own robots, gadgets, or projects, you're in the right place. This book will show you how to go from idea to reality quickly.

Theory, skills, and basics are useful—as long as they serve your creativity. Feel free to experiment with your ideas, and have the courage to publish your results on the Web.

Each chapter presents a mini project to show how you can combine different technologies. For example, you'll build a wooden box that you open with a fingerprint and a color-changing chameleon dome. These are fun projects, but also good starting points for things you invent later yourself.

The skills you learn with Arduino are easily applicable to real-life projects. For example, we used Arduino to build the sun sensor prototype for Finland's first satellite (Figure P-1).

Figure P-1. *Finland launches its first satellite in 2014. We designed and built the sun sensor prototype with Arduino.*

How to Read This Book

When you get an idea, you can quickly build your first prototype with the help of this book. Instead of spending hours with component data sheets, you can simply pick a sensor and use ready-made breadboard diagrams and code. You can use sensors as building blocks for your project, but unlike construction kits such as Meccano or Lego, the possibilities with Arduino and Raspberry Pi are nearly endless.

If you know what you want to measure, you can easily find a sensor for it. The book is arranged by the real-life phenomena you can measure:

- Distance (Chapter 3)
- Smoke and gas (Chapter 4)
- Touch (Chapter 5)
- Movement (Chapter 6)
- Light (Chapter 7)

- Acceleration and angular momentum (Chapter 8)
- Identity (Chapter 9)
- Electricity (Chapter 10)
- Sound (Chapter 11)
- Weather and climate (Chapter 12)

You can also use *Make: Sensors* as a maker's coffee-table book: browse it to get ideas of what's available, and look for inspiration for new projects.

If you want to understand how sensors are connected to Arduino and Raspberry Pi, you'll enjoy the in-depth explanations. All the sensor code examples are fully self-contained, completely showing the interaction with the sensor. Understanding the sensors in the book helps you apply your skills to new sensors, even ones that aren't on the market yet.

When we chose the sensors for you, we picked a variety of useful and interesting sensors. We didn't just pick easy or difficult ones. This means you'll get to see solutions to the wide variety of challenges involved in connecting sensors to Arduino and Raspberry Pi.

In each chapter you'll find experiments, environmental experiments, and a test project:

1. Experiments give you quick instructions on how to use a single sensor with Arduino and Raspberry Pi. You can easily use these as building blocks for your own projects or just to see how the sensor works.
2. Environmental experiments let you play with sensors and monitor changes in the surrounding environment. This gives you insight into how sensors see the world and how they really work.
3. Sensors are more fun when you actually do something with the readings they give you. In test projects you'll build a device or gadget around one sensor. You'll learn how to use different outputs such as RGB LEDs, e-paper, and servo motors. Test projects also work as quick starting points for your own innovations.

Input, Processing, Output

Any robot or gadget you build must handle three things: *input*, *processing*, and *output*.

1. Because most of the devices you build won't have a keyboard or a mouse, sensors are your inputs. Take a quick look at the table of contents, and keep in mind that this is just a fraction of what's out there. There are countless sensors to measure everything you could imagine.
2. Processing happens in your program, running in Arduino or Raspberry Pi. In your program, you get to decide what happens next.

3. Outputs affect the world around the device. You could light an LED, turn on a servo motor, or play a sound. Those are three of the most common types of output, but there are others (for example, haptic feedback such as vibration, displaying something on an e-paper screen, or turning on a household appliance).

Protocols

A *protocol* defines how a sensor talks to the microcontroller board, such as Arduino or Raspberry Pi. The protocol defines how the wires should be connected and how your code should ask for measurements.

Even though there are a staggering amount of different sensors, there are a limited number of popular protocols. You'll learn each of the protocols as you work through experiments and projects, but here's an overview of what you'll be seeing.

You can get an overview of common sensor protocols in Table P-1.

Digital resistance
: Some sensors work like a button and have two states, on or off. These sensors are easy to read. The on state is represented when a voltage referred to as *HIGH* is applied to the microcontroller input pin. This is usually either 3.3 volts or 5 volts depending on the microcontroller board you're using.

Analog resistance
: Analog resistance sensors change their resistance in response to a physical change (such as turning the knob of a dial). Arduino and Raspberry Pi measure the changes in resistance by measuring the voltage level that passes through the sensor. For example, you can turn a potentiometer to make its resistance larger or smaller. These analog resistance sensors are very easy to make with Arduino. Raspberry Pi needs an external chip for measuring analog values. You'll learn to use the MCP3002 analog-to-digital converter to measure resistance with Raspberry Pi in "Experiment: Follow Movement with Infrared (IR Compound Eye)" on page 50. Most analog input sensors report their value using resistance, so they are analog resistance sensors.

Pulse width
: Some sensors report their value with a pulse width, or the period of time in which the pin is held HIGH. You use functions like `pulseIn()` or `gpio.pulseInHigh()` to read the length of the pulse. Because this is handled by a function, you don't have to get into low-level microcontroller operations such as *interrupts*; it is all handled by a library.

Serial port
: A *serial port* sends text characters between two devices. It's the same technique your computer uses when talking to Arduino over USB. You'll become quite familiar with the serial port when you print some messages to the Arduino serial monitor in various projects.

I2C
: I2C is a popular industry standard protocol. It is commonly found inside computers and well known from Wii Nunchuk joysticks. I2C allows 128 devices to be connected to the same wires. In *Make: Sensors*, you'll get ready-made code and circuits for two sensors using I2C.

SPI
: SPI is another industry standard protocol. You'll find it easy to use the code in this book for using an analog-to-digital converter on the Raspberry Pi. But creating your own code from scratch for new devices using SPI will be a bit more work.

Bit-banging
: Sometimes, a sensor is unusual enough that a standard protocol won't work with it. In those cases, you need to craft up your own code to talk to that sensor. This is often called *bit-banging*, because you're manipulating the signal from the sensor, often at the bit level. You'll see an example of that later in the book in "Experiment: Is It Humid in Here?" on page 332.

As you play with the sensors, you'll get much more familiar with these protocols. Or, if you're in a hurry to put new sensors in your robots and innovative devices, you can just use the code in this book and look at the details later.

Table P-1. Sensor protocols, easiest first

Protocol	Example value	Arduino	Raspberry Pi Python	Example sensors
Digital resistance	1 or 0	digitalRead()	botbook_gpio.read()	Button, IR sensor switch, tilt sensor, passive infrared movement
Analog resistance	5%, 10%, 23 C	analogRead()	botbook_mcp3002.readAnalog(), chip	Potentiometer, light-dependent resistor, MQ-3 alcohol, MQ X gas family (smoke, hydrocarbon, CO…), FlexiForce pressure, KY-026 flame, HDJD-S822-QR999 color, LM35 temperature, soil humidity
Pulse length	20 milliseconds	pulseIn()	gpio.pulseInHigh()	Ping and HC-SR04 ultrasonic distance, MX2125 acceleration
Serial port	A9B3C5B3C5	Serial.read()	pySerial.read()	GT-511C3 Fingerprint scanner, ELB149C5M RFID identity
I2C	(2.11 g, 0.0 g, 0.1g), very precise values	Wire.h	smbus	Wii Nunchuk, MPU 6050 accelerometer and gyro combination, GY65 atmospheric pressure
SPI	57 deg, very precise values	Bit-banging	spidev	MCP3002 analog-to-digital converter
Bits encoded to very short pulses	53%	Bit-banging	Bit-banging	DHT11 humidity

Building Things Your Way

Most users won't find raw circuit boards and components compelling to play with. Making an attractive package for your gadget or robot makes a huge difference.

This book gives you one example for each project, but there's no need to follow our instructions blindly. Try different materials and use different tools.

How about using cardboard (Figure P-2), fabric (Figure P-3) or 3D printing (Figure P-4)?

Figure P-2. *Cardboard model. Photo from Ars Electronica in Linz (not made by us)*

Figure P-3. *Fabric robot. Photo from Ars Electronica in Linz (not made by us)*

Figure P-4. *3D Bender. Photo from Ars Electronica in Linz (not made by us)*

Trying out and learning new techniques makes the process of work more interesting, such as welding or making something out of clay between all the soldering (see Figure P-5).

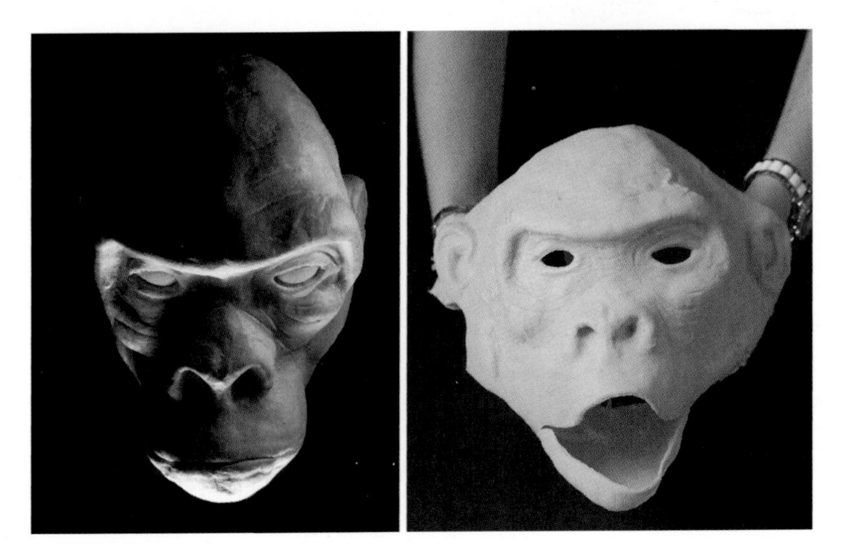

Figure P-5. *Base model for animatronic gorilla head and latex skin made from it.*

We also use a lot of recycled materials in our own projects. Obviously they are cheap (free!) but they also give a unique look to a project.

Buying Components

If you need high quality components without fuss, pick a well-known shop, preferably in the Western world. If you want cheap components, look to Asia.

Quality shops mainly selling to makers include Maker Shed, SparkFun, Parallax, and Adafruit. Maker Shed is the shop from the publisher of this book. SparkFun sells a lot of breakout boards, which require you to solder in headers. Parallax created Basic Stamp, the previous generation of microcontroller boards for makers. Adafruit has a lot of parts, many designed by them. The SparkFun and Adafruit websites have a lot of information about their components, including tutorials.

These days, even big-name distributors like Element14 and RS electronics have broken into the Maker market. Finding parts from their huge catalogs is becoming easier, as they've started providing clear areas for Arduino and Raspberry Pi.

For some special parts and sometimes very cheap prices, Asia is the continent to go to. DealExtreme (*http://dx.com*) is very popular at the moment. Its shipping is slow and quality varies, but the prices are low and the assortment is wide. AliExpress (*http://www.aliexpress.com*) is another Asian shop worth checking out.

Conventions Used in This Book

The following typographical conventions are used in this book:

Italic
: Indicates new terms, URLs, email addresses, filenames, and file extensions.

`Constant width`
: Used for program listings, as well as within paragraphs to refer to program elements such as variable or function names, databases, data types, environment variables, statements, and keywords.

`Constant width bold`
: Shows commands or other text that should be typed literally by the user.

`Constant width italic`
: Shows text that should be replaced with user-supplied values or by values determined by context.

This icon signifies a tip, warning, or general note.

Using Code Examples

You can download all the source code for this book from *http://makesensors.botbook.com*.

You can extract the ZIP package by double-clicking it, or by right-clicking and selecting "Extract" from the pop-up menu.

This book is here to help you get your job done. In general, you may use the code in this book in your programs and documentation. You do not need to contact us for permission unless you're reproducing a significant portion of the code. For example, writing a program that uses several chunks of code from this book does not require permission. Selling or distributing a CD-ROM of examples from MAKE books does require permission. Answering a question by citing this book and quoting example code does not require permission. Incorporating a significant amount of example code from this book into your product's documentation does require permission.

We appreciate, but do not require, attribution. An attribution usually includes the title, author, publisher, and ISBN. For example: "*Make: Sensors* by Tero Karvinen, Kimmo Karvinen, and Ville Valtokari. Copyright 2014 Tero Karvinen, Kimmo Karvinen, and Ville Valtokari, 978-1-449-36810-4."

If you feel your use of code examples falls outside fair use or the permission given here, feel free to contact us at *bookpermissions@makermedia.com*.

Safari® Books Online

Safari Books Online is an on-demand digital library that delivers expert content in both book and video form from the world's leading authors in technology and business.

Technology professionals, software developers, web designers, and business and creative professionals use Safari Books Online as their primary resource for research, problem solving, learning, and certification training.

Safari Books Online offers a range of product mixes and pricing programs for organizations, government agencies, and individuals. Subscribers have access to thousands of books, training videos, and prepublication manuscripts in one fully searchable database from publishers like O'Reilly Media, Prentice Hall Professional, Addison-Wesley Professional, Microsoft Press, Sams, Que, Peachpit Press, Focal Press, Cisco Press, John Wiley & Sons, Syngress, Morgan Kaufmann, IBM Redbooks, Packt, Adobe Press, FT Press, Apress, Manning, New Riders, McGraw-Hill, Jones & Bartlett, Course Technology, and dozens more. For more information about Safari Books Online, please visit us online.

Maker Media has uploaded this book to the Safari Books Online service. To have full digital access to this book and others on similar topics from MAKE and other publishers, sign up for free at *http://my.safaribooksonline.com*.

How to Contact Us

Please address comments and questions concerning this book to the publisher:

MAKE
1005 Gravenstein Highway North
Sebastopol, CA 95472
800-998-9938 (in the United States or Canada)
707-829-0515 (international or local)
707-829-0104 (fax)

MAKE unites, inspires, informs, and entertains a growing community of resourceful people who undertake amazing projects in their backyards, basements, and garages. MAKE celebrates your right to tweak, hack, and bend any technology to your will. The MAKE audience continues to be a growing culture and community that believes in bettering ourselves, our environment, our educational system—our entire world. This is much more than an audience, it's a worldwide movement that Make is leading—we call it the Maker Movement.

For more information about MAKE, visit us online:

MAKE magazine: *http://makezine.com/magazine/*
Maker Faire: *http://makerfaire.com*
Makezine.com: *http://makezine.com*
Maker Shed: *http://makershed.com/*

We have a web page for this book, where we list errata, examples, and any additional information. You can access this page at:

http://bit.ly/make-sensors

To comment or ask technical questions about this book, send email to:

bookquestions@oreilly.com

Acknowledgments

The authors would like to thank Hipsu, Marianna, Nina, Paavo Leinonen, and Valtteri.

Raspberry Pi

1

We recommend you start with the Raspberry Pi Model B, which includes wired Ethernet and enough USB ports for a mouse and keyboard. This makes it much easier to get started.

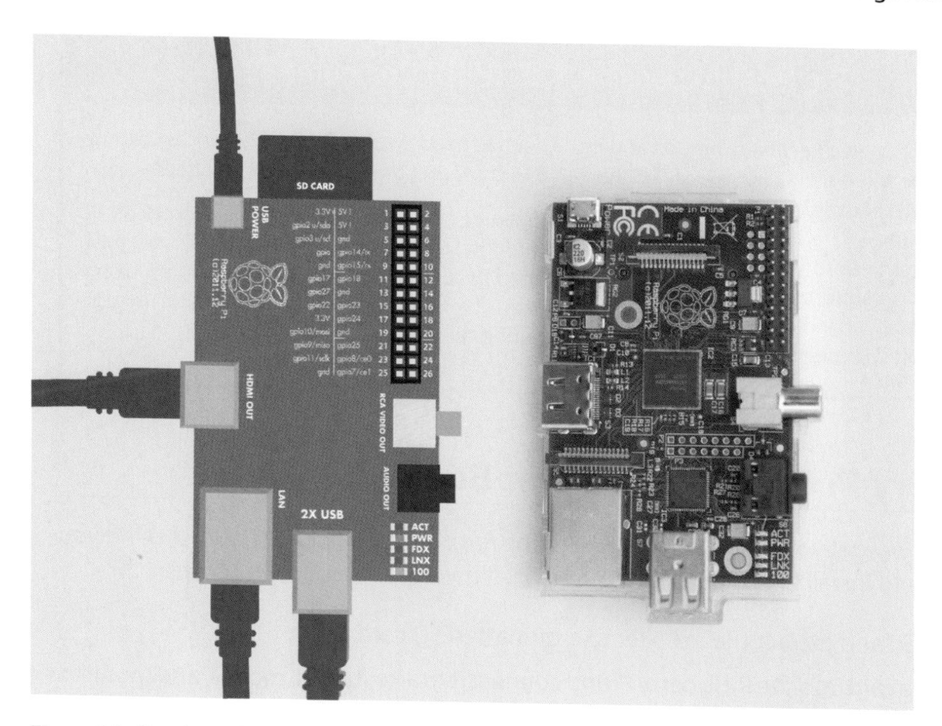

Figure 1-1. *Raspberry Pi peripheral connections*

Unless you buy your Raspberry Pi as part of a kit, it probably didn't come with an enclosure, but you can just put the bare board on your table for extra geek credibility. Or, if you have access to a 3D printer, CNC, or laser cutter, you can find many enclosures to fabricate on *http://www.thingiverse.com*.

A 4 GB SD memory card is big enough to fit the operating system. A bigger card may be less susceptible to wearing out over time (more storage to allocate to wear-leveling), so if you have an 8 GB or bigger card, even better.

The Raspberry Pi can drive a full high-def display, and can even send sound over HDMI. Most likely, an HD television will work nicely as a display for your Pi.

Having a keyboard and a mouse will make it easy to get started. Raspberry Pi Model B has exactly two USB ports, just enough for the mouse and the keyboard.

If you want to add a USB WLAN adapter, you need a powered USB hub. See http://elinux.org/RPi_USB_Wi-Fi_Adapters for a list of WiFi adapters that are known to work with the Raspberry Pi. You'll be able to configure WiFi on your Pi by double-clicking the WiFi Config icon on the desktop after you install the operating system and boot to the graphical desktop environment.

The Most Expensive $35 (USD) Computer?

Buying all the cables, keyboard, mouse, and display can cost more than a couple of Raspberry Pis. If you don't already have all those parts gathering dust somewhere, it can be quite a lot for a tiny computer. Even so, it saves time (== money) to establish a comfortable development environment. Later, when your project is working, you can easily trim down the system to just the needed parts. As they say, Raspberry Pi is the only $35 computer that costs a hundred bucks.

If you decide to interact with your Raspberry Pi through SSH or VNC over the network, you only need to connect network and power and won't need the keyboard, mouse, or monitor except during the initial setup.

Raspberry Pi from Zero to First Boot

This chapter will get you up and running with the Raspberry Pi quickly. The first thing you need to do is to install Linux on the Raspberry Pi. It involves the following steps:

- Download and extract the installer to a formatted SD card.
- Insert the card into the Raspberry Pi and connect it to a keyboard, mouse, and monitor.
- Turn it on, choose what to install, and wait.

Once that's done, you are ready to boot the Pi into a graphical Linux desktop.

You'll need the following parts:

- Raspberry Pi Model B
- Micro USB cable and USB charger (or computer)
- 4 GB SD card
- Display with HDMI port
- HDMI cable
- USB mouse
- USB keyboard

Extract NOOBS*.zip

Download *NOOBS_vX_Y_Z.zip* (as of this writing, it was *NOOBS_v1_3_4.zip* but the filename may be different by the time you read this) from *http://raspberrypi.org/downloads*.

> *You can also find all the important links mentioned in this book on http://botbook.com, along with mirrored copies of some files.*

Insert the SD card into your computer. Most SD cards are FAT32 formatted at the factory, so unless you're using an SD card that you've formatted yourself, extracting the NOOBS zip to the SD card is enough. After you unzip the file, make sure that the *bootcode.bin* file is in the root (top-level) directory of the SD card.

> *If you need to format the SD card, use the formatting tool from the SD Card Association (https://www.sdcard.org/downloads/formatter_4/).*

In modern versions of Linux, Windows, and Mac you can just double-click or right-click the NOOBS zip file to extract it. For older versions of Windows, you can install 7zip (*http://www.7-zip.org*) to let you extract zip files.

Connect Cables

Connecting the cables is easy, because each cable will fit only its correct socket. Plug the mouse and the keyboard into the Raspberry Pi's USB ports. If you're using an HDMI monitor, connect an HDMI cable between the monitor and Raspberry Pi. If you're using an NTSC or PAL monitor, use a composite video cable to connect the yellow plug on the Raspberry Pi to the monitor.

Next, connect the micro USB cable to Raspberry to supply power. Plug that cable into either a computer's USB port or a 5 volt USB charger that provides at least 700 mA.

Boot and Install Raspbian

As soon as you connect power to the Raspberry Pi, it boots. No power switch is needed.

> *If nothing appears on the screen, you may need to select the right output mode for the Raspberry Pi. The default output mode is HDMI, but if you are connected via HDMI and see nothing, try pressing 2 on the keyboard connected to your Raspberry Pi to select HDMI Safe Mode. If you are connected via the composite (yellow) connector, press 3 for a PAL monitor or television, or 4 for an NTSC monitor or television.*

You are greeted with a graphical menu of different operating systems as well as language and keyboard type. Choose "Raspbian [RECOMMENDED]" (Figure 1-2) and select your language and type of keyboard you'll be using.

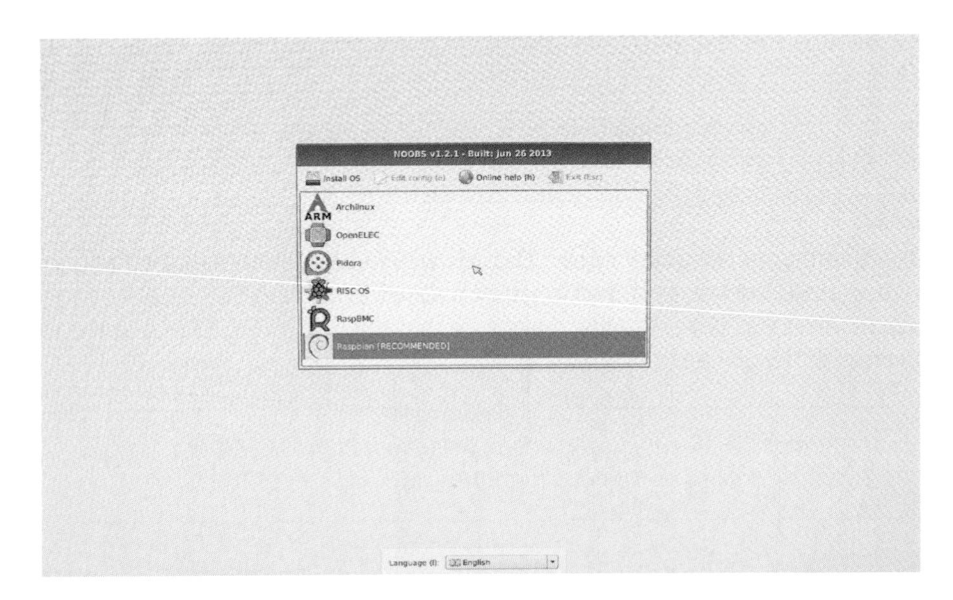

Figure 1-2. *Choosing an operating system*

If you know any Debian, Mint or Ubuntu, you will feel at home with this choice; if you don't, read on and you'll still feel at home! Raspbian takes a few minutes to finish installing (Figure 1-3). After the installer completes, it will indicate that it installed the operating system successfully. Press Enter or click OK to reboot.

Figure 1-3. *Raspbian installs*

The Raspberry Pi configuration utility opens. Use arrow keys and Tab to navigate, and press Enter/Return to select an option, as shown in Figure 1-4.

Figure 1-4. *Changing your password*

You'll want to enable the Boot to Desktop option. When you have finished changing settings, use Tab to select Finish and reboot when asked.

After the Raspberry Pi reboots, it will start up in a graphical desktop and will log you in automatically.

If you have chosen not to enable Boot to Desktop, you'll always start in the command-line interface. Log in as "raspberry" with password "pi" (unless you changed the password). After you log in, type `startx` to start the X Window System, which is the graphical desktop.

Welcome to Linux! You have now installed Raspbian on Raspberry Pi (Figure 1-5).

Figure 1-5. *Welcome to Linux*

> *To turn off your Raspberry Pi, double-click the Shutdown icon on the desktop. After it finishes the power down process, you should unplug the power.*

Troubleshooting Your Raspberry Pi Installation

Here are some solutions to common problems.

Is your card FAT32 formatted?
: If you're having trouble booting from the SD card, it might not be formatted correctly. On Linux, use the built-in graphical partition editor (type `sudo gparted` to run it). Format the entire disk to FAT. You can run another tool with the command `sudo palimpsest` (or `sudo gnome-disks`) and for advanced users, `sudo parted` will get you into a classic command-line partitioning tool. On Windows and Mac, use the SD Association formatter (*https://www.sdcard.org/downloads/formatter_4/*). On Windows, choose the "format size adjustment" option in the formatter. On Mac, use "Overwrite Format."

Red power LED (PWR) not lit?
: Power LED dim or blinking? Power LED briefly on but then goes off? Raspberry Pi is not getting enough power. Connect a USB power supply that can provide 5 volts and 1 amp or more. If you're powering the Pi via a laptop USB, switch to a desktop computer's USB or use a powerful mobile phone or tablet charger.

Black screen (but red PWR LED lit)?
: It's possible that the Raspberry Pi can't read the bootloader from the SD card. Power down, remove the SD card, and reinsert it firmly. Verify that the first file used in the boot sequence, *bootcode.bin*, is on the top level of the SD card. If the problem persists, format the SD card and extract the NOOBS zip on it again. If that doesn't help, try a different SD card.

Four colored boxes on the screen?
: The bootloader was read from the SD card, but the operating system *kernel.img* failed to boot. Format the SD card and extract the NOOBS zip again, or try another SD card.

Boot fails with some error messages?
: Try disconnecting all USB devices, such as keyboard, mouse, and WiFi adapter (if you connected one). Leave only the SD card, display, and power. Remove and insert the SD card to ensure that it makes proper contact. If the problem persists, reformat the SD card and extract the NOOBS zip on it again.

Messed up your operating system?
: If normal commands don't work, your screen is full of garbage, or Raspberry Pi stops working suddenly, don't worry. Hold down the Shift key on boot, and choose the option to reinstall Raspbian. This is quite fast and easy, but it deletes all data on the memory card. If it doesn't help, reformat the card, extract the NOOBS zip to it, and go through the installation process again.

No Internet?
: If you have connected an Ethernet cable before boot, it should work like a charm in a typical network. Check that the link present (LNK) light is lit on the Raspberry Pi. If LNK is not lit, this means the Raspberry Pi thinks that one end of the Ethernet cable is not connected. Typically, you should see 100 (indicating a 100 Mbit/s or faster connection), and FDX (full duplex) should be lit. If LNK is lit and you still have problems, you may need some more advanced troubleshooting commands like the following: `ifconfig` (displays network adapter configuration), `route -n` (displays the routing table for the network connection), `cat /etc/resolv.conf` (shows the nameserver in use), and `ping -c 1 google.com` (tells you whether you can reach Google over the network).

If you get any error messages not mentioned here, try typing the exact error message into a Google search. Be sure to use the exact error message shown. If the message appears as the system is booting, take a photo of the screen with your camera or cell phone and transcribe it.

Feeling at Home in Linux

Raspberry Pi *is* Linux. Well, it's built on Linux. The name "Linux" is used to describe the operating system kernel and the operating system itself. The Linux operating system is composed of the kernel as well as thousands of utilities and applications from various sources.

Raspberry Pi is not a workstation-class device. In terms of computing power, it's more comparable to an entry-level portable tablet or a mobile phone. So even if you boot to its graphical desktop, don't expect to dump your laptop or desktop computer just yet. Extremely low computing power combined with little memory means that applications like LibreOffice and Mozilla Firefox won't be usable.

Command-Line Interface is Everywhere, Forever

Are you ready to feel the power of the $ prompt?

The command line has stood the test of time, and it might be something you'll teach your grandchildren. The commands you use all the time, like *pwd*, *ls*, or *cat*, existed long before Linux was invented. (Linux was written by a Finnish student in Helsinki, just five kilometers from the botanical garden where I'm writing this.) Power users on both OS X and Windows dive down into the command line when they need to do something they can't do with a mouse alone.

Most of the commands you'll use with Raspberry Pi are the same you'd use on a Mac or Linux computer, and are similar even to the command-line tools on Windows.

As you may know, most of the world's servers run Linux. Google, Facebook, Amazon, and most supercomputers run Linux. Web servers don't run a graphical user interface, so that's why most programmers and system administrators must know how to use the command-line interface (CLI). You can use these commands on a Linux desktop or laptop, too.

The CLI Really Is Everywhere

If your smartphone is like many smartphones, it runs Linux, and you can run some of these commands on your phone. Android supports a limited subset of commands even without jailbreaking or rooting (you'll need a shell environment app such as ZShaolin (*http://www.dyne.org/software/zshaolin/*)). Like Apple's computers, the iPhone and iPad are based on a distant cousin of Linux, BSD, and you can access the CLI on a jailbroken iPhone.

The CLI is easy to automate. Whatever commands you can type on the command prompt (also known as the *shell*), you can also turn into a program. You just put each command into a text file, one command per line (we show you how to do this in the next section). Then you can run that program with `dash` ***`filename`*** (on Raspberry Pi, the shell is named *dash*). Whenever you realize you're typing the same 10 commands over and over, it's time to turn them into a script.

Even though CLI scripts are so easy to write that they are good for quick-and-dirty one-off scripts, they are also suitable for serious, mission-critical applications. For example, Linux uses scripts for booting and controlling daemons (server applications that run in the background).

Looking Around

After you boot your Raspberry Pi into the graphical user interface, you can start up the command-line interface by double-clicking the LXTerminal icon on your desktop.

> *Besides being great for browsing the Web, the Midori web browser is useful if you want to copy and paste some sample code from http://botbook.com.*

The prompt *$* means Linux is waiting for your command. Type *pwd* to print your current working directory (where you are in the file system). Linux answers with a path, such as */home/pi/*.

To list the files in the working directory, type *ls* and press Return or Enter. These are the files you can readily manipulate. Whenever you get an error like *No such file or directory*, just use *pwd* to see where you are and *ls* to see what files are in the working directory.

> *You will always need to type Return or Enter after you type a shell command such as* `ls`*. Before you press Return/Enter, you can use the Backspace and arrow keys to edit the command.*

To edit (or create) a file called *foo.txt*, use `nano foo.txt`. Type some text, then press Control-X to save it (when prompted whether to save, type y. When asked for the filename, just press Enter or Return). To edit the file some more, type the command `nano foo.txt` again.

Text Files for Configuration

In Linux, most things are text files. With just the commands listed in the previous section, along with the *sudo* command, a skillful hacker can change many system settings. All configuration in Linux is stored in text files. System-wide configuration files are in the */etc/* directory and per user configuration is in the user's home directory, */home/pi/*.

Even the Pi's input and output pins can be manipulated by editing text files under the */sys/* directory. Later in this book, you will learn to connect sensors and LEDs to Raspberry using GPIO pins.

If you connected the Pi to the network with an Ethernet cable before booting, your Raspberry Pi should already be connected to the Internet. Test this by typing `ping www.google.com`. You'll see a result every second. Kill the command after a while by typing Control-C. If the network is working well, it should report a packet loss of 0%. You can also download whole web pages with commands like `curl botbook.com` and `wget botbook.com`.

sudo Make Me a Sandwich

Linux is well known for its robust security model. The *separation of user privileges* is one of its key features. Normal users can make changes only to files that affect their working environment. This means that they can modify files only in their home directory (*/home/pi/*) and in temporary working directories such as */tmp/*.

The super user, or root user, is all-powerful and can change any file on the system. To use root's privileges, put `sudo` in front of a command. Putting `sudo` in front of the command runs that commands under the privileges of the root user.

For example, Raspbian's package manager makes it very easy to install additional software, but you need to use root's privileges to install anything. Before you install new software, you need to update the list of what's available with `sudo apt-get update`. This requires a network connection because all the software packages are on a file server.

Many Linux and Unix systems (such as OS X) are configured such that you need to type your user password when you use sudo. This is an extra safety step. By default, Raspberry Pi's Raspbian operating system does not ask for this. Be careful using sudo, because you can easily make mistakes that would render the operating system unbootable.

You can install any program from a repository by specifying its package name. To install ipython (an interactive tool for experimenting with the Python language and data visualization), use `sudo apt-get -y install ipython`. The `-y` parameter tells the package manager to assume a "yes" answer to any questions it asks. The package manager (apt) does everything for you.

After a moment, you can run the newly installed package by typing `ipython`. Any python command will work here, but `exit()` will get you back to the command prompt ("$"), where you can type shell commands.

Different prompts are one way of indicating which program you're talking to. When you see the shell, or command prompt ($), you can type the name of built-in shell commands as well as programs installed on your Raspberry Pi. When you see another prompt, such as the ipython prompt (`In [1]:`), you can type Python commands.

Installing daemons (also known as servers) is just as easy. Try installing the most popular web server in the planet, Apache (*http://www.apache.org*), with this command: `sudo apt-get -y install apache2`. When it finishes, you can pull down your web server's raw home page with `curl localhost`. To browse your Apache web server from another computer, determine your IP address with the output of the `ifconfig` command and type that address into a web browser on another computer connected to your network.

You will see at least two adapters listed in the output of `ifconfig`. Use the Ethernet adapter's (`eth0`) address, or if you're using WiFi, use that adapter's address.

Phew! That was a lot of command line if you are (were) a beginner. To give yourself a well-earned break, power off the Raspberry Pi by typing the command `sudo shutdown -P now`. Just like you do on your workstation, you must shut down properly so that data gets actually written to disk before you power off. Once it shuts down, you can unplug it from the USB power source. When you're ready to use it again, plug it back in.

Still wondering what we're talking about with the sudo sandwich? Try searching the Web for "xkcd sudo sandwich." You can use the rest of your break to read some of the other comics there.

See Appendix A for a cheat sheet of Linux commands.

One Handed Wonder

We sometimes ask Linux students to guess how long it takes me to type "supercalifragilisticexpialidocious.foo.bar.txt.botbook.com.pdf.cc" with my left hand. After some guesses, we'll test it. As a preparation, I create the file with this command (type it all on one line with no space between `foo.` and `bar`):

```
$ nano
supercalifragilisticexpialidocious.foo
.bar.txt.botbook.com.pdf.cc
```

I count down to three, and type this in less than five seconds:

```
$ ls
supercalifragilisticexpialidocious.foo
.bar.txt.botbook.com.pdf.cc
```

In reality, I just need to type `ls s` and press Tab. The shell will complete the word you're typing after you press Tab. You should use this feature all the time. It's weird how well Tab works. It can guess not only directories and files but also network servers after commands like `ssh` and `ping`. Because Tab completes only correct filenames, you won't be making many typos.

Connecting Electronics to Raspberry Pi Pins

The Pi's GPIO (general-purpose input and output) pins let you connect electronic components directly to the Raspberry Pi. These pins are called general purpose because you can decide what purpose they serve, and you can even configure the same pin to be an input or an output at different times. Throughout this book, you will learn the following:

- Digital output (turn an LED on and off)
- Digital resistance (detect whether a button is pressed or a sensor is active)

- Digital input for very short pulses (used by sensors such as a distance sensor)
- Analog resistance (analog resistance sensors for pressure, light, temperature)
- Industry standard protocols, such as I2C and SPI (used by the Wii Nunchuk and analog-to-digital converters)

Unlike most other tutorials available at the time of writing, we'll teach you how to use digital input and output without having to invoke root privileges all the time. This provides security and stability benefits.

For digital input, you'll learn to use the Pi's internal pull-up resistor, so that your circuit uses a minimum number of components.

For measuring analog resistance, you'll use an external analog-digital converter chip.

Many components in everyday products communicate over industry standard protocols, such as I2C and SPI. You will see examples of both protocols later in the book.

But first, let's show you how to use the most basic form of digital output.

Hello GPIO, Blink an LED

In this "Hello GPIO World" example, you'll attach a new LED to Raspberry Pi and blink it.

We start all of our projects with a "Hello World" on any platform, on any language. So whenever you are about to build something more complicated, it's a good idea to build this "Hello GPIO World" first. This lets you confirm that the hardware and software are functioning at the most basic level. If your "Hello World" example doesn't work, you need to fix it before trying something more complicated.

Parts needed:

- Raspberry Pi
- Female-to-male jumper wires, black and green or yellow color
- Solderless breadboard
- 470 Ohm resistor (yellow-violet-brown stripes)
- An LED

If you don't have all the parts at hand, see "Troubleshooting" on page 17 for suggestions. For wires, a female connector is one that has a receptacle and a male connector is one that has an extending pin. You will often see female abbreviated as F and male abbreviated as M, along with other characteristics of the connector, such as its length. The type of female-to-male jumper wires you need are typically sold as "male to female breadboard jumper wires."

GPIO pins are not protected against overcurrent (Figure 1-6). Unlike Arduino, Raspberry Pi is not forgiving on user mistakes. Data pins can take only 3.3 V. Connecting a +5 V pin to a data pin can easily break your Raspberry Pi, or at a minimum, render that pin unusable. Double-check anything that you build on a breadboard, and be very careful where you place any test probes if you use a multimeter with the GPIO pins.

Figure 1-6. *GPIO header*

Building the Circuit

The circuit is simply an LED with a current limiting resistor, connected in series between GPIO pin 27 and ground (Figure 1-7). Connect the short (negative) lead of the LED to the black wire, and the long (positive) lead to the resistor. You should make these connections while the Raspberry Pi is shut down and unplugged.

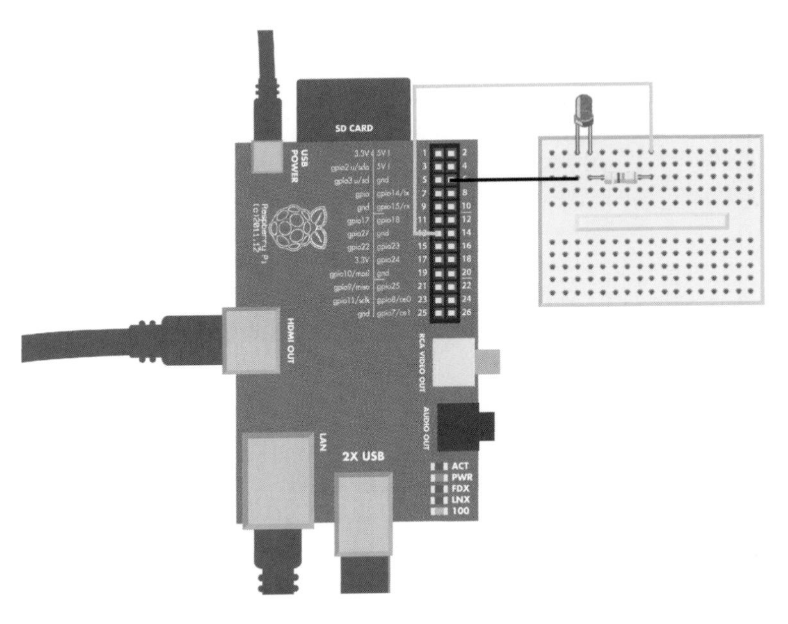

Figure 1-7. *LED hello breadboard*

Build the circuit on the breadboard (Figure 1-8). Double-check all connections to avoid breaking your Raspberry Pi. After you're sure you've connected the wires correctly, you can power up your Pi.

To learn about how the Raspberry Pi's pins are numbered, read on.

Figure 1-8. *Hello GPIO!*

Two Numbering Systems: Purpose and Location

Each GPIO pin has two numbers: purpose and physical location. To find the correct pin in the GPIO header, you should learn to convert between the two numbering systems.

Locate the GPIO pin header on Raspberry Pi (Figure 1-6). When converting between purpose and physical location numbers, use the numbering diagram (Figure 1-9).

Figure 1-9. *GPIO numbering*

The left side of the numbering diagram shows the purpose of the pins (GND, GPIO 27). The right side of the numbering diagram shows the physical location (1 to 26).

The physical pin header has a running number, from 1 to 26. There is a tiny white box drawn near pin one, and it is also labeled P1. This physical pin header number tells you where to insert the jumper wire. These physical numbers are also known as *board numbers*.

There is also another, purpose-based numbering. The code you'll be writing uses GPIO numbers, such as GPIO 27, which corresponds to physical pin 13 (which is the pin you connected to the resistor with a wire). Circuit diagrams are likely to refer to ground (GND), +5 V, and +3.3 V, and Figure 1-9 will help you find them. Some pins can have multiple purposes. For example, GPIO 10 can be used for the SPI bus. These purpose-based numbers are also known as *BCM numbers*.

To help you get started, the two pins used here are listed in Table 1-1. For future projects, learn to find these numbers in the numbering diagram (Figure 1-9).

Table 1-1. Pins used in Hello GPIO

GPIO pin (BCM, used in code)	Physical location (Board)
GPIO 27	13
GND	14

Controlling GPIO Pins from the CLI

Let's see how to use the command-line interface to control the GPIO pins we just connected. First you'll try it out as root, then advance to using it without needing to invoke sudo privileges each time.

Text files control everything in Linux. The kernel GPIO driver (a piece of software that controls how Linux talks to GPIO pins) makes the GPIO pins available to you through the virtual */sys/* file system. To control GPIO, you simply edit or otherwise make changes to these text files.

You don't need a graphical user interface at this point, so we can do everything from the LXTerminal command-line interface. Double-click its icon to launch it.

If you configured your Raspberry Pi to not start in the graphical interface, or if you've used SSH to connect to it remotely from another computer, you don't need to start up the graphical desktop system to try this out.

To turn on the pin, you first *export* it, configure it for "out" mode, and write the number "1" to it. All this is done by editing text files.

Writing to Files Without an Editor

Earlier, you saw how to modify the contents of a file with the nano text editor. Here's how you can modify files without needing the editor.

First, let's look at how to display text. You can print text to the terminal using this command (don't type the $; it indicates the shell prompt):

```
$ echo "Hello BotBook"
```

Any text you display this way can also be redirected to overwrite a file (be careful using this, and don't overwrite anything important):

```
$ echo "Hello BotBook" > foo.txt
```

If the file *foo.txt*, did not exist, it will be created. If it existed, it was overwritten without warning. You can use the > (redirection) operator to send the output of almost any command to a text file. You can see what's inside that file with the `cat` command:

```
$ cat foo.txt

Hello BotBook
```

Couldn't you just use *nano foo.txt*? Yes, of course you could. But it requires more typing, and it's not as easy to automate.

Light Up the LED

We'll use sudo the first time through lighting an LED. Does it feel wrong to use root for non-administrative tasks? If not, it should; if you mistype a command with sudo, you can potentially render your operating system unusable and will need to reinstall it.

Don't worry, you'll use sudo just to initially try it. Later, you'll fix Linux's file permissions and interact with the files that control GPIOs as a normal user.

Type `sudo -i` to get a *root shell*. Use root shell only as long as required by this task, and type `exit` when you're done. You'll notice that your prompt changes to a hash mark, #. Be careful what you type as root, as mistakes can break your operating system.

After you start the root shell, export GPIO pin 27 to be able to manipulate it (remember, just type the text to the right of the # shell prompt):

```
# echo "27">/sys/class/gpio/export
```

This creates the new virtual file you'll use to blink the LED. Next, set pin 27 to *out* mode, so that you can turn it on and off.

```
# echo "out" > /sys/class/gpio/gpio27/direction
```

Now turn on the pin:

```
# echo "1" > /sys/class/gpio/gpio27/value
```

Your LED should light up now. Once you have enjoyed the light for a while, turn it off:

```
# echo "0" > /sys/class/gpio/gpio27/value
```

Did your LED light up? Hello, GPIO world!

Once you are done with playing root, remember to type `exit`. Your prompt should return to the dollar sign symbol $, indicating that you are working with the shell as a normal user.

Troubleshooting

If you had trouble setting things up, check out the following:

Lacking a 470 Ohm resistor?
: Use any resistor in the hundreds of ohms range to take only a minor risk with your board. The resistor is just used for limiting current through the LED, to avoid frying the LED or the pins on the Raspberry Pi. If you don't mind the LED being a bit too dim or slightly overworked, you can use any resistor value between 100 Ohm and just under 1 kOhm. (Such resistors will have a brown third stripe; see "The Third Stripe Rule" on page 19.)

Can't find female-male jumper wires?
: You can use an old 40-wire IDE hard drive cable (not 80-wire). Put the female side of the cable into GPIO header. Cut the other end. Pay attention to wire numbers. If you don't need all the wires, don't strip them. If you strip them all, at least use tape to cover +5 V wire ends. See *http://bit.ly/1f0GWfV* for a tutorial on using an old IDE cable. There are also many ready-made connectors available such as this one from Adafruit (*http://bit.ly/POg1rj*). MAKE's

Raspberry Pi Starter Kit (*http://bit.ly/1hlLDew*), which includes a Raspberry Pi, includes the cable and a breadboard-ready breakout board.

My LED does not light
: Did you insert it the right way? LEDs have polarity, and if you insert them backward, they won't light up. The long positive leg of the LED goes to GPIO 27 (through the resistor). The negative side of the LED has a flattened area in the plastic body of the lamp, and a shorter lead than the positive lead. The negative side of the LED goes to ground. The arrow in Figure 1-10 is pointing to the cathode. You may need to look closely to see the flattened part at the base of the LED package. If you have trouble seeing it, grab an LED and check it out yourself.

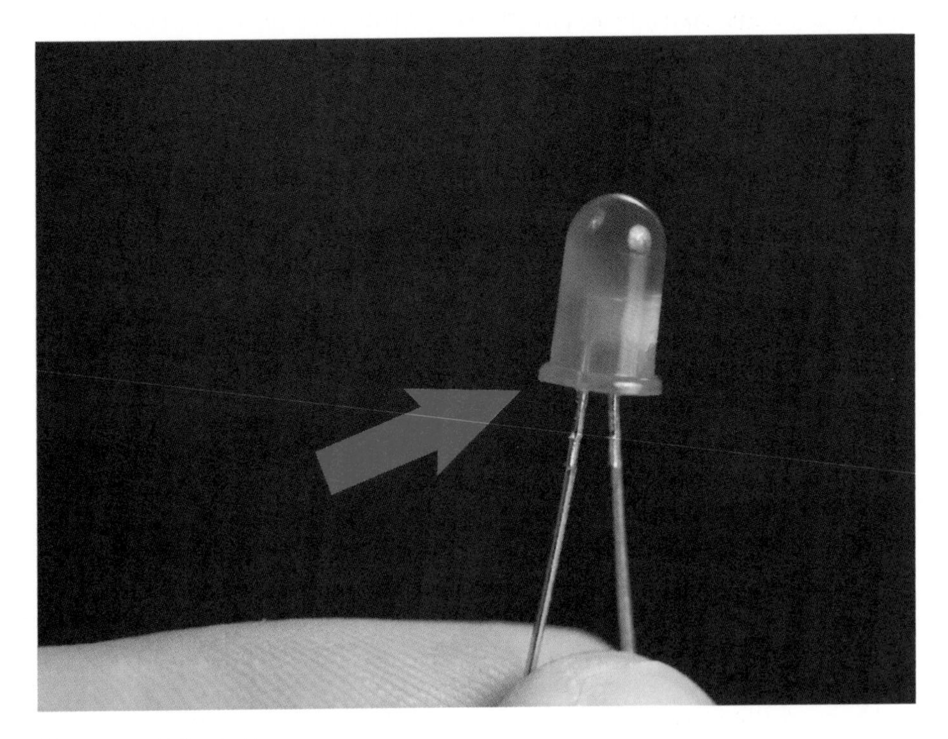

Figure 1-10. *LED polarity*

Nothing happens when I type root commands
: Don't type the prompt (#); that is shown in the previous examples to show you what you'll see on your Raspberry Pi screen. The shell shows that prompt when it is waiting for you to type a command while in a root shell. You wouldn't type the normal user prompt "$" either, would you? The hash symbol has special meaning to the shell: it means that the rest of the line is a comment (a human-readable remark you leave in a program), and is ignored by the shell.

I can't access the shell
: Review the instructions in "Looking Around" on page 9.

I get an error message

Copy and paste the exact error message into a search engine like Google. Also try putting quotes around your error message. You can try "Raspbian" or "Raspberry Pi" as additional search terms. When you solve your problem, it's a good idea to blog about it and include the exact error message in your blog post so other people searching can find it.

The Third Stripe Rule

The third stripe trick identifies a resistor easily. You can quickly find a resistor with approximately the value you are looking for by looking at the multiplier ring. The multiplier ring is usually the third one (with five band resistors, the multiplier is the fourth ring). In many connections, the exact value of resistor is not important as long as it's near enough.

For example, 10 kOhm resistor has kilo (thousand, 10^3) multiplier, 3: orange. A resistor from 10 MOhm to 50 MOhm has mega (million, 10^6) multiplier, so the third stripe is 6: blue.

For calculating all the color bands, search the Web for "resistor color code calculator." To double-check your resistor identification, measure the resistance with a multimeter.

GPIO Without Root

Avoiding root privileges will make the system more secure and more stable. For example, think about a program that serves sensor data to web. Would you run a program that strangers can connect to as root?

Before you begin, make sure that you can turn the LED on and off ("Light Up the LED" on page 17).

In modern versions of Linux, devices attached to your system are controlled by *udev*. Udev is a rule-based system that can run scripts when devices are plugged in. If you have developed apps for Android under Linux, you may have created a udev rule to modify permissions when your cell phone is plugged into the computer. If you have developed with Arduino on Linux, you have probably added yourself to the *dialout* group to get access to serial over USB.

Normally, the GPIO files in */sys/class/gpio/* are owned by the user "root" and the group "root." You'll see how to write a udev rule to change the group to "dialout." You'll then allow that group to read and write the files under */sys/class/gpio/*. Finally, you'll make the folders' group *sticky*, so that any newly created files and folders under it will also be owned by the "dialout" group.

All system-wide configuration in Linux is under */etc/*. Not surprisingly, udev configuration is in */etc/udev/*. First, open an editor with sudoedit so you can create a new rule file:

```
$ sudoedit /etc/udev/rules.d/88-gpio-without-root.rules
```

Add the text shown in Example 1-1 to the file. Be sure to type each line as shown (don't type the numeric symbols; those are there to explain to you what is going on in this file). Udev rules are very sensitive to typos.

Example 1-1. 88-gpio-without-root.rules

```
# /etc/udev/rules.d/88-gpio-without-root.rules - GPIO without root on Raspberry Pi  # ❶
# Copyright 2013 http://BotBook.com
# ❷
SUBSYSTEM=="gpio", RUN+="/bin/chown -R root.dialout /sys/class/gpio/"
SUBSYSTEM=="gpio", RUN+="/bin/chown -R root.dialout /sys/devices/virtual/gpio/"
# ❸
SUBSYSTEM=="gpio", RUN+="/bin/chmod g+s /sys/class/gpio/"
SUBSYSTEM=="gpio", RUN+="/bin/chmod g+s /sys/devices/virtual/gpio/"
# ❹
SUBSYSTEM=="gpio", RUN+="/bin/chmod -R ug+rw /sys/class/gpio/"
SUBSYSTEM=="gpio", RUN+="/bin/chmod -R ug+rw /sys/devices/virtual/gpio/"
```

❶ This comment explains the purpose of the file.

❷ Sets the owner of the two directories to be root, and the group to be dialout.

❸ Sets the *sticky bit* flag on these two directories.

❹ Configures the permissions on the directories to give members of the dialout group read and write permission.

The rules are processed in numeric order, but this is probably the only rule affecting the GPIO directories, so the number does not matter. In Morse code (CW), 88 is short for hugs and kisses. We prefer it over the often-picked number 99, which means "get lost."

To avoid typing and inevitable typos, you can download a copy of the *88-gpio-without-root.rules* file from *http://botbook.com*.

Save the file (Control-X, press y, and then press Enter).

To use your new rules, restart the udev daemon and trigger your new rule with these commands:

```
$ sudo service udev restart
$ sudo udevadm trigger --subsystem-match=gpio
```

In modern Linux, all daemons ("servers") are controlled with scripts. So you could just as easily tell the Apache web server to reread its configuration with `sudo service apache2 reload`. Or you can make SSH server restart (stop and start) with `sudo service ssh restart`.

Next, check whether the ownership is correct:

```
$ ls -lR /sys/class/gpio/
```

The listing should mention the "dialout" group many times. The parameter `-l` means to display a long listing (with owner, group, permissions), and `-R` means recursively list directory contents, too.

Let's try GPIO without root now. Notice that our prompt is a dollar sign (`$`), indicating that we're running as a normal user.

```
$ echo "27" > /sys/class/gpio/unexport
$ echo "27" > /sys/class/gpio/export
$ echo "out" > /sys/class/gpio/gpio27/direction
$ echo "1" > /sys/class/gpio/gpio27/value
```

The first command unexports the GPIO so that the export command doesn't get an error.

The LED should be lit now. You can turn it off with this:

```
$ echo "0" > /sys/class/gpio/gpio27/value
```

Can you control the LED as a normal user? Well done!

This paves the way to use GPIO from almost any programming language.

Troubleshooting GPIO

File permissions have not changed

If `ls -lR /sys/class/gpio/` does not show "dialout," the most likely cause is a typo in the rule file */etc/udev/rules.d/88-gpio-without-root.rules*. Download the file from *http://botbook.com* and move it to the right place (for example, `sudo mv 88-gpio-without-root.rules /etc/udev/rules.d/`). To make Linux use the new rules, use `sudo service udev restart` or reboot with `sudo shutdown -r now`.

The LED is not lit

Try it as root first ("Light Up the LED" on page 17) to make sure it works.

I get an error message

Write it down and search the Web for it.

GPIO in Python

You can use GPIO from Python by just writing and reading files in */sys/*. This is the same method you used earlier with the shell.

Hello Python

As always, start with a "Hello World" to test your environment. Using a text editor, create the file:

```
$ nano hello.py
```

If you can't save the file, you might have used the cd command to navigate to part of the file system where you don't have permissions. Type cd ~ to return to your home directory and try it again.

The file needs only one line:

```
print "Hello world!"
```

Save the file (in nano, you save with Control-X, then type y and press Enter/Return).

Now run your program:

```
$ python hello.py

Hello world!
```

Running the python command without any parameters starts an interactive console. See also "The Python Console" on page 221.

Python GPIO

Let's blink the LED connected to GPIO pin 27. Wire up the LED if you don't still have it set up from the earlier experiment (Figure 1-7). Save the code in Example 1-2 to a file called *led_hello.py* on your Raspberry Pi and run it:

```
$ python led_hello.py

Blinking LED on GPIO 27 once...
```

Did your LED light up for two seconds? Hello, GPIO!

Any problems? Read on to code explanation and troubleshooting. "Hello World" is the right place to solve these problems. Whenever you have problems with more complicated code, make sure that you can still run this "Hello World."

Example 1-2. *led_hello.py*

```
# led_hello.py - light a LED using Raspberry Pi GPIO
# (c) BotBook.com - Karvinen, Karvinen, Valtokari

import time     # 1
import os

def writeFile(filename, contents):       # 2
        with open(filename, 'w') as f:   # 3
                f.write(contents)
```

```
# main

print "Blinking LED on GPIO 27 once..."          # ❹

if not os.path.isfile("/sys/class/gpio/gpio27/direction"):      # ❺
        writeFile("/sys/class/gpio/export", "27")       # ❻

time.sleep(0.1)
writeFile("/sys/class/gpio/gpio27/direction", "out")    # ❼

writeFile("/sys/class/gpio/gpio27/value", "1")  # ❽
time.sleep(2)   # seconds       # ❾
writeFile("/sys/class/gpio/gpio27/value", "0") # ❿
```

❶ Import the libraries you need. Each library has *namespace* with the same name as the library, so all commands you use from `time` start with that word, like `time.sleep(2)`.

❷ Define a new helper function for writing files. The function only runs later, when it's called.

❸ The modern way to access files in Python is the "with" syntax. That will automatically handle closing the file in case of any exceptions. In open(), filename could be */sys/class/gpio/gpio27/direction* and `w` means to open it for writing. This creates a new file handle `f` that we use for the actual file operations.

❹ Even though you want to blink an LED, it's a good idea to print something on the screen to confirm that the program runs. In Python this is not critical, as Python can print quite good *traceback* error messages that explain what went wrong.

❺ Check that the pin is not already exported. Otherwise, the second run would result in "IOError: [Errno 16] Device or resource busy." This style of checking the conditions first is sometimes called "asking for permission." Another style, not used here, is to "ask for forgiveness" and use `try...except`. We chose "asking for permission" here so that we can use an if-clause and keep the program simple.

❻ Export the pin. This creates all the files for controlling the pin, like *direction* and *value*.

❼ Because you want to write to the pin to set it on or off, you set the direction to "out." If you want to read the value, you'd set the direction to "in." If you are familiar with Arduino, you might remember a similar Arduino command `pinMode()`.

❽ Set the value to "1" to light the LED. The value "1" means that the pin is set to 3.3 V, the HIGH level of a Raspberry Pi GPIO pin.

❾ Wait for two seconds. It's a good idea to put units in comments at least once for each function or variable so that others reading your code know what you were planning. During this time, the pins are left in their current state. In this case, GPIO 27 is "1" and lights the LED.

❿ Set the value to "0" to turn off the LED.

Troubleshooting

You receive a permission denied error

If you get an error like `IOError: [Errno 13] Permission denied` or `IOError: [Errno 2] No such file`, you can live dangerously and try running as root:

```
$ sudo python led_hello.py # for testing only
```

If it works correctly as root, good! You can now fix the permissions on the GPIO virtual files ("GPIO Without Root" on page 19). If you still have the same problem, reboot Raspberry Pi by shutting down, unplugging power, and plugging it back in. If all goes well, you can run it the correct way, as a normal user:

```
$ python led_hello.py
```

LED does not light, but the program doesn't give any errors

Check the LED polarity and connections. Check that you have connected the jumper wires to the correct pins on the GPIO header (Figure 1-9). If that doesn't help, you can use a multimeter to verify that you used the correct resistor (e.g., 470 Ohm). You can test the LED with a circuit that just has a battery, the resistor, and the LED in series. This will tell you whether the LED works.

What's Next?

You have now set up your own $35 Linux environment. And you can even wire it directly to hardware. This means you have combined the power of Linux and electronics.

The system administration techniques you have been practicing are leading you in the right direction. It's always good to use minimum (non-root) privileges whenever possible.

To keep playing with your Raspberry Pi, you can now move ahead to the sensor projects in this book. The individual sensor examples teach you how to use digital input, analog input, industry standard protocols, and to measure pulse length with an interrupt.

You can apply your new system administration and GPIO skills with all the sensors in this book. Raspberry Pi usually shines with sensors that use more advanced protocols. This advantage is clear with I2C sensors such as Wii Nunchuk, MPU 6050 accelerometer-gyro, and the GY65 atmospheric pressure sensor.

Welcome to embedded Linux!

Arduino

2

Arduino is a simple and robust development board (Figure 2-1). It's one of the simplest options available for making the electronics world programmable, and it's extremely reliable as well.

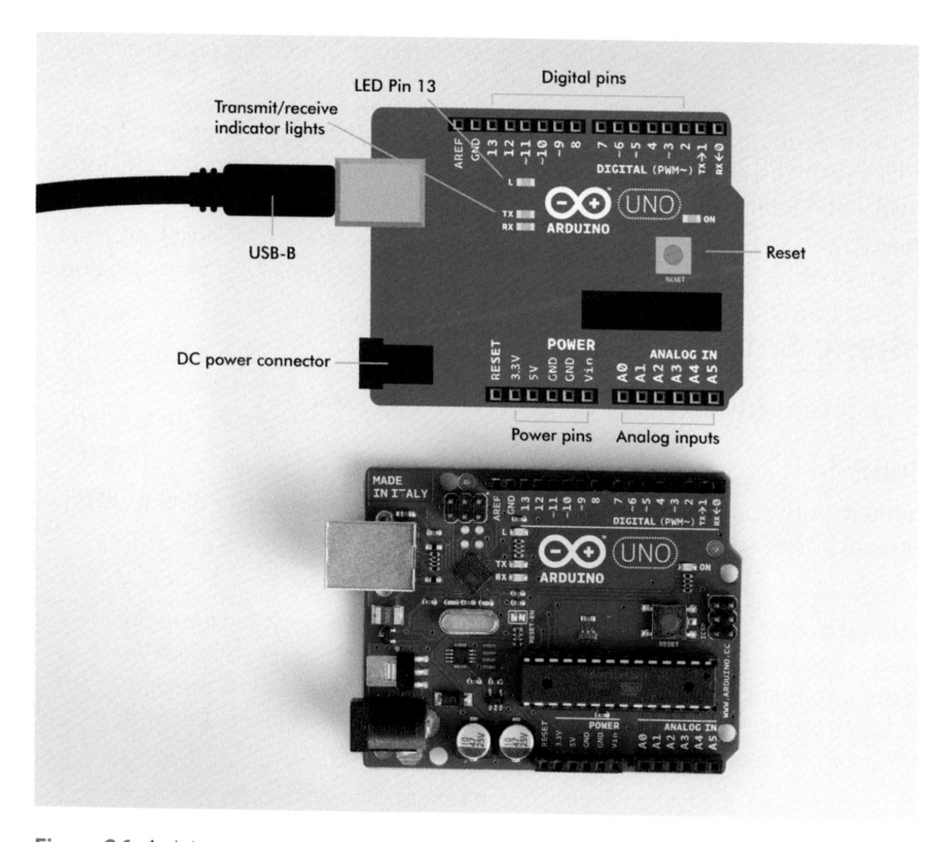

Figure 2-1. *Arduino connections*

It doesn't take much to get started with Arduino. To make something interesting happen, you just need an Arduino Uno and a USB cable; together, it shouldn't cost more than $35 or $40 USD. The software is free (the source code is available for people to use, study, modify, and share with others).

First, we'll show you how to install the Arduino development environment (often called IDE, or integrated development environment) on your computer. After that, you'll plug in a USB cable and upload your first program (called a *sketch* in Arduino parlance). There's only one program you install on the Arduino—the sketch that you're running. Aside from that, there's nothing else to maintain because, unlike with Raspberry Pi, Arduino has no operating system. It's just you, your program, and the bare metal.

There's one more piece, actually. Arduino has a bootloader that occupies a small amount of the chip's available storage. The bootloader is a small program that runs briefly when you power up or reset the board, and lets you load programs over USB without the need for a separate hardware programmer device.

The Arduino Uno is robust. It's unlikely to suffer damage even if you were to connect a wire the wrong way (but don't get too careless because, with enough abuse, it is possible to fry a pin on the Arduino).

It's very easy to learn Arduino. Beginners can accomplish a lot of things just by turning pins on and off. Unlike with Raspberry Pi, you can plug analog resistance sensors directly into the Arduino without needing external hardware, because Arduino has a built-in analog-to-digital converter.

Basic Arduino Setup

Here's how to get set up with Arduino on Linux, Windows, and Mac.

Ubuntu Linux

Connect Arduino to your computer with a USB cable. Arduino draws power directly from USB, so no external power supply is needed. Start the terminal application.

You can start the command-line terminal in many ways. You can open it in the main menu with Applications→Accessories→Terminal (on Xubuntu and other XFCE-based distributions such as Debian with XFCE). Super-T, also known as the ugly key *or* Windows key*, works on many desktops. If you are using Unity in the standard Ubuntu distribution, search for "Terminal" in Dash (top-left corner).*

To install the Arduino IDE, install the `arduino` package. Here's how you'd do it on Ubuntu Linux:

```
$ sudo apt-get update
$ sudo apt-get -y install arduino
```

Give yourself the permission to access the serial over USB port (this is required by the Arduino development environment to function). The first command adds you to the dialout group, and the second command switches you into that group without you needing to log out and back in again:

```
$ sudo adduser $(whoami) dialout
$ newgrp dialout
```

Start Arduino:

```
$ arduino
```

The Arduino IDE opens.

After you have logged out and back in, you can also start Arduino IDE from the menus.

Now you're ready to test your installation. See "Hello World" on page 28.

Windows 7 and Windows 8

Download the latest version of the Arduino Software from *http://arduino.cc/en/Main/Software*. Unzip the file you downloaded to any location that you find suitable (your *Desktop* or *Downloads* directory for example).

Connect your Arduino Uno to your computer with a USB cable. Arduino draws power directly from USB, so no external power supply is needed. Windows will start an automatic installation process for the Arduino drivers. It may fail after a while and display an error dialog.

If it fails to install the driver:

1. Open Windows Explorer, right-click Computer, and choose Manage.
2. From Computer Management, choose Device Manager on the left. Locate Arduino Uno in the device list, right-click it, and choose Update Driver Software.
3. Choose "Browse my computer for driver software." Navigate to the Arduino folder you extracted, open the *drivers* directory, choose *arduino.inf*, and click Next.
4. Windows will now install the driver.

Launch the Arduino IDE by double-clicking the Arduino icon inside the folder you unzipped.

Time to test your installation; see "Hello World" on page 28.

OS X

Download the latest version of the Arduino Software from *http://arduino.cc/en/Main/Software*. Unzip the file you downloaded, and copy it to your */Applications* folder.

Connect your Arduino Uno to your computer with a USB cable. Arduino draws power directly from USB, so no external power supply is needed. You don't need to install a driver for OS X.

Launch the Arduino IDE by double-clicking the Arduino icon in the */Applications* folder.

Time to test your installation; see "Hello World" on page 28.

Hello World

Now that you have Arduino the IDE open, you can run the Arduino equivalent of "Hello World."

First, confirm that you have the correct board selected. The Arduino Uno is the default. If you have another board, such as a Mega or a Leonardo, choose it from the Tools→Board menu.

Now you need to load the Blink test program. Choose File→Examples→1.Basics→Blink. Click the Upload button (or choose File→Upload) to compile and upload your program to Arduino.

The first time you do this, Arduino may display an error popup: "Serial port COM1 not found." That's because you haven't chosen which serial port to use (the connection between your computer and Arduino is represented as a USB serial port). Select your serial port from the drop-down menu. On Linux, it's probably */dev/ttyACM0*. On Mac, it may be something like */dev/usbmodem1234*, and on Windows, it's one of the COM ports.

> *If you see a different error message instead of a request to choose a serial port, choose your serial port from Tools→Port. If you can't figure out which port Arduino is connected to, pay attention to the ports listed, unplug the Arduino, and make a note as to which port went away. That's the Arduino port. OS X lists each port twice, for example as /dev/cu.usbmodem1234 and /dev/tty.usbmodem1234. Either one will work.*

While the program is uploading, Arduino's TX and RX (transmit and receive) lights blink rapidly. Finally, when the program is running, the tiny light labeled "L" is blinking.

The L LED blinking means that everything was successfully installed, and you just got your first sketch running.

Congratulations! Remember this simple procedure: if you ever get so stuck you are wondering whether Arduino is even running your code at all, return to this "Hello World" example. Whenever you start a new program, start with a "Hello World" to make sure everything is working.

Anatomy of an Arduino Program

An Arduino program starts by executing the code inside the `setup()` function once. After that, the code inside `loop()` is repeated forever (or until you disconnect the power). See Example 2-1.

Example 2-1. blink.ino

```
// blink.ino - blink L LED to test development environment
// (c) BotBook.com - Karvinen, Karvinen, Valtokari
void setup() {  // ❶
  pinMode(13, OUTPUT);  // ❷
}

void loop() {   // ❸
  digitalWrite(13, HIGH);        // ❹
  delay(1000);  // ms   // ❺
  digitalWrite(13, LOW);
  delay(1000);
}
```

❶ When Arduino boots, it executes `setup()` once.

❷ Configures digital pin D13 to be in OUTPUT mode, so that you can control it from your program.

❸ After `setup()` has finished, Arduino calls `loop()`. After `loop()` finishes, it's called again. And again. Forever.

❹ Sets D13 to be HIGH, which indicates Arduino is giving the pin +5 V.

❺ During the delay, the pins stay as they are. Here, D13 stays HIGH, so the Arduino's built-in L LED stays lit. During the next delay, it's LOW, so it's off. On for one second (1000 milliseconds), and off for one second. Forever.

Shields Make It Easy and Robust

Shields are boards that attach on top of Arduino and extend its features or make it more usable (Figure 2-2). There are many different shields available, from simple prototyping shields to more complex shields such as an Ethernet or WiFi shield. One of the best things about shields is how they reduce the need for extra wires; this is because they stack on top of the Arduino and use pin-to-pin connections instead of jumper wires. Of course there won't always be a shield for your needs, but they are one good option to keep in mind.

Some shields don't have any electronics on them, but are simply designed to help you prototype: these usually bring out the Arduino header pins so they are adjacent to a solderless breadboard so you can easily connect jumper wires. Our all-time favorite is the priceless ScrewShield, which adds "wings" with terminal blocks to both sides of Arduino. This eliminates loose wires, which is likely the most annoying thing about building prototypes.

You can also consider building your own shields to make easy-to-use and robust Arduino add-ons Figure 2-3. Just solder pin headers to a circuit board so that they match the pin layout of Arduino.

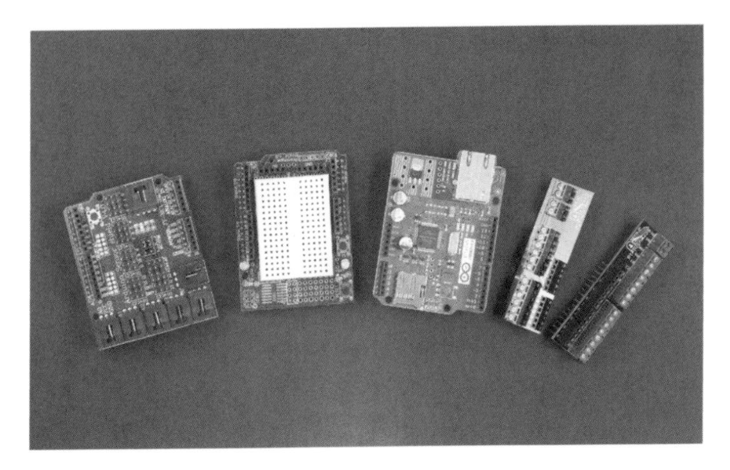

Figure 2-2. *Shields*

Figure 2-3. *Shields made by Andreas Zingerle*

3 Distance

How far is it? An ultrasonic distance sensor is one of the most popular sensors in the embedded courses we teach. A robot must know when an obstacle is near if it is to navigate around it. And isn't it more convenient to just wave your hand in the air instead of clicking a physical button? A burglar alarm can detect an intruder by noticing change in distance or heat pattern. Your home, office, or school probably has an alarm like that.

The two most common ways to measure distance are sound echoes and light reflection. To avoid annoying people with constant beeping and blinking, the sound frequency is usually so high that humans can't hear it, and the light frequency is so low humans can't *see* it. The high-frequency sound is *ultrasonic*, and the low-frequency light is *infrared*.

Even though infrared is invisible, we'll show you how you can observe it with some common household items.

An ultrasonic sensor can provide exact distance readings. For example, it could tell you that the distance to an object is 36 cm.

To detect the *proximity* of humans and other living things, sensors can detect the heat they radiate. This lets you detect the presence of hot things in the measured area, but not their exact distance. There are many ways for heat to move: conduction, convection, and radiation. A *passive infrared sensor* measures radiated heat in the form of infrared light.

In contrast to passive infrared sensors, an active *infrared distance sensor* sends invisible light and tests whether it reflects back. It can tell if something is closer than a given distance. For example, an active infrared sensor could tell you that there is an object closer than 30 cm, but it wouldn't know if it's 5 cm or 29 cm away. As a rare exception, some sensors estimate distance from reflected infrared light.

A common use for active infrared is an automatic faucet and automatic hand dryer in a public toilet. Some automatic trashes open their lids when you go near them. Infrared light makes things more hygienic, as you don't have to touch objects that many other people have touched.

Long-distance range finders can use a laser beam to measure distance. Most of them are based on factoring in the speed of light and the time it takes for a beam to be reflected. Because light is very fast, the circuit must be able to do very precise timing. This makes them quite expensive (prices start from $100 USD). They are far less commonly used for prototyping with Arduino or Raspberry Pi than sound and IR.

Experiment: Measure Distance with Ultrasonic Sound (PING)

Ping, 1, 2, 3... pong. An ultrasonic sensor sends a sound, and then measures the time for the echo to return. Because you know that sound moves at about 330 meters per second, your program can calculate the distance.

Nowadays, there are many cheap ultrasonic sensors inspired by the Ping sensor from Parallax (Figure 3-1). Later in this chapter, you'll see some code for one of these cheap sensors, the HC-SR04 ("HC-SR04 Ultrasonic Sensor" on page 38). To better understand all the other ultrasonic sensors similar to Ping, it's useful to be familiar with the original, so we'll show you some code for that next. Also, many universities, hackerspaces, and makerspaces already have Ping sensors in their collections, so it's a good one to know.

To understand how an ultrasonic sensor measures distance, see "Echo Calculations Explained" on page 42.

Ping is an older, popular sensor by Parallax. Compared with the alternatives, it's a bit expensive, about $30 USD. If you need a lot of distance sensors, you might want something cheaper, but if you're just buying one, Ping is a great choice. The similar HC-SR04 costs only a couple of dollars, and the only difference in configuration between the Ping and HC-SR04 is one pin. (HC-SR04 uses one pin to trigger sending a pulse and another to read the echo.) The sensors have almost identical code.

Figure 3-1. *Ping sensor*

Ping Code and Connections for Arduino

Figure 3-2 shows the wiring diagram for the Ping sensor and Arduino. Build the circuit, and then compile and upload the code using the Arduino IDE.

You can download the example code from http://makesensors.botbook.com.

To see the readings, use the serial monitor (Arduino IDE→Tools→Serial Monitor). If you get gibberish instead of text, make sure that you specify the same speed (bit/s or "baud") in both your code (`Serial.begin`) and the Arduino Serial Monitor.

Even though there is a lot of code in this example, it's easy to reuse it so you can measure distance in your own projects. Just copy the supporting parts of the code (the `distanceCm()` function and global variables) and paste them into your own code. You can then measure distance with this line of code:

```
int d=distanceCm();
```

Because Ping works by listening to the echo of sound, its placement is quite important. If you always get the same reading (such as 2 cm), make sure that the wide beam of sound isn't bouncing off of something, like the edge of your breadboard or a table. If you put Ping onto the edge of a breadboard, you're not going to get reflections from it.

You can easily put Ping farther away from Arduino by using a servo extension cable, of the male-female type. Ping has just three pins, so it fits this type of cable perfectly.

Example 3-1 shows the complete code for reading a Ping ultrasonic distance sensor.

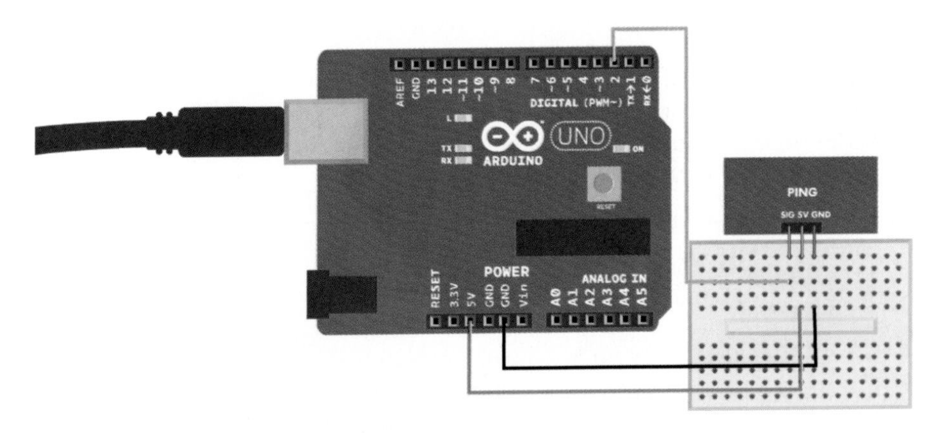

Figure 3-2. *Ping sensor circuit for Arduino*

Example 3-1. distance_ping.ino

```
// distance_ping.ino - distance using ultrasonic ping sensor
// (c) BotBook.com - Karvinen, Karvinen, Valtokari

int pingPin = 2;
float v=331.5+0.6*20; // m/s  // ❶

void setup()
{
  Serial.begin(115200);
}

float distanceCm(){
  // send sound pulse
  pinMode(pingPin, OUTPUT); // ❷
  digitalWrite(pingPin, LOW);
  delayMicroseconds(3); // ❸
  digitalWrite(pingPin, HIGH);
  delayMicroseconds(5); // ❹
  digitalWrite(pingPin, LOW);

  // listen for echo
  pinMode(pingPin, INPUT);
  float tUs = pulseIn(pingPin, HIGH); // microseconds  // ❺
  float t = tUs / 1000.0 / 1000.0 / 2; // s  // ❻
  float d = t*v; // m  // ❼
```

```
  return d*100; // cm
}

void loop()
{
  int d=distanceCm();   // ❽
  Serial.println(d, DEC);        // ❾
  delay(200); // ms     // ❿
}
```

❶ Calculate the speed of sound v for temperature 20 C (if your ambient temperature is significantly different, change 20 to the ambient temperature in C). The speed is about 340 meters per second or 1200 km/h.

❷ Ping uses the same pin for input and output.

❸ Wait for the pin to settle. 1 µs == 1 millionth of a second, or 1e-6 s == 0.000001 s

❹ Send a very short beep. 5 µs, or 5e-6 s

❺ Measure how long it takes for pingPin (D2) to go LOW, in microseconds.

❻ Convert to SI (Système Internationale, metric) base units, seconds (see *http://en.wikipedia.org/wiki/SI_base_unit*). Notice we're using a floating point divider (1000.0) instead of integer 1000 so that we get a floating point result. This one-way time is half of the round trip.

❼ Distance is time multiplied by speed.

❽ Measure distance and save it to a new variable, d. This is how you'd use it in your own code.

❾ Print the value of d to the Serial Monitor.

❿ Always have some delay in your loops. If you run your sketch without pausing, you'll be taxing the Arduino CPU and wasting power (doing anything as fast as possible can take 100% of power on any single-core CPU).

Ping Code and Connections for Raspberry Pi

Build the circuit for Ping in Raspberry Pi as shown in Figure 3-3, and then run the code listed in Example 3-2.

Be careful when connecting anything to the GPIO header. A wrong connection can easily damage (at best) one pin or (at worst) your whole Raspberry Pi. You can avoid problems by disconnecting power when making or changing connections, and double-checking connections to the pins before powering up.

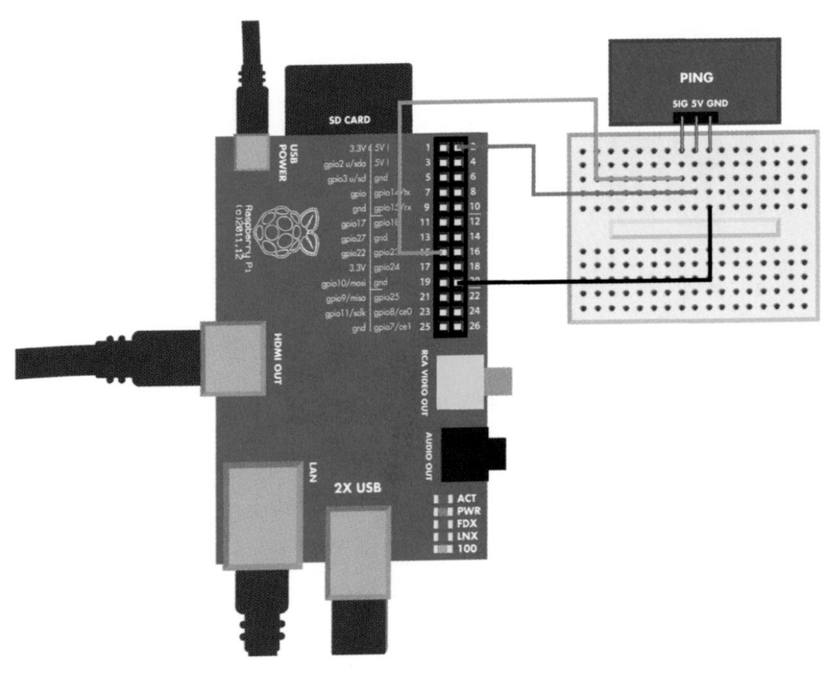

Figure 3-3. *Ping sensor circuit for Raspberry Pi*

Example 3-2. distance_ping.py

```
# distance_ping.py - print distance
# (c) BotBook.com - Karvinen, Karvinen, Valtokari
import time # ❶
import botbook_gpio as gpio # ❷

def readDistanceCm():
        sigPin=22
        v=(331.5+0.6*20)

        gpio.interruptMode(sigPin, "both")      # ❸

        gpio.mode(sigPin, "out")        # ❹
        gpio.write(sigPin, gpio.LOW)    # ❺
        time.sleep(0.5) # s

        gpio.write(sigPin, gpio.HIGH)   # ❻
        time.sleep(1/1000.0/1000.0)     # ❼
        gpio.mode(sigPin, "in") # ❽

        #Read high pulse width
        t = gpio.pulseInHigh(sigPin) # s        # ❾
        d = t*v
        d = d/2 # ❿
        return d*100    # cm
```

```
def main():
        d = readDistanceCm()    # ⓫
        print "Distance is %.2f cm" % d # ⓬
        time.sleep(0.5)

if __name__ == "__main__":
        main()
```

❶ Importing the time library creates a *namespace* of the same name (time) that contains the library's function, so this line lets you invoke `time.sleep(1)` later in your code.

❷ To import your own libraries, they must be in the same directory. So make sure that *botbook_gpio.py* is in the same directory as *distance_ping.py*. You can find this directory in the sample code available from *http://makesensors.botbook.com*. (See "GPIO Without Root" on page 19 for information on configuring your Raspberry Pi for GPIO access.)

❸ The interrupt mode `both` means that `pulseInHigh()` will measure a whole pulse from the signal's rising edge (from 0 to 1) to the falling edge (from 1 back to 0).

❹ With the Ping sensor, we switch the same pin between "out" and "in" mode as needed. Other sensors, such as the HC-SR04, use separate pins for each function.

❺ Turn off the pin and wait for the pin to settle. Half a second is a safe amount of time.

❻ Start the pulse (rising edge). This is where the time-critical code starts.

❼ Wait for a microsecond (1e-6 s), or one millionth of a second.

❽ Set the pin to "in" mode. This has the side effect of turning off the pulse, creating the falling edge of the short pulse.

❾ Read the pulse width in seconds. `gpio.pulseInHigh()` measures the length of the whole pulse, from start (rising edge) to finish (falling edge). Raspbian runs a whole operating system, so timing is not as precise as with Arduino. Other programs running on the system can affect the timing.

❿ One-way distance is half of round trip.

⓫ This is the line you need for measuring distance in your own programs.

⓬ Print the distance to the terminal window that this program is running in. The "%.2f" is part of the format string. It marks a place for the variable d. "%f" is floating point (decimal), and ".2" means to show two decimal places. If you just used "print d," you would get a very long decimal number.

HC-SR04 Ultrasonic Sensor

The HC-SR04 is just like the Ping but is available at a fraction of the cost. The code for this sensor is almost the same as Ping code, except the HC-SR04 uses separate pins for triggering the sound and listening for the echo. For detailed code explanations, see "Ping Code and Connections for Arduino" on page 33 and "Ping Code and Connections for Raspberry Pi" on page 35; the explanations in this section will focus on the differences.

> *To understand how an ultrasonic sensor measures distance, jump ahead to "Echo Calculations Explained" on page 42.*

Figure 3-4. *HC-SR04 ultrasonic sensor*

HC-SR04 Code and Connection for Arduino

Build the circuit as shown in Figure 3-5 and upload the code.

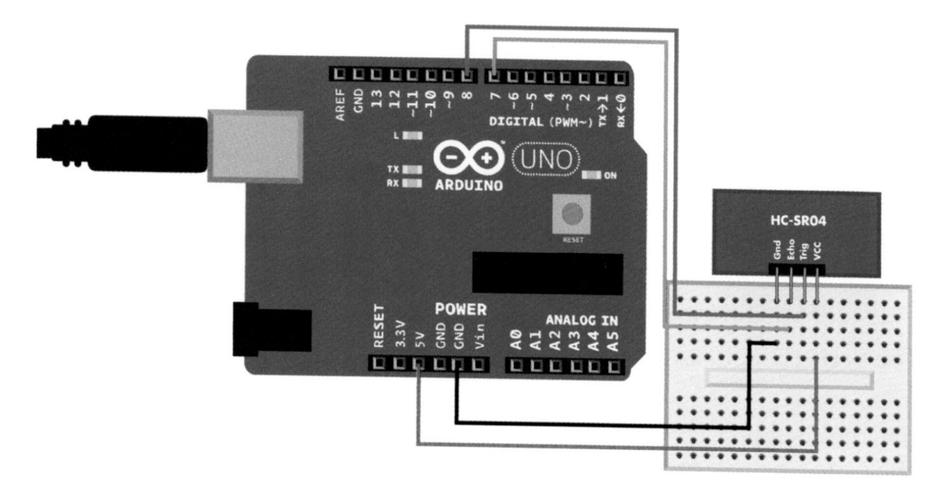

Figure 3-5. *HC-SR04 sensor circuit for Arduino*

Example 3-3. **hc-sr04.ino**

```
// hc_sr04.ino - print distance to serial
// (c) BotBook.com - Karvinen, Karvinen, Valtokari

int trigPin = 8;
int echoPin = 7;
float v=331.5+0.6*20; // m/s

void setup()
{
  Serial.begin(115200);
  pinMode(trigPin, OUTPUT);       // ❶
  pinMode(echoPin, INPUT);        // ❷
}

float distanceM(){
  // send sound pulse
  digitalWrite(trigPin, LOW);
  delayMicroseconds(3);
  digitalWrite(trigPin, HIGH);
  delayMicroseconds(5);
  digitalWrite(trigPin, LOW);

  // listen for echo
  float tUs = pulseIn(echoPin, HIGH); // microseconds
  float t = tUs / 1000.0 / 1000.0 / 2; // s
  float d = t*v; // m
  return d*100; // cm
}

void loop()      // ❸
{
```

```
  int d=distanceM();
  Serial.println(d, DEC);
  delay(200); // ms
}
```

❶ With the Ping, we didn't follow the normal practice of setting pin modes in the `setup()` function because we had to keep changing them (the Ping has one pin used for both triggering a pulse and reading the reflection). HC-SR04 uses a pin labeled Trig for triggering the sound.

❷ The pin labeled Echo returns the time it took to read the reflected echo as a pulse length.

❸ Other than the changes to the setup, using HC-SR04 in your main program looks a lot like the code you used for the Ping sensor.

HC-SR04 Code and Connections for Raspberry Pi

Build the circuit (Figure 3-6) and upload the code shown in Example 3-4. Take notice that in addition to jumper wires, you also need to add two 10 kOhm resistors. (To identify resistors, you can use the method described in "The Third Stripe Rule" on page 19.) The code is very similar to Ping.

Example 3-4. hc-sr04.py

```
# hc-sr04.py - print distance to object in cm
# (c) BotBook.com - Karvinen, Karvinen, Valtokari
import time
import botbook_gpio as gpio

def readDistanceCm():
        triggerPin = 22 # ❶
        echoPin = 27

        v=(331.5+0.6*20) # m/s

        gpio.mode(triggerPin, "out")

        gpio.mode(echoPin, "in")
        gpio.interruptMode(echoPin, "both")

        gpio.write(triggerPin, gpio.LOW)
        time.sleep(0.5)

        gpio.write(triggerPin, gpio.HIGH)
        time.sleep(1/1000.0/1000.0)
        gpio.write(triggerPin, gpio.LOW)

        t = gpio.pulseInHigh(echoPin) # s
```

```
        d = t*v
        d = d/2
        return d*100    # cm

def main():
        d = readDistanceCm()    # 2
        print "Distance is %.2f cm" % d
        time.sleep(0.5)

if __name__ == "__main__":
        main()
```

1. The only difference from the Ping example is that HC-SR04 uses two pins (Trig and Echo).
2. You can read a result from the HC-SR04 just as you would with the Ping.

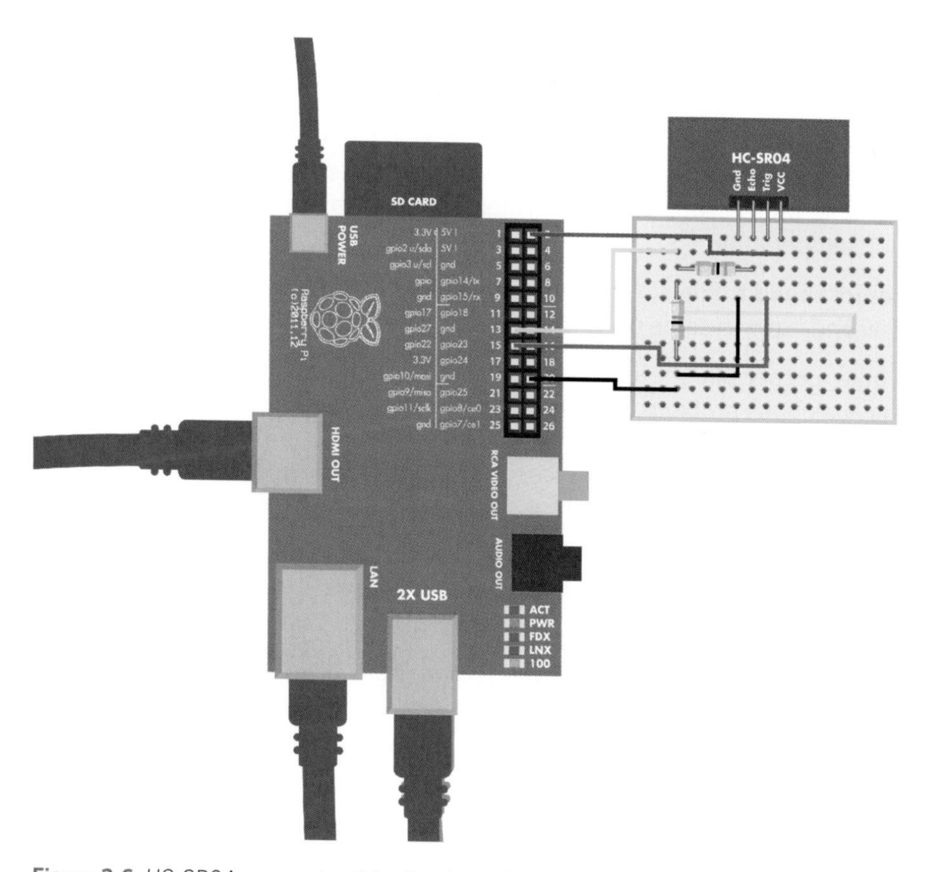

Figure 3-6. *HC-SR04 sensor circuit for Raspberry Pi*

Why does the HC-SR04 need a resistor, but the Ping doesn't? According to the HC-SR04 data sheet, its output is TTL level, which means +5 V. The data sheet for Ping promises compatibility with +3.3 V logic level. We verified these maximum values by measuring the output. The Raspberry Pi's GPIO pins' maximum voltage is +3.3 V, and sending +5 V to it would damage the pins or the Raspberry Pi.

Echo Calculations Explained

Thunderstorm nearby? You can estimate the distance to where lightning is striking by comparing the time between seeing lightning and hearing the thunder. Count the number of seconds after you see a lightning flash. Each second corresponds to 330 meters to the location of the strike (the actual number differs depending on the air temperature; we'll get into this shortly).

To play with the math, 330 meters/second * 3 seconds = 990 meters/second, so the sound moves roughly one kilometer in 3 seconds (and roughly one mile in 5 seconds). We see light nearly instantly, but the sound takes more time to reach us.

An ultrasonic sensor typically measures distances from 3 cm to 6 m. If you buy a distance measuring tool ("ultrasonic tape measure") from a hardware store, it can go farther, about 20 m (it uses a cone to project sound and a thermometer to calibrate the speed of sound for the current air temperature).

The time it takes for sound to travel 1 centimeter is very short, just 30 microseconds: 30 millionths of a second. How to come up with this number?

For a difficult problem, especially one involving very small or very large numbers, it's helpful to model it as an analogous problem using familiar, everyday quantities. For example, if I drive for two hours (t) at the speed (v) of 50 km per hour, isn't that an annoyingly slow trip? But with t and v, you can calculate the distance (d):

```
t = 2 h

v = 50 km/h

d = t*v = 2 h * 50 km / h =
    2*50 km * h/h = 100 km
```

That seemed easy. Now we know that two hours is equivalent to 100 km in this system. Let's try the exact same formula with a very short time (3.33 milliseconds), much faster speed, and base units of meters and seconds. (A milli- prefix means one thousandth, so a millisecond is one thousandth of a second.)

```
t = 3.33 ms = 0.00333 s

v = 330 m/s

d = t*v = 0.00333 s * 330 m/s = 1.10 m
```

This is how you can measure distance in your program with the ultrasonic ping sensor: if it takes 3.33 milliseconds for the reflected sound to return to you, it's traveled 1.1 m.

If you read code written by others, you might also see someone count a pace for sound. Instead of noting meters per second, many examples count the inverse, seconds per meter, expressed as *milli*seconds/meter here:

```
1/v = 1/(330 m/s) = 0.00303 s/m = 3.03 ms/m
```

It takes about 3 milliseconds for sound to move a meter.

Sound moves faster when it's warm. Sound is the vibration of air, and the vibrations move better if air molecules are already vibrating with heat. If you live in a warm place, we envy you because you probably need less calibration. In the north of Finland, it might be +22 C inside and -40 C outside, resulting in over 60 C difference in temperature. A change this big will clearly affect measurements. Temperature (T) affects the speed of sound (r) according to the formula

```
v = (331.3+0.606*T) m/s
```

This formula gives the speed of sound in practice (343 m/s at 20 C). If you start getting fancy with it, you could be calibrating for many factors. If you climb to a mountain or live in a submarine, you must also take into account the change in air pressure. If you go from the Sahara to a laundry room, calibrate for air humidity, too. That said, common commercial ultrasonic distance measuring tools tend to calibrate for temperature only.

When you use these calculations in your code, just put them in the beginning of the code. Arduino or Raspberry Pi can calculate them in an instant, and calculations outside `loop()` are performed only once anyway. Be sure to comment these calculations, and you will thank yourself when you can understand your code a week later.

Environment Experiment: Invisible Objects

You can easily fool an ultrasonic sensor so that it thinks there is nothing in front of it. Attach the sensor to a helping hand tool (aka third hand tool) and point it at a solid, flat object. Upload the code and open the serial monitor as you did earlier in this chapter. Now you should get a normal distance reading.

Next, try putting a soft pillow or similar plush object between the sensor and the solid object (see Figure 3-7). Check the serial monitor again. Is the solid object still there?

Inclined planes are another Achilles' heel of ultrasonic sound sensors. Remove the soft object and start tilting the solid flat object that is facing the sensor. Keep checking the serial monitor as you tilt the object to a steeper angle.

Why does this happen? Soft objects (like our Monty Python killer rabbit in Figure 3-7) absorb so much sound that there's not enough echo. On the other hand, an inclined plane echoes the sound, but in the wrong direction (not back at the sensor). This is similar to how a stealth aircraft fools radar.

Figure 3-7. *Testing ping sensor with a soft object*

Experiment: Detect Obstacles With Infrared (IR Distance Sensor)

An infrared switch (Figure 3-8) is more reliable than an ultrasonic one, but less versatile. You can't fool it as easily as you fooled ultrasound in the experiment you did earlier. But an infrared switch can tell you only if there is something present, not the distance to it. And because the sun is a great big source of infrared light, it's strong enough to blind an infrared switch.

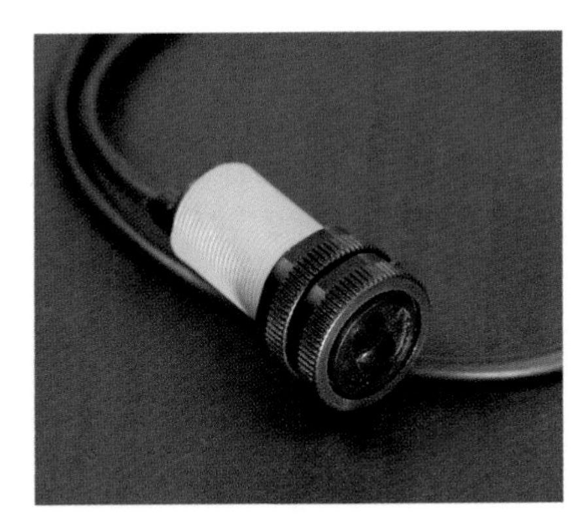

Figure 3-8. *Image of infrared sensor switch*

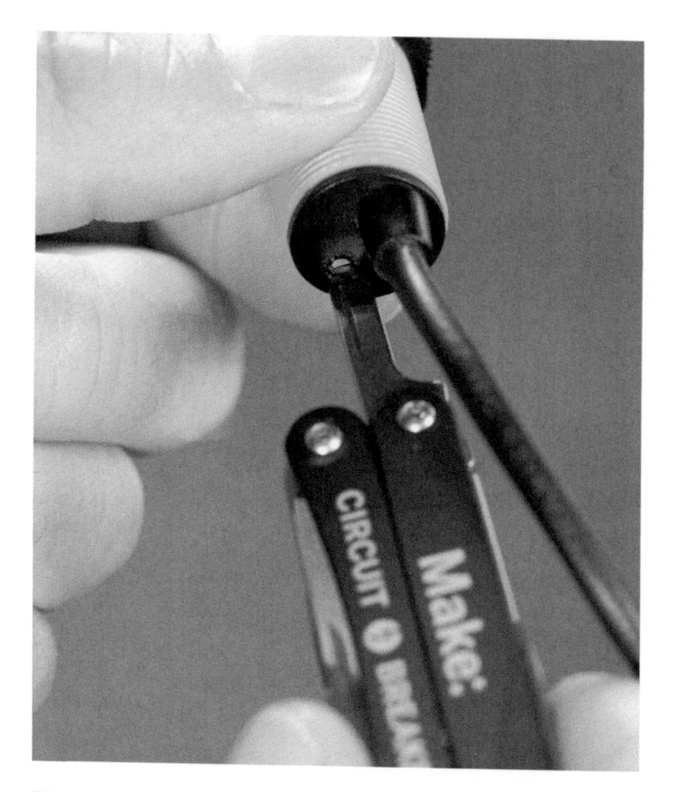

Figure 3-9. *You can adjust the distance to which the sensor detects obstacles*

IR Switch Code and Connections for Arduino

Figure 3-10 shows how to connect Arduino to the infrared sensor. The sketch is shown in Example 3-5.

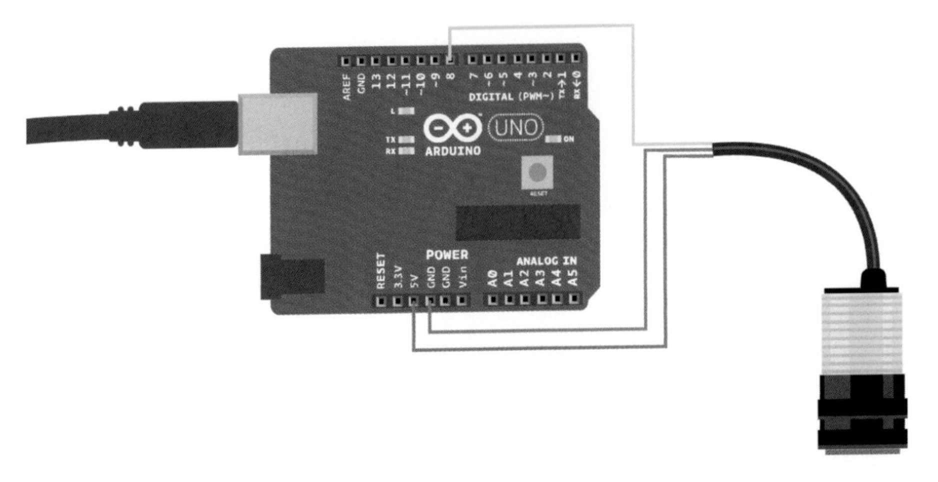

Figure 3-10. *Infrared sensor connections on Arduino*

Example 3-5. *adjustable_infrared_sensor_switch.ino*

```
// adjustable_infrared_sensor_switch.ino - print detection to serial and light LED.
// (c) BotBook.com - Karvinen, Karvinen, Valtokari

const int sensorPin = 8;
const int ledPin = 13;

//Sensor value
int switchState = 0;

void setup() {
  Serial.begin(115200); // ❶
  pinMode(sensorPin, INPUT);
  pinMode(ledPin, OUTPUT);
}

void loop() {
  switchState = digitalRead(sensorPin); // ❷
  Serial.println(switchState);  // ❸
  if(switchState == 0) {
    digitalWrite(ledPin, HIGH);
    Serial.println("Object detected!"); // ❹
  } else {
    digitalWrite(ledPin, LOW);
  }
  delay(10); // ms      // ❺
}
```

❶ Open the Arduino Serial Monitor (Tools→Serial Monitor). You must set the same speed in your code and in the serial monitor. The fastest is 115,200 bit/second. If you have an unreliable (or very long) USB cable, change this to 9,600 bit/second.

❷ An infrared sensor switch is just like a button. This is the line that reads the sensor.

❸ Print the sensor pin state for debugging purposes.

❹ A state of 0 means that an object is detected. We turn Arduino's built-in LED on to indicate that an object was detected.

❺ You should always have some, even tiny, delay in `loop()`. This will prevent the sketch from using 100% of the Arduino's CPU all the time.

IR Switch Code and Connections for Raspberry Pi

Figure 3-11 shows the wiring diagram for Raspberry Pi and the switch. The corresponding Python code is in Example 3-6.

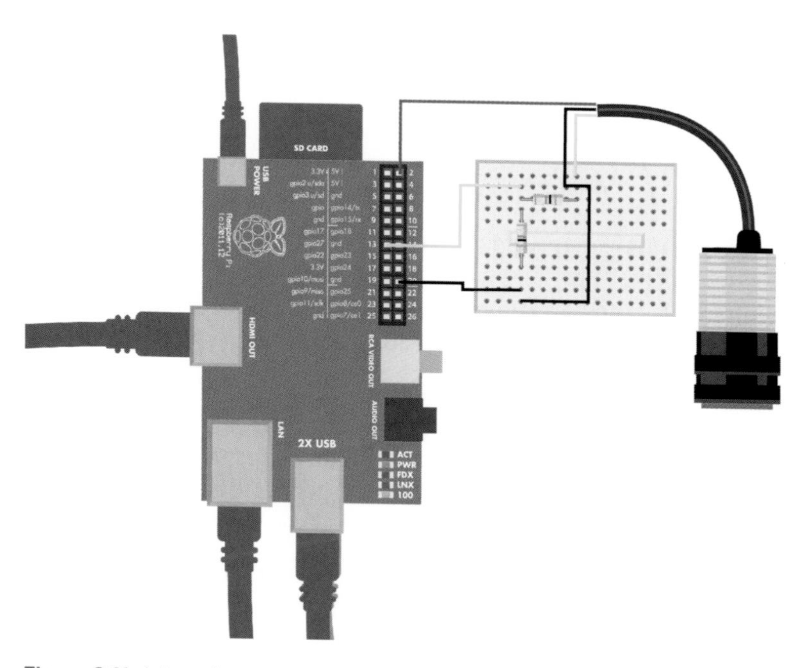

Figure 3-11. *Infrared sensor connections on Raspberry Pi*

Example 3-6. *adjustable-infrared-sensor-switch.py*

```
# adjustable-infrared-sensor-switch.py - read infrared switch
# (c) BotBook.com - Karvinen, Karvinen, Valtokari
import time
import botbook_gpio as gpio     # ❶

def main():
        switchPin = 27
        gpio.mode(switchPin, "in")       # ❷
        x = gpio.read(switchPin)         # ❸
        if( x == gpio.LOW ):    # ❹
                print "Something is inside detection range"
```

```
        else:
                print "There is nothing inside detection range"
        time.sleep(0.1)

if __name__ == "__main__":
        main()
```

❶ Import the *botbook_gpio* library. It must be in the same directory as this code, so make sure that *botbook_gpio.py* is in the same directory as *adjustable-infrared-sensor-switch.py*. You can find this directory in the sample code available from *http://makesensors.botbook.com*. (See "GPIO Without Root" on page 19 for information on configuring your Raspberry Pi for GPIO access.)

❷ Configure the pin that the switch is connected to; this puts it into input mode.

❸ Read the state of the pin, and store it in the variable x.

❹ If the pin is low, it means an object was detected in range.

Environment Experiment: How to See Infrared

As we mentioned earlier, infrared is outside the range of visible light. How could you see it if you were determined to do so? You could use night-vision goggles or, if you don't have any of those handy, any cheap digital camera.

Try looking at an IR sensor through your smartphone's camera. You should be able to see a violet glow on the IR emitter Figure 3-12. Dim the lights and close the curtains to make it even more visible. It's not as cool as the night vision, but it is a quick and easy way to see that the emitter is working.

What if you want to try this with an expensive SLR camera? Be aware that some cameras have strong infrared filters that prevent unwanted wavelength from being part of your photos (it works with cheaper cameras because the infrared filters aren't as good, so some infrared light manages to excite some of the sensors in your camera). If this is the case, you can go to a dark room and set your camera to a very slow shutter speed (possibly many seconds) and take a picture using a tripod. As the IR is the only light in the room, it will eventually appear in the picture.

If you happen to have night-vision goggles, they are perfect for observing how IR sensors work. Night-vision goggles amplify visible light, but they are especially greedy for infrared spectrum. The cheapest models actually rely solely on an IR beam they produce themselves. If the model you're using has an IR emitter, be sure to turn it off (or tape it over) in order to see the weaker light from your IR distance sensor. The fun in using night-vision goggles is not just seeing that your sensor works, but also the reflections that IR light creates off of other objects (Figure 3-13).

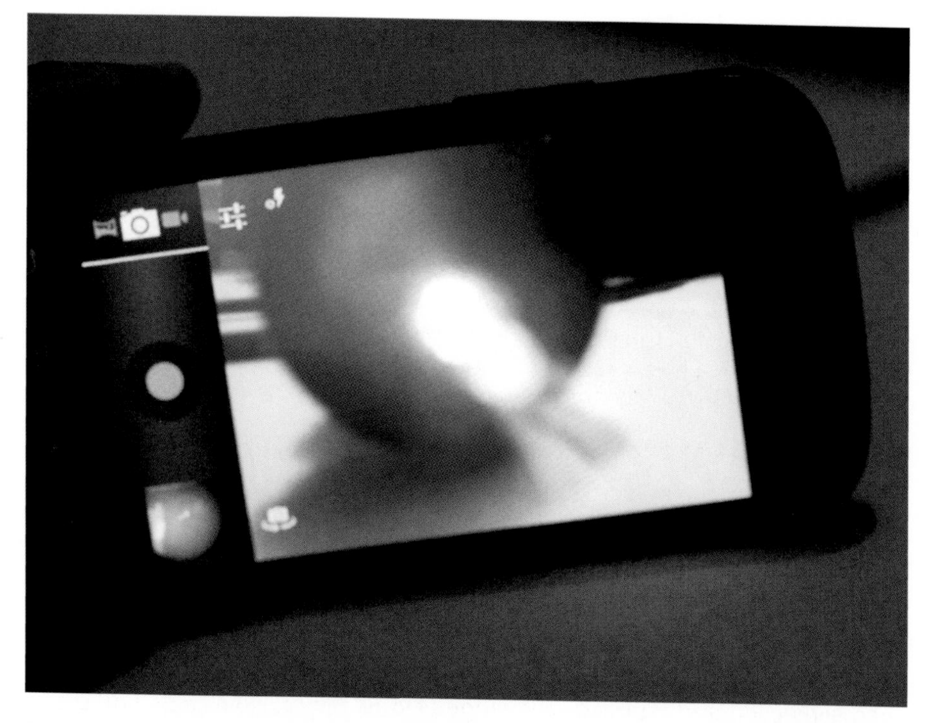

Figure 3-12. *How a phone's camera sees IR sensors*

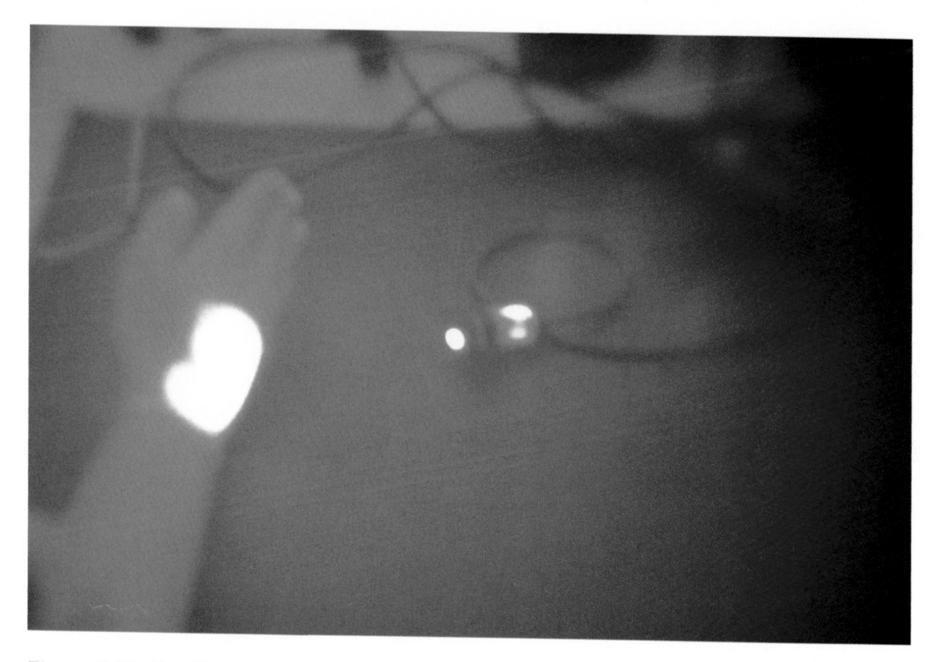

Figure 3-13. *The IR sensor seen through night-vision goggles*

Experiment: Follow Movement with Infrared (IR Compound Eye)

A compound eye has many infrared-sensitive transistors and LEDs. It can track movement within 20 centimeters. Even though it's one sensor, each of the infrared (IR) light-sensitive transistors can be read separately. Ambient light correction is done by turning off the IR LEDs and comparing values.

Figure 3-14. *Image of IR Compound Eye*

The exact name of this sensor is "IR Compound Eye." If you ever start inventing and selling sensors, please give them a unique code name in addition to such a general name. A unique code will make it much easier to search for the component.

Figure 3-15. *Real compound eye*

If you want to improve readings from your compound eye, you must calibrate it. Wait for the night when there is no ambient IR light. Even closed window shades can pass IR light, so if you can't wait for the night, go to your cellar or a windowless room. Build the circuit shown in Figure 3-16 so that you can measure values. Put paper in front of the sensor (about 20 cm away) and see how much the values differ for each pin. Values should be almost the same (+/- 100) with the paper. If one of the values is too high, you can block some of the IR light with opaque tape or shrink wrap. If the value is too low, block some IR light going to other sensors.

When you use this sensor, you're measuring analog resistance. For the simplest example of reading analog resistance, see "Experiment: Potentiometer (Variable Resistor, Pot)" on page 98.

Compound Eye Code and Connection for Arduino

Figure 3-16 shows the wiring diagram for Arduino and the compound eye. The Arduino sketch is shown in Example 3-7.

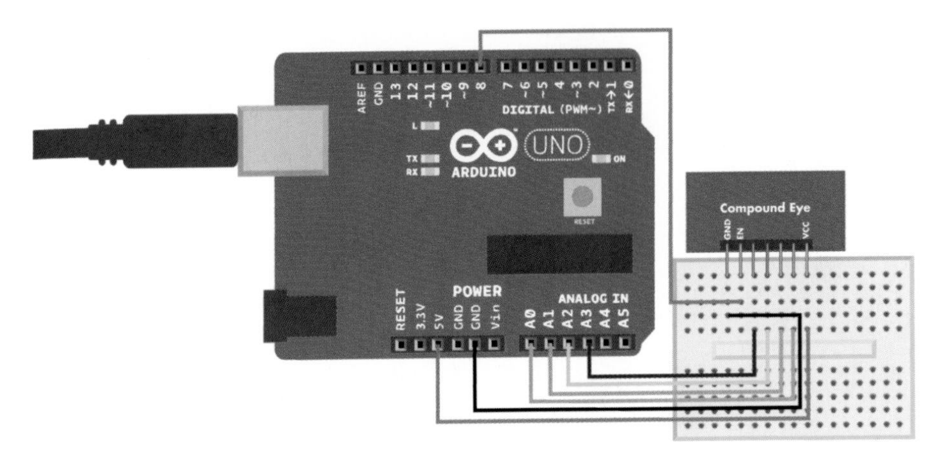

Figure 3-16. *Compound eye connections on Arduino*

Example 3-7. **compound_eye.ino**

```
// compound_eye.ino - print distance and direction values to serial
// (c) BotBook.com - Karvinen, Karvinen, Valtokari

const int irEnablePin = 8;      // ❶
const int irUpPin = 0;
const int irDownPin = 2;
const int irLeftPin = 1;
const int irRightPin = 3;

int distance = 0;        // ❷
int irUpValue = 0;
int irDownValue = 0;
int irLeftValue = 0;
int irRightValue = 0;

void setup() {
  Serial.begin(115200);
  pinMode(irEnablePin, OUTPUT);
}

void loop() {
  readSensor(); // ❸
  Serial.print("Values: "); // ❹
  Serial.print("irUpValue"); Serial.print(irUpValue); Serial.print(",");
  Serial.print("irDownValue"); Serial.print(irDownValue); Serial.print(",");
  Serial.print("irLeftValue"); Serial.print(irLeftValue); Serial.print(",");
  Serial.print("irRightValue"); Serial.print(irRightValue); Serial.print(",");
  Serial.print("distance"); Serial.println(distance);
  delay(100);
}

void readSensor() {
  digitalWrite(irEnablePin, HIGH);      // ❺
```

```
    delay(5); // ms         // ❻
    irUpValue = analogRead(irUpPin);
    irDownValue = analogRead(irDownPin);
    irLeftValue = analogRead(irLeftPin);
    irRightValue = analogRead(irRightPin);

    int ambientLight = 0; // ❼
    digitalWrite(irEnablePin, LOW);         // ❽
    delay(5);
    ambientLight = analogRead(irUpPin);    // ❾
    irUpValue = irUpValue - ambientLight; // ❿

    ambientLight = analogRead(irDownPin);
    irDownValue = irDownValue - ambientLight;

    ambientLight = analogRead(irLeftPin);
    irLeftValue = irLeftValue - ambientLight;

    ambientLight = analogRead(irRightPin);
    irRightValue = irRightValue - ambientLight;

    distance = (irUpValue+irDownValue+irLeftValue+irRightValue) / 4;        // ⓫
}
```

❶ Pin numbers are declared constant (`const`), so you won't be able to change them elsewhere in your sketch (if you try to write code that reassigns them, even—perhaps especially—by accident, you'll get an error when you try to verify or upload the sketch).

❷ Sensor values will be stored in global variables. Globals are available in all functions. In C and C++ (which Arduino is based on), it's good practice to initialize variables at the same time you declare them (for example, `int foo = 0;`).

❸ `readSensor()` will not return any values. Instead, it modifies some global variables. This way, it can modify multiple values when you run it.

❹ Print the results. `Serial.print()` doesn't add a newline (but `Serial.println()` will).

❺ Turn on the IR LED to illuminate the target for measurement.

❻ Wait for the inputs to settle.

❼ Start measuring ambient light (for example, the invisible IR light from the sun).

❽ Turn off the IR LED. All light detected now is from ambient sources.

❾ Use each of the IR-sensitive transistors to measure ambient light.

❿ Remove the ambient light value from each of the sensors.

⓫ Calculate the average distance over all four IR-sensitive transistors.

Compound Eye Code and Connections for Raspberry Pi

The IR compound eye contains eight infrared-sensitive sensors, which are connected in pairs, so there is a total of four sensors you can read. Each IR sensor is read as an analog resistance sensor.

The Raspberry Pi requires an external ADC (analog-to-digital converter) to read the IR sensors. One MCP3002 chip can read two analog inputs. Because we need to read four sensors, we use two MCP3002 chips.

This circuit (shown in Figure 3-17) has many things in it, but the principle is simple: there are four analog resistance sensors, and you read them one by one. Build the circuit as shown, and then run the code shown in Example 3-8.

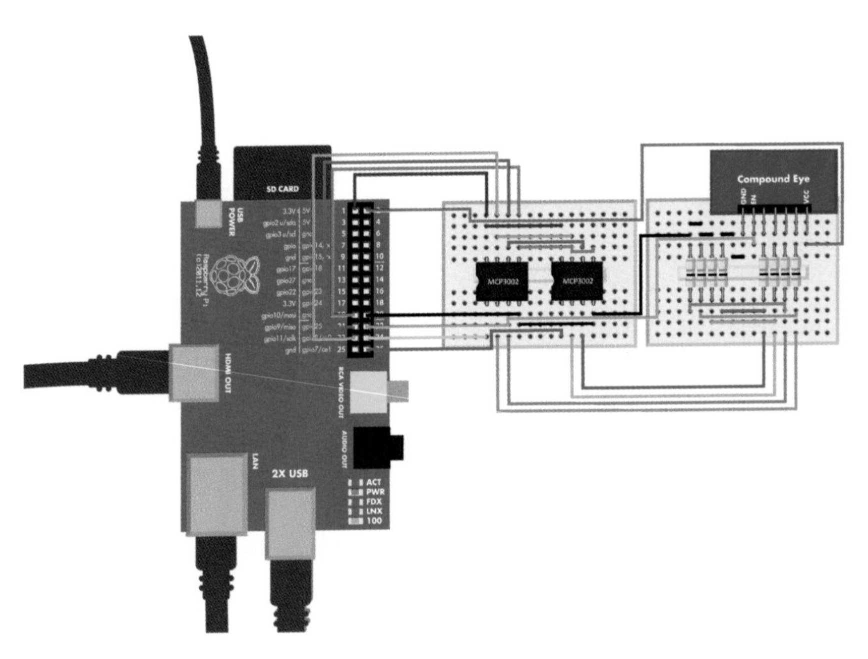

Figure 3-17. *Compound eye connections on Raspberry Pi*

Example 3-8. compound_eye.py

```
# compound_eye.py - read distance and direction.
# (c) BotBook.com - Karvinen, Karvinen, Valtokari

import time
import botbook_gpio as gpio      # ❶
import botbook_mcp3002 as mcp    # ❷

irUpValue = 0   # ❸
irDownValue = 0
irLeftValue = 0
irRightValue = 0
```

```
distance = 0

def readCompoundEye():
        global irUpValue,irDownValue,irLeftValue,irRightValue,distance  # ❹
        ledPin = 25
        gpio.mode(ledPin, "out")        # ❺
        gpio.write(ledPin, gpio.HIGH)
        #Wait for sensors to get ready
        time.sleep(0.05)        # ❻

        irUpValue = mcp.readAnalog(0, 0)        # ❼
        irDownValue = mcp.readAnalog(1, 0)
        irLeftValue = mcp.readAnalog(0, 1)
        irRightValue = mcp.readAnalog(1, 1)

        ambientLight = 0
        gpio.write(ledPin, gpio.LOW)        # ❽
        time.sleep(0.05)
        ambientLight = mcp.readAnalog(0, 0)        # ❾
        irUpValue = irUpValue - ambientLight        # ❿
        ambientLight = mcp.readAnalog(1, 0)        # ⓫
        irDownValue = irDownValue - ambientLight
        ambientLight = mcp.readAnalog(0, 1)
        irLeftValue = irLeftValue - ambientLight
        ambientLight = mcp.readAnalog(1, 1)
        irRightValue = irRightValue - ambientLight

        distance = (irUpValue+irDownValue+irLeftValue+irRightValue)/4   # ⓬

def main():
        global irUpValue,irDownValue,irLeftValue,irRightValue,distance
        while True:     # ⓭
                readCompoundEye()       # ⓮
                print "Values:"
                print "Up: %f" % irUpValue
                print "Down: %f" % irDownValue
                print "Left: %f" % irLeftValue
                print "Right: %f" % irRightValue
                print "Distance: %f" % distance
                time.sleep(0.5) # s     # ⓯

if __name__ == "__main__":
        main()
```

❶ Import `gpio` for turning digital pins (in this case, gpio25) on and off. The *botbook_gpio.py* file from the book's sample code must be in the same directory as this program (*compound_eye.py*).

❷ Import the `mcp3002` library for reading analog sensor values using MCP3002 analog-to-digital converter chip (ADC). You'll use it for reading values of each of the IR light sensitive transistors. The *botbook_mcp3002.py* library file must be in the same directory as *compound_eye.py*. You must also install the `spidev` library, which is imported by *botbook_mcp3002.py*. See the comments in the beginning of *botbook_mcp3002/botbook_mcp3002.py* or "Installing SpiDev" on page 56.

❸ Declare global variables.

❹ To use global variables inside a function, they must be listed at the beginning of the function.

❺ Turn on the gpio25 pin connected to IR LEDs. This will illuminate the target area for measurement.

❻ Wait for the pins to settle.

❼ Read each of the IR-sensitive transistor values. As Raspberry Pi has no built-in analog-digital conversion, we use the external MCP3002 chip.

❽ To remove the effect of ambient light, we turn the IR LED off.

❾ ⓫ The value from each IR-sensitive transistor is read again.

❿ The ambient light value is removed from the actual measurement (which was made with IR illumination).

⓬ The distance is the average of measurements from all four IR-sensitive transistors.

⓭ In embedded applications, `while(True)` is a common method to keep performing the same action forever. Most embedded devices are supposed to keep doing their thing, and are not expected to exit the program and stop functioning after a while. To kill a program running in a `while(True)` loop, press Control-C in the same terminal session you started it from.

⓮ The `readSensor()` function doesn't need to return any values, as it modifies global variables. In Python, you could alternatively return multiple values, as in `a,b,c=foo()`.

⓯ When running a loop, a small delay ensures that this trivial loop doesn't take 100% of the CPU time.

The compound eye has quite a few connections by itself. Combined with the ADCs, this means a lot of jumper wires.

Installing SpiDev

The MCP3002 analog-to-digital converter uses the SPI protocol. SPI is quite a complicated protocol, but you can install the SpiDev library to handle the details.

The SpiDev library is required by all code that uses `import spidev`. This includes the potentiometer code for Raspberry Pi, and every analog resistance sensor used in this book, because SpiDev is imported by *botbook_mcp3002*.

On your Raspberry Pi, open a terminal. First, install prerequisites:

```
$ sudo apt-get update
$ sudo apt-get -y install git python-dev
```

Download the latest version of SpiDev from its version control site:

```
$ git clone https://github.com/doceme/py-spidev.git
$ cd py-spidev/
```

And install it to your system:

```
$ sudo python setup.py install
```

Next, you need to enable the SPI module on the Raspberry Pi. First, make sure it is not disabled. Edit the */etc/modprobe.d/raspi-blacklist.conf* with the command `sudoedit /etc/modprobe.d/raspi-blacklist.conf` and *delete* this line:

```
blacklist spi-bcm2708
```

Save the file: press Control-X, type y, and then press Enter or Return.

To allow access to SPI without root, copy the udev file from Example 3-9 (or from the example code in the *botbook_mcp3002* directory) into place:

```
$ sudo cp 99-spi.rules /etc/udev/rules.d/99-spi.rules
```

Example 3-9. 99-spi.rules

```
# /etc/udev/rules.d/99-spi.rules - SPI without root on Raspberry Pi
# Copyright 2013 http://BotBook.com

SUBSYSTEM=="spidev", MODE="0666"
```

Reboot your Raspberry Pi, open LXTerminal, and confirm that you can see the SPI devices and that the ownership is correct:

```
$ ls -l /dev/spi*
```

The listing should show two files, and they should list permissions of `crw-rw-rwT`. If not, go over the preceding steps again.

Now you can use MCP3002 chip and other SPI devices with Raspberry Pi.

Alternative Circuits for Raspberry Pi

The Raspberry Pi circuit for IR compound eye is quite complicated. Even though it's not hard to understand, it has a lot of wires to connect. To build a simpler system, you could either use a another ADC or an Arduino (see "Pi + Arduino" on page 58).

To get by with just one ADC chip, you could use MCP3008 ADC, which has eight inputs. That would require modifying the *botbook_mcp3002* library, though. You could also look into using Adafruit's MCP3008 code (see *https://github.com/adafruit/Adafruit-Raspberry-Pi-Python-Code*).

Pi + Arduino

If you find the amount of jumpers excessive, an alternative is to read the sensor from Arduino, and send the result to the Raspberry Pi over a USB-serial connection. Another option is the Pi a la Mode (*http://bit.ly/1icPd0z*), an Arduino-compatible board that piggybacks on top of a Raspberry Pi to create the ultimate hybrid.

See "Talking to Arduino from Raspberry Pi" on page 337 for more details.

Test Project: Posture Alarm (Arduino)

Every geek knows this problem: the more intensely you work, the closer your head moves to the computer screen. This is not quite how Mother Nature intended our posture. By combining an IR distance sensor and a piezo beeper, you can easily build a gadget that will warn you when you are too close to the screen (Figure 3-18).

Figure 3-18. *Ready posture alarm*

What You'll Learn

In the *Posture Alarm* project, you'll learn how to:

- Combine input, processing, and output.
- Play tones with a piezo beeper.
- Enclose your project.

Piezo Beeper

A piezoelectric crystal changes shape when you apply voltage to it. By using alternating current (AC) or a simple on-off square wave, you can make the piezo crystal vibrate. This makes the air vibrate, and air vibration is sound. In this case, an annoying sound.

Figure 3-19. *A piezo speaker*

The most common piezo element beeps when you send it a square wave (Figure 3-19). A square wave varies between HIGH (5 V for Arduino) and LOW (0 V).

You can easily create a square wave by repeatedly turning a data pin on and off with `digitalWrite`. Alternatively, you could use the built-in `tone()` function that uses a more advanced and complicated implementation to produce the same wave.

> *The piezoelectric phenomena also works the other way: you can generate electricity by squeezing a piezoelectric crystal. Electric lighters often use the piezoelectric effect to create a spark.*

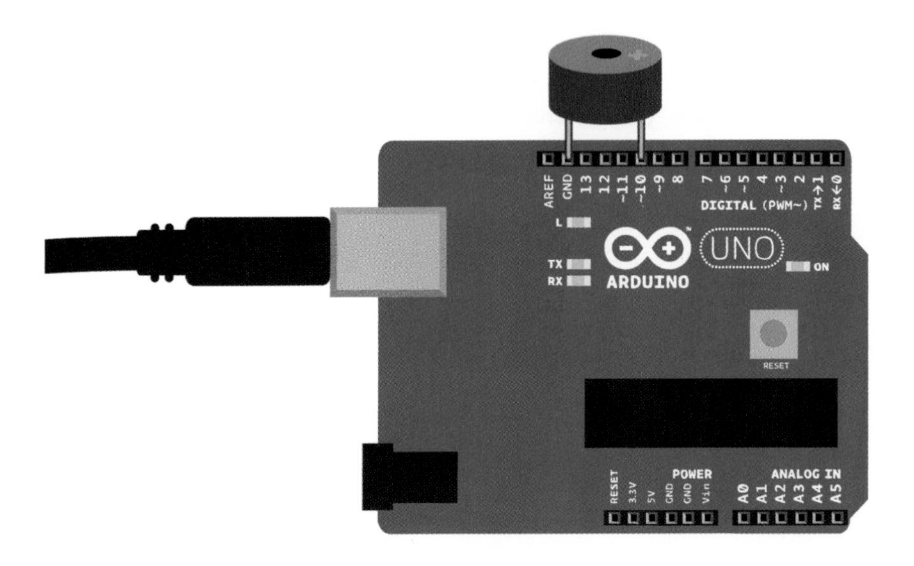

Figure 3-20. *Piezo beeper connected*

Example 3-10. *piezo_beep.ino*

```
// piezo_beep.ino - beep on a given frequency with a piezo speaker
// (c) BotBook.com - Karvinen, Karvinen, Valtokari

int speakerPin = 10;

void wave(int pin, float frequency, int duration)       // ❶
{
  float period=1/frequency*1000*1000; // microseconds (us)      // ❷
  long int startTime=millis();  // ❸
  while(millis()-startTime < duration) {        // ❹
    digitalWrite(pin, HIGH);    // ❺
    delayMicroseconds(period/2);
    digitalWrite(pin, LOW);
    delayMicroseconds(period/2);
  }
}

void setup()
{
 pinMode(speakerPin, OUTPUT);
}

void loop()
{
  wave(speakerPin, 440, 500);   // ❻
  delay(500);
}
```

❶ To use the wave function in your own projects, you just need to specify pin, note frequency, and duration in milliseconds. The contents of the function are implementation details: educational, but not required for using the function yourself.

❷ Calculate the period *T*: how long does one wave (one HIGH + one LOW) take? It is the inverse of frequency *f*: `T = 1/f`. For example, if you have 2 waves in one second (2 hertz, which is `2 * 1/s`), one wave will take half a second (`1 / 2 Hz`, or `1/(2 * 1/s)`, which works out to `1/2 s`). Note that hertz (number of cycles per second) is abbreviated Hz.

❸ This is a common pattern used in code to perform something for set time: first, save the starting time into a variable (`millis` gives the number of milliseconds since the Arduino powered up)...

❹ ...and wait until it's been more than the specified duration since the starting time.

❺ Create one whole wave. First create the high part, then the low part.

❻ This call is all you need for creating a beep in your own main program.

Alarm, Alarm!

It's time to advance from a beep to a whole new world of notes: the alarm.

Example 3-11. piezo_alarmtone.ino

```
// piezo_alarmtone.ino - use piezospeaker to sound alarm sound
// (c) BotBook.com - Karvinen, Karvinen, Valtokari

int speakerPin = 10;

void wave(int pin, float frequency, int duration)       // ❶
{
  float period=1 / frequency * 1000 * 1000; // microseconds (us)
  long int startTime=millis();
  while(millis()-startTime < duration) {
    digitalWrite(pin, HIGH);
    delayMicroseconds(period/2);
    digitalWrite(pin, LOW);
    delayMicroseconds(period/2);
  }
}

void setup()
{
 pinMode(speakerPin, OUTPUT);
}

void loop()
{
  wave(speakerPin, 440, 40);    // ❷
  delay(25);
  wave(speakerPin, 300, 20);
  wave(speakerPin, 540, 40);
```

```
  delay(25);
  wave(speakerPin, 440, 20);
  wave(speakerPin, 640, 40);
  delay(25);
  wave(speakerPin, 540, 20);
}
```

❶ Use the exact same `wave()` function you used to generate a simple beep.

❷ Play a note, wait for a very short time, play another… repeat forever.

Combining Piezo and IR Sensor

Build the posture alarm circuit (Figure 3-21). Upload the code and prepare for a life with better posture—or endless alarm.

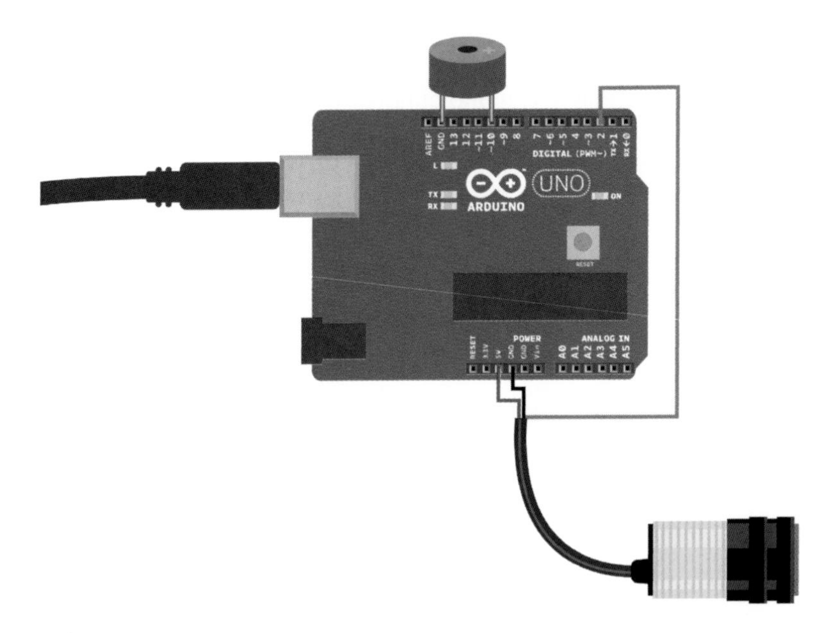

Figure 3-21. *Posture alarm build*

Example 3-12. *posture_alarm.ino*

```
// posture_alarm.ino - sound an alarm when IR switch detects bad posture
// (c) BotBook.com - Karvinen, Karvinen, Valtokari

int speakerPin = 10;
const int sensorPin = 2;
int switchState = 0;

void wave(int pin, float frequency, int duration)        // ❶
```

```
{
  float period=1/frequency*1000*1000; // microseconds (us)
  long int startTime=millis();
  while(millis()-startTime < duration) {
    digitalWrite(pin, HIGH);
    delayMicroseconds(period/2);
    digitalWrite(pin, LOW);
    delayMicroseconds(period/2);
  }
}

void alarm() // ❷
{
  wave(speakerPin, 440, 40);
  delay(25);
  wave(speakerPin, 300, 20);
  wave(speakerPin, 540, 40);
  delay(25);
  wave(speakerPin, 440, 20);
  wave(speakerPin, 640, 40);
  delay(25);
  wave(speakerPin, 540, 20);
}

void setup()
{
  pinMode(speakerPin, OUTPUT);
  Serial.begin(115200);
  pinMode(sensorPin, INPUT);
}

void loop()
{
  switchState = digitalRead(sensorPin);
  Serial.println(switchState,BIN);
  if (switchState==0) { // ❸
    alarm(); // ❹
  }
  delay(10);
}
```

❶ Use the same `wave()` function you've used before for creating beep and alarm.

❷ Use the same `alarm()` you used earlier.

❸ Use the `digitalRead()` function to get a value from the infrared distance switch, like you did before.

❹ If something is detected in the monitor's area, play the alarm. In practice, if you put your head too close to the monitor, the alarm will sound.

Putting Everything in a Neat Package

Prototypes are more impressive and robust when you put them inside an enclosure. We used a box for Arduino made by SmartProjects, because it happened to be an exact fit for this project. First we sprayed the whole thing black and made a hole for the infrared sensor with a 19 mm drill bit (see Figure 3-22).

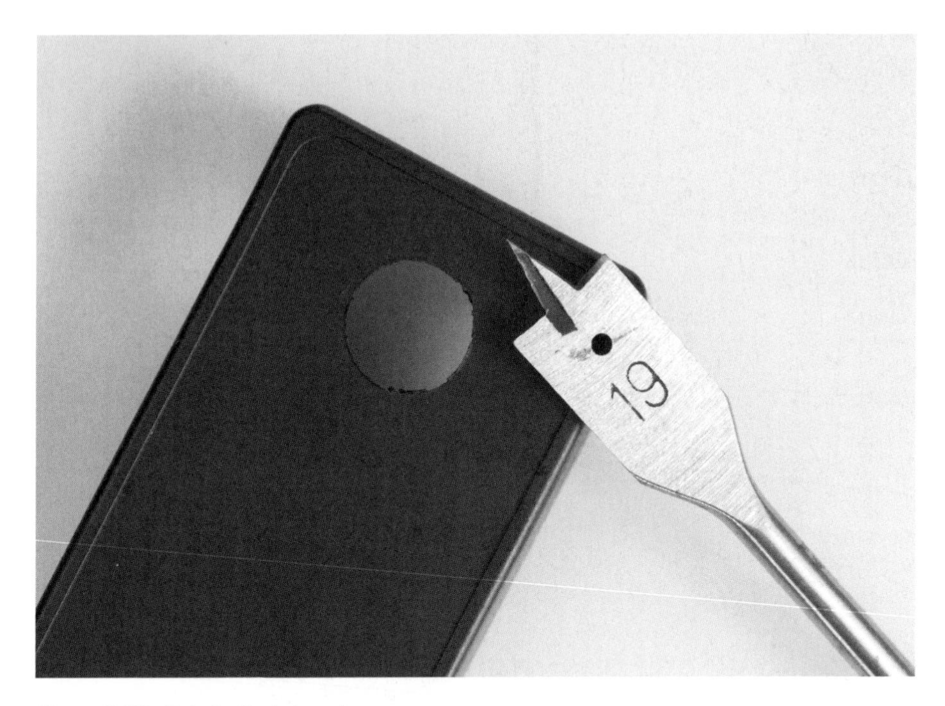

Figure 3-22. *Hole for the infrared sensor*

We removed the little hatch from the back of the chassis to make the sensor fit inside better and to be able to adjust the distance screw later (see Figure 3-23).

We attached the sensor in place with the plastic nuts that are part of it. You can slide the Arduino through the posts in the chassis, and it will stay put nicely, as shown in Figure 3-24.

Figure 3-23. *From the back of the chassis, you can adjust the distance screw*

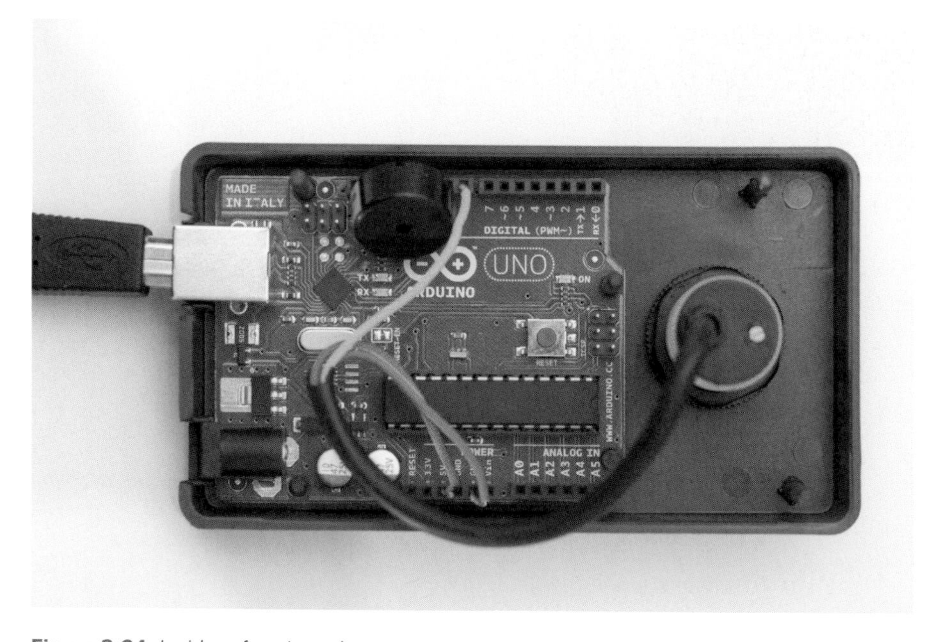

Figure 3-24. *Insides of posture alarm*

Close the enclosure by pushing the top and bottom half together, and you're ready to start protecting your posture (see Figure 3-18). This box has a ready-made hole for the USB cable, so now you actually have a neat-looking USB gadget to decorate your workspace.

You have now learned to measure distance using multiple methods. Your projects can know if something is near or how far it is to objects nearby. With more than one sensor, you can create more sophisticated behaviors. For example, combine two IR sensors with a servo motor and make it turn toward the direction of the nearest detected object. This way you would have a simple hand follower. Two IR receivers in a rover robot would enable it to follow flame. How will you use distance sensors in your projects?

Smoke and Gas 4

BEEP BEEP BEEP! The loud noise of a fire alarm has saved many lives, waking up residents before carbon monoxide lulls them into a permanent sleep. Another gas sensor, an alcometer, has kept many drunk drivers off the road and avoided lethal consequences.

When you start yawning at work or in an otherwise interesting class, the culprit could be carbon dioxide (CO2). It's the gas all animals (including humans) exhale. Sensors in a building's ventilation/air conditioning system could notice an elevated CO2 level and send you some needed fresh air.

The fire department can measure if there is *hydrocarbon vapor* in the air to avoid explosive surprises. There is also a gas sensor inside your car engine, which measures the fuel-air ratio. A correct fuel-air ratio ensures that all the gasoline burns in the cylinder—it wouldn't be good if there was gasoline dripping out of the exhaust pipe.

> *It's easy and fun to prototype with inexpensive smoke and gas sensors. But there are strict requirements, testing protocols, and certification programs for safety and security products. The projects in this chapter are no substitute for such products.*

MQ is a series of inexpensive gas sensors. There are sensors for many gases, some of which are listed in Table 4-1.

Table 4-1. Some MQ gas sensors

MQ Sensor	Gases detected
MQ-2	Flammable gas and smoke
MQ-3, MQ-303A	Alcohol (ethanol)
MQ-4	Methane (CH4)
MQ-7	Carbon monoxide
MQ-8	Hydrogen
MQ-9	Carbon monoxide, methane, LPG (propane or butane)

Experiment: Detect Smoke (Analog Gas Sensor)

An MQ-2 smoke sensor reports smoke by the voltage level it puts out. The more smoke there is, the higher the voltage. The MQ-2 we used has a built-in potentiometer for adjusting sensitivity (see Figure 4-1).

Figure 4-1. *Analog gas sensor*

What is the most dangerous poison gas? If you measure danger by the number of victims, it's carbon monoxide (CO). In a fire, most victims die from smoke inhalation before the flames get them.

Lethal CO is different from the mostly harmless carbon dioxide, CO2. The human body creates CO2 when it produces energy, and blood is made for carrying CO2 to the lungs where it's exhaled.

CO is poisonous because it doesn't want to leave. It sticks to blood hemoglobin, preventing some blood from carrying oxygen and CO2 for hours. If all hemoglobin is stuck with CO, tissues are not able to get oxygen, and death follows.

As the MQ-2 has three leads, it doesn't need any pull-up or pull-down resistors. From the Arduino point of view, the MQ-2 is like a potentiometer in a three-lead configuration.

The MQ-2 takes ground (black, 0 V) and +5 V (red). It measures smoke and sets its S pin voltage higher (nearer to +5 V) when it detects smoke.

MQ-2 Code and Connection for Arduino

Arduino has a built-in analog-to-digital converter, so you can read the MQ-2 with a call to `analogRead()`. Use the potentiometer on the breakout board to adjust its sensitivity.

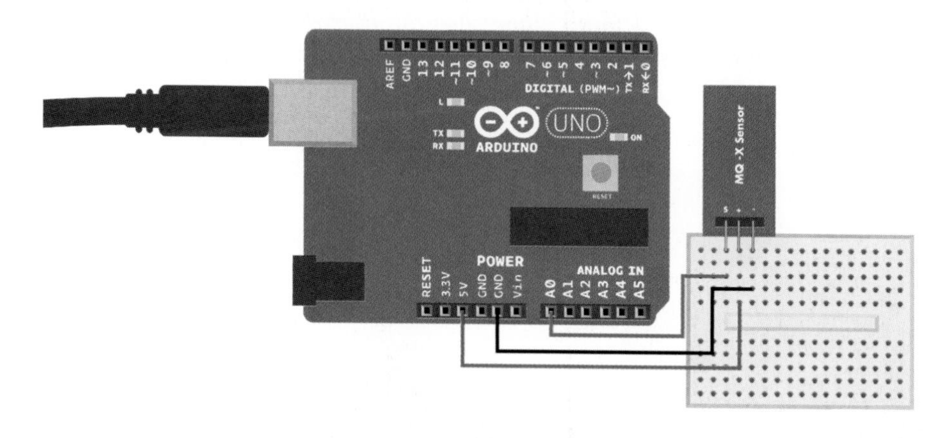

Figure 4-2. *MQ-2 sensor circuit for Arduino*

Example 4-1. *mq_x_smoke_sensor.ino*

```
// mq_x_smoke_sensor.ino - print smoke level to serial
// (c) BotBook.com - Karvinen, Karvinen, Valtokari

const int sensorPin = A0;
int smoke_level = -1;    // ❶
```

```
void setup() {
  Serial.begin(115200); // bit/s
  pinMode(sensorPin, INPUT);
}

void loop() {
  smoke_level = analogRead(sensorPin);  // ❷
  Serial.println(smoke_level);
  if(smoke_level > 120) {        // ❸
    Serial.println("Smoke detected");
  }
  delay(100); // ms
}
```

❶ Initialize smoke level to an impossible value to help debugging. If you ever see -1 in the output, you know it has never been replaced by a value read by `analogRead()`.

❷ The MQ-2 is a simple analog resistance sensor, so you can read the value with `analogRead()`. This returns a value between 0 (0 V) and 1023 (+5 V).

❸ Even though there is a cutoff value in the code, it's best to adjust sensitivity with the potentiometer on the sensor itself.

A fire alarm doesn't have to be a dull box, as the Lento alarm designed by Paola Suhonen proves (see Figure 4-3). A moth shape gives this everyday item a whole new look. Keep this in mind while designing your own gadgets.

Figure 4-3. *Lento alarm designed by Paola Suhonen*

MQ-2 Code and Connection for Raspberry Pi

The Raspberry Pi needs an external analog-to-digital converter (ADC) to read the MQ-2. Similar to other analog resistance sensors (see "Compound Eye Code and Connections for Raspberry Pi" on page 54), you can use a cheap MCP3002 chip for this conversion.

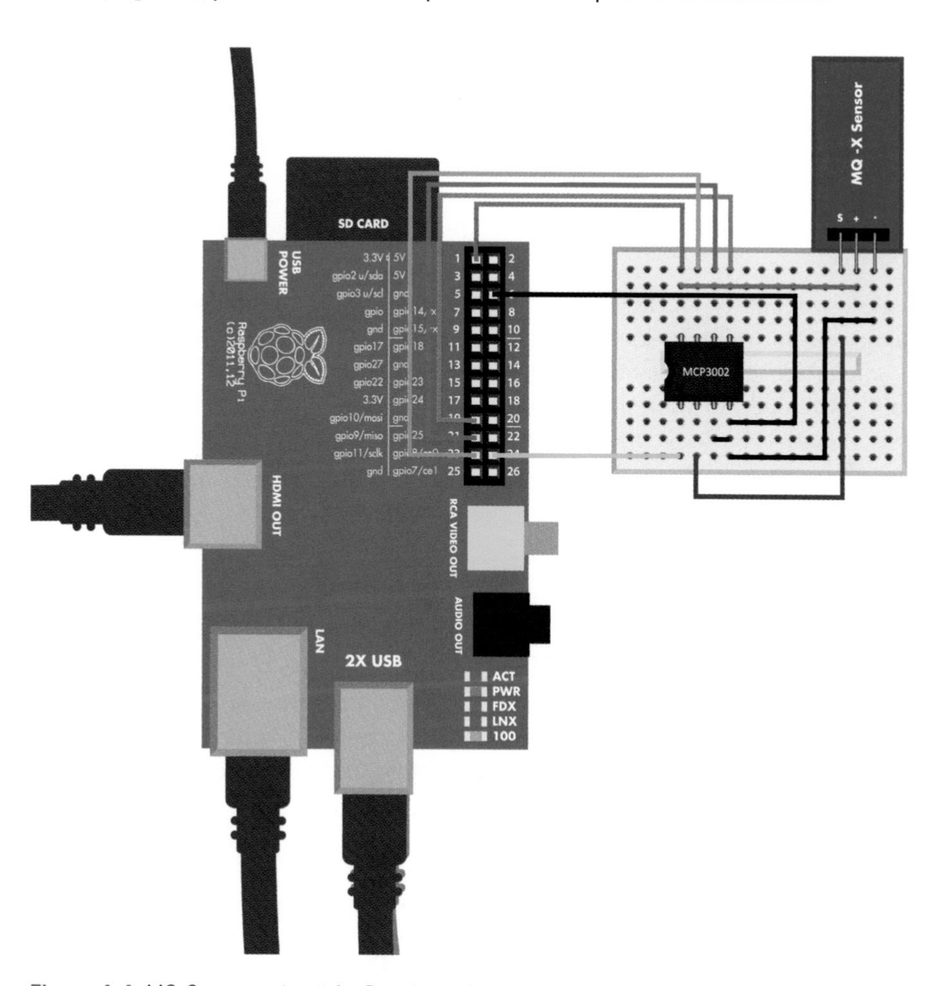

Figure 4-4. *MQ-2 sensor circuit for Raspberry Pi*

Example 4-2. mq_x_smoke_sensor.py

```
# mq_x_smoke_sensor.py - print smoke level
# (c) BotBook.com - Karvinen, Karvinen, Valtokari
import time
import botbook_mcp3002 as mcp    # ❶

smokeLevel = 0

def readSmokeLevel():
```

```
        global smokeLevel
        smokeLevel = mcp.readAnalog()   # ❷

def main():
  while True:    # ❸
    readSmokeLevel()    # ❹
    print("Current smoke level is %i " % smokeLevel)    # ❺
    if smokeLevel > 120:
      print("Smoke detected")
    time.sleep(0.5) # s

if __name__ == "__main__":
  main()
```

❶ The botbook.com library for MCP3002 saves a lot of coding and makes it easy to read analog values. The library (*botbook_mcp3002.py*) must be in the same directory as *mq_x_smoke_sensor.py*. You must also install the *spidev* library, which is imported by *botbook_mcp3002*. For more information, see the comments in the beginning of *botbook_mcp3002/botbook_mcp3002.py* or "Installing SpiDev" on page 56.

❷ Read the voltage of the first device connected to the MCP3002 ADC. The parameters in `readAnalog(device=0, channel=0)` are default values, so you could also leave them out.

❸ Repeating a program forever is common for embedded devices. When you use `while True`, you should always add some delay at the end of the loop. You can kill the program with Control-C when you are done.

❹ It's practical to wrap the primary task of your program inside its own function. The purpose of `readSmokeLevel()` is obvious from reading its name, which helps when you want to use it as a building block when you build bigger projects with many sensors and outputs.

❺ Create the string to be printed with a format string. The value of *smokeLevel* is treated as an integer and replaces `%i` in the format string.

Environment Experiment: Smoke Goes Up

There is a good reason why fire alarms are attached to the ceiling instead of the floor. Smoke is usually created by fire, so it's warmer than the air around it. Warming makes gases less dense or, in other words, lighter.

Why is warm air lighter than cool air? Warmth is movement: molecules are shaking and bouncing against one another. The warmer it gets, the more molecules shake and bounce. As they bounce, they push one another farther away. This makes the gas less dense. Having fewer molecules in a liter of gas means it's lighter. And surrounded by heavier air, the lighter gas goes up.

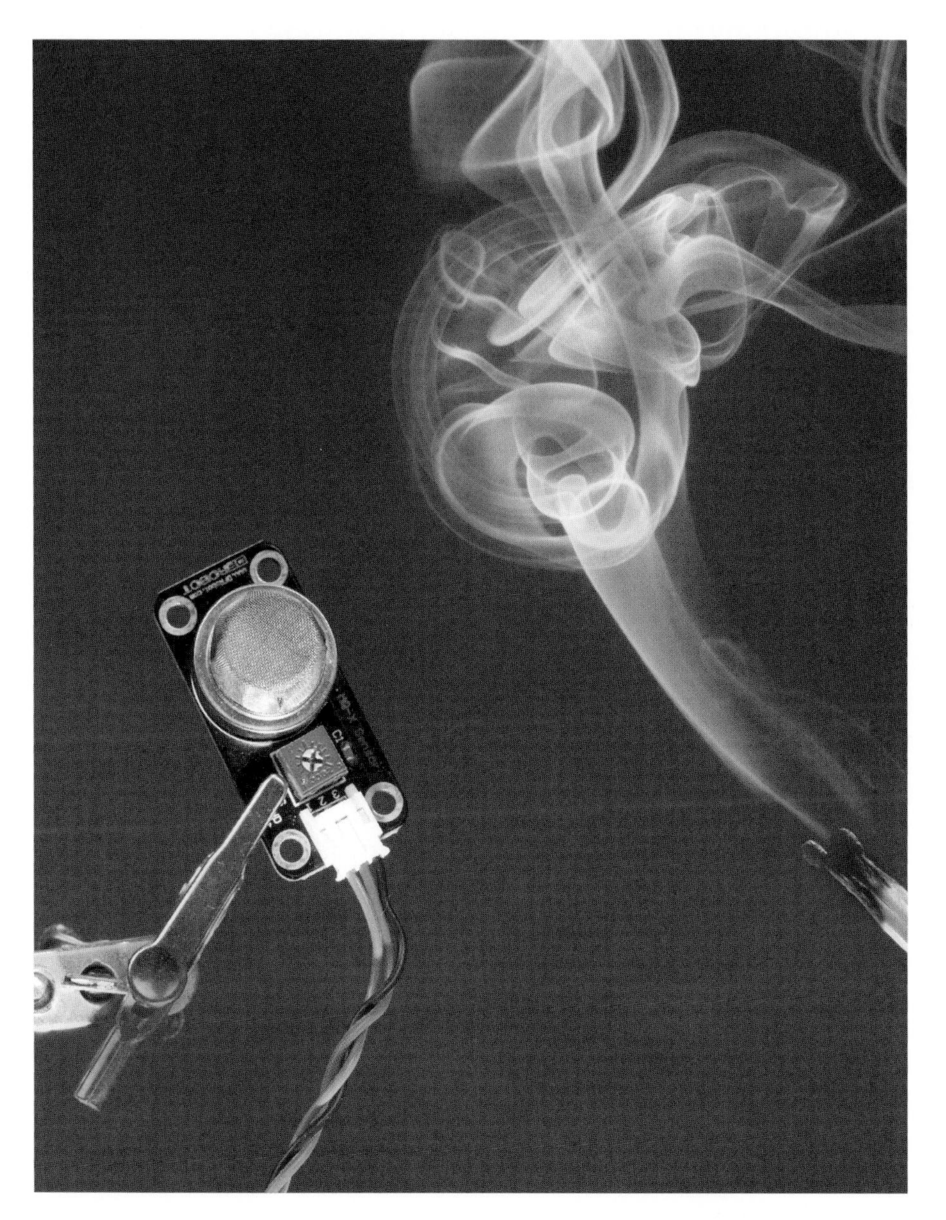

Figure 4-5. *Smoke goes up*

Use the smoke and gas sensor code and extinguish a match next to it. How high can you hold the extinguished match and still get a reading? When you reach the same level with the sensor you won't likely see any difference to normal air. As the experiment title says: smoke goes up. This is why there is nothing to measure if you let the warm smoke escape to heights above the sensor.

Experiment: Breathalyzer (Alcohol Sensor MQ-303A)

A Breathalyzer is used for checking whether a person has alcohol in his blood. More specifically, ethanol, the alcohol found in wine, beer, and liquor.

The MQ-3 alcohol sensor looks a lot like the other MQ series gas sensors. (see Figure 4-6).

Figure 4-6. *MQ-3 alcohol sensor*

Before you call it a day and pour a glass of Castillo Montroy, at least have a look at "Environment Experiment: Try It Without Drinking" on page 77.

Just as the gas exchange in lungs brings in oxygen and gets rid of carbon dioxide, some blood alcohol is released in the air you exhale. This is the alcohol measured by an alcometer.

The more ethanol in your blood, the more there is in the air you exhale. Blood alcohol content gives a good indication how drunk a person is.

Even though more alcohol in blood makes the same person more drunk, the drunkenness differs from person to person. For making laws, a typical value is chosen as the limit. For example, the limit for DUI charges in Finland is 0.5%.

Officially accepted alcometers are calibrated periodically to get reliable readings. Alcometers use a built-in formula to estimate blood alcohol content from exhaled air alcohol content.

Figure 4-7 shows the Arduino circuit for the MQ-3, and the Arduino sketch is shown in Example 4-3. You can find the Raspberry Pi circuit in Figure 4-8 and the Python code in Example 4-4.

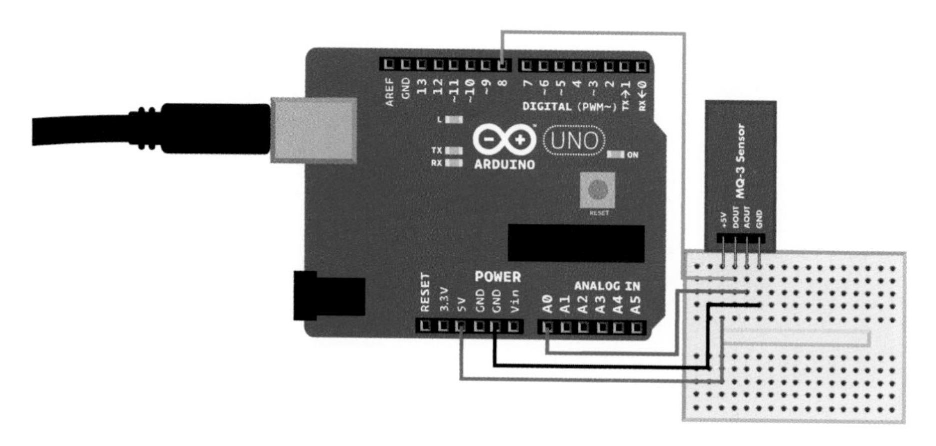

Figure 4-7. *MQ-3 sensor circuit for Arduino*

Example 4-3. mq_3_alcohol_sensor.ino

```
// mq_3_alcohol_sensor.ino - print alcohol value and limit digital info.
// (c) BotBook.com - Karvinen, Karvinen, Valtokari


const int analogPin = A0;
const int digitalPin = 8;

int limit = -1;
int value = 0;

void setup() {
  Serial.begin(115200);
  pinMode(digitalPin,INPUT);
}

void loop()
{
  //Read analog value
  value = analogRead(analogPin);
  //Check if alcohol limit is breached
  limit = digitalRead(digitalPin);
  Serial.print("Alcohol value: ");
  Serial.print(value);
  Serial.print(" Limit: ");
  Serial.println(limit);
  delay(100);
}
```

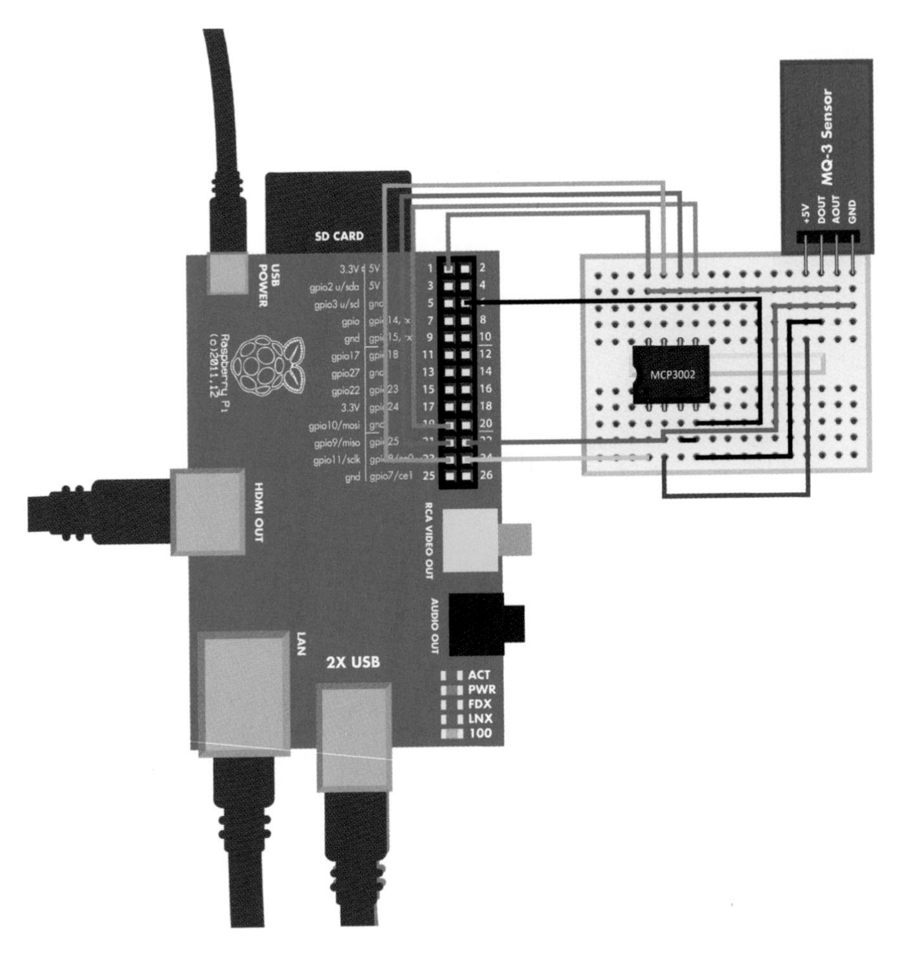

Figure 4-8. *MQ-3 sensor circuit for Raspberry Pi*

Example 4-4. *mq_3_alcohol_sensor.py*

```
# mq_3_alcohol_sensor.py - read digital output from alcohol sensor
# (c) BotBook.com - Karvinen, Karvinen, Valtokari

import time
import botbook_gpio as gpio

def readLimit():
  limitPin = 23
  gpio.setMode(limitPin,"in")
  return gpio.read(limitPin)


def main():
  while True:
```

```
        if readLimit() == gpio.HIGH:
            print("Limit breached!")
        time.sleep(0.5)

if __name__ == "__main__":
    main()
```

Environment Experiment: Try It Without Drinking

Drinking alcohol while trying to learn electronics would be slightly counterproductive. The good news is that you don't need a bottle of whiskey to test your sensor. You can get impressive readings by using other products that have alcohol in them. Mouthwash and liquor candy worked really well for us.

Use the Breathalyzer code and open the serial monitor. Chew some candy or take a sip of mouthwash (don't swallow it) and breathe into the sensor. The readings should instantly jump up. You can still get higher-than-normal levels from readings you take several minutes later.

Test Project: Emailing Smoke Alarm

Get an email when your gadget detects smoke.

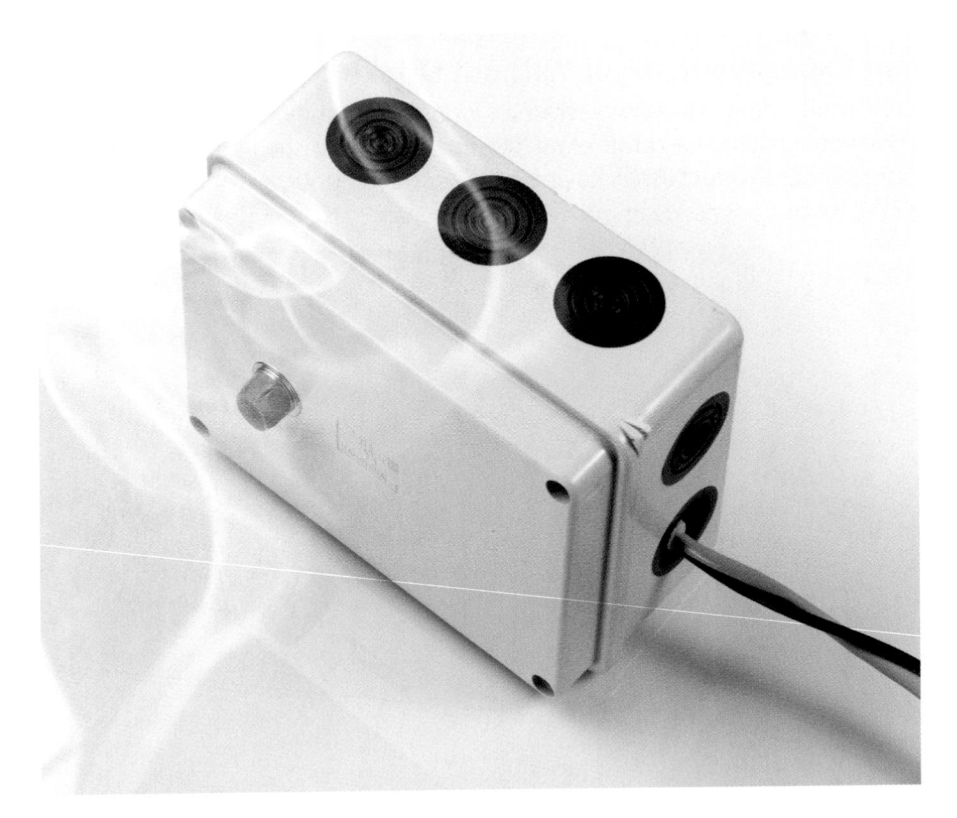

Figure 4-9. *Ready emailing smoke alarm*

What You'll Learn

In this project, you'll learn:

- How to react to environment changes with actions: if there's smoke, send an email warning.
- How to automatically send email from Raspberry Pi.
- The basics of sending email.

Python for Email and Social Media

This is an example of a project that's easy with Raspberry Pi. Sending email automatically from Raspberry Pi is no different from automatically emailing on any Linux system.

Python is known for having "batteries included," in that there is a library for everything. Email sending is trivial using existing libraries. Similarly, there are libraries for sending and receiving data over the Web using the HTTP protocol.

What about social media? You could adapt the program to send messages to Twitter, Facebook, or similar "social media" services. But all of these use protocols that though they are built on open technologies, are proprietary and create lock-in to their services. Those giants have rules on how they run their service—rules that could unilaterally change any day. For example, Twitter could change the number of requests you're permitted each day or remove the functionality you've grown to depend on. Don't build your sand castle on someone else's private beach. Or at least know the score before you make the decision.

Building It

Connect the MQ-2 smoke sensor to Raspberry Pi, as shown in "MQ-2 Code and Connection for Raspberry Pi" on page 71. It's a good idea to test the smoke sensor with just the sensor code (Example 4-2) first.

Once you're sure it works, you can advance to sending emails in your code. Test your email credentials normally before using them in this program. For example, you could enter them into a regular mail user agent (MUA) such as Thunderbird, KMail, Claws, or, in the worst case, Outlook. Even though the example uses a Gmail account, any mail account will work.

Don't use your personal email account for this kind of testing. If you make a mistake in your code and send out too many email messages, your email server may interpret this as an attempt at sending spam. It's best to create an email account just for these purposes. However, services like SendGrid (http://sendgrid.com/) and Amazon SES (http://aws.amazon.com/ses/) are designed for exactly this sort of scenario, so consider using them if you need to send many messages.

How Does Email Work?

Do you still remember how email works? In the time of Gmail and other webmail clients, it's too easy to forget the basics.

Here we'll describe the workings of email from the point of a mail user agent running on your local computer, not webmail. To keep things simple, we'll just show key points in an example case.

It's essential to remember that sending and receiving email involves two different servers (the sender's and the recipient's). It's common that the sending and receiving server are in different networks, even on different continents.

When you click Send in your mail user agent, your computer contacts your *SMTP server*. The SMTP can be in your local network, run by your ISP (such as `smtp.verizon.net` or `smtp.comcast.net`) and may accept all connections originating within its network without password or login. Or it could be run by your email provider (such as `smtp.gmail.com`, `mail.gmx.com`), and require TLS/SSL encryption and login. When your email is accepted, the SMTP server will deliver your message to the recipient.

When your recipient checks her mail, her mail user agent contacts an *IMAP server*. This is run by her email provider (such as Gmail or GMX) or her ISP. Reading private email of course requires a login and password, and common sense dictates that one should use encryption (SSL/TLS) when connecting (most email providers require encryption). From the IMAP server, a mail user agent can download message headers and full copies of messages. And when someone sends you an email, the reverse happens: *her* mail user agent contacts *her* SMTP server, delivers it to your mail server, and when you check your mail, you retrieve your copy of the email from your IMAP server.

Messages are typically left on IMAP server, which is why you can check your mail on your computer and mobile phone and see the same list of messages on both.

Could Arduino Send Email? Not Easily

A typical way to build a similar project with Arduino would use an external computer. The Arduino would read the smoke sensor and communicate this value to a desktop computer using serial over USB. On the desktop computer, a Python program could read serial (with the pySerial API) and send the email just as shown in the next example.

But even with an Ethernet or WiFi shield, it would be difficult for you to send mail from Arduino. Although the email protocol is, on its surface, fairly simple, there are enough possible surprises when sending an email that it would be hard to write a reliable SMTP library for Arduino (also, many servers require SSL, and the Arduino Uno and similar models don't support this).

Obviously, it is easier to do this project with Raspberry Pi, or a similarly capable microcontroller board such as the BeagleBone (or the Arduino/Linux hybrid Arduino Yún).

Code for Raspberry Pi

Example 4-5. *smoke_alarm.py*

```
# smoke_alarm.py - send email every 5 minutes when smoke is detected
# (c) BotBook.com - Karvinen, Karvinen, Valtokari

import time
import botbook_mcp3002 as mcp
import smtplib  # ❶
from email.mime.text import MIMEText    # ❷
```

```
# Email addresses
email_to = 'example@gmail.com'  # ❸
email_from = 'example@gmail.com'         # ❹

# SMTP email server settings
server = 'smtp.gmail.com'        # ❺
mail_port = 587
user = 'example@gmail.com'       # ❻
password = 'password'    # ❼
gracePeriod = 5 * 60 # seconds  # ❽

def sendEmail(subject, msg):    # ❾
        msg = MIMEText(msg)     # ❿
        msg['Subject'] = subject        # ⓫
        msg['To'] = email_to    # ⓬
        msg['From'] = email_from

        smtp = smtplib.SMTP(server,mail_port)   # ⓭
        smtp.starttls()
        smtp.login(user, password)      # ⓮
        smtp.sendmail(email_from, email_to, msg.as_string())     # ⓯
        smtp.quit()     # ⓰

def main():
        while True:
                smokeLevel = mcp.readAnalog()
                print("Current smoke level is %i " % smokeLevel)
                if smokeLevel > 120:
                        print("Smoke detected")
                        sendEmail("Smoke","Smoke level was %i" % smokeLevel) # ⓱
                        time.sleep(gracePeriod) # ⓲
                time.sleep(0.5) # s     # ⓳

if __name__ == "__main__":
        main()
```

❶ "Batteries included": Python has a built-in library for SMTP communication. You don't have to do low-level socket programming on your own.

❷ Email used to be plain text. Headers one per line, an empty line, and then the body. But nowadays we expect localized, non-ASCII characters to work. That's why an email body is often MIME encoded, just as attachments are. Learn funny Finnish: ä (a with two dots) is pronounced like a and e at the same time. Päivää!

❸ The *To* address is the recipient of your email. You should probably use your own address here to receive the smoke warning messages. These are global variables, visible to all functions.

❹ The *From* address in the email could theoretically be anything. But in the time of spam filters, choosing a weird address could affect how well your mail goes through. You should use your email address, because its domain name will usually match that of the SMTP server you're using.

❺ The SMTP server is the sending server. As this program only sends email and never receives any, the SMTP server is the only server needed.

❻ Your SMTP login name. It could be your email address (*example@gmail.com*) or a shorter login name (jdoe).

❼ Your SMTP password should be the same you use for logging into webmail or connecting from an email client.

❽ Even when smoke is detected, you probably don't want to get 60 emails every minute. Trying to send email that fast would probably get your email blacklisted by the server. That's why you should add a grace period: the time program should wait after sending an email. The grace period, 5 minutes, is converted to seconds by multiplying by 60 seconds per minute. (Write simple calculations into code instead of calculating magic numbers on paper. Avoid the temptation to repeat the value in comment.)

❾ The code that sends email is in its own function for clarity. Topic (email subject) and msg (email body) come from function parameters.

❿ The body of the message is MIME encoded to reliably handle non-ASCII characters. As you can see from the capitalized first letter, `MIMEText` is a class. The constructor `MIMEText()` returns a new object that's stored into variable `msg`. `MIMEText` is just for creating the message; it does not take care of connecting anywhere or sending anything.

⓫ To modify email headers, you can use the MIMEText object `msg` just like a dictionary data type (*http://bit.ly/1jFV3V8*).

⓬ Recipient and sender come from global variables.

⓭ Create a new object of class SMTP. This will be used for connecting to the SMTP server. The beginning part (smtplib) is the *namespace*, automatically created with "import smtplib."

⓮ Connect to the SMTP server, using the username and password from global variables.

⓯ Send the email, using the `sendmail()` method of the `smtp` object of the class `smtplib.SMTP`. The `smtp.sendmail()` method expects to see string, so the `msg` object is dumped to string with its built-in method.

⓰ Close the connection and clean up. The `quit()` function is a method of the `smtp` object.

⓱ Use a format string to create the `msg`, replacing "%i" with the value of `smokeLevel`.

⓲ Wait for 5 minutes after sending email before you send another email.

⓳ Avoid running the loop at maximum speed by sleeping half a second.

Packaging

We used a general-purpose enclosure for packaging our email smoke alarm (see Figure 4-10). These boxes look ascetic but are available widely, and their rubber covered ready-made holes make modifications fast and easy.

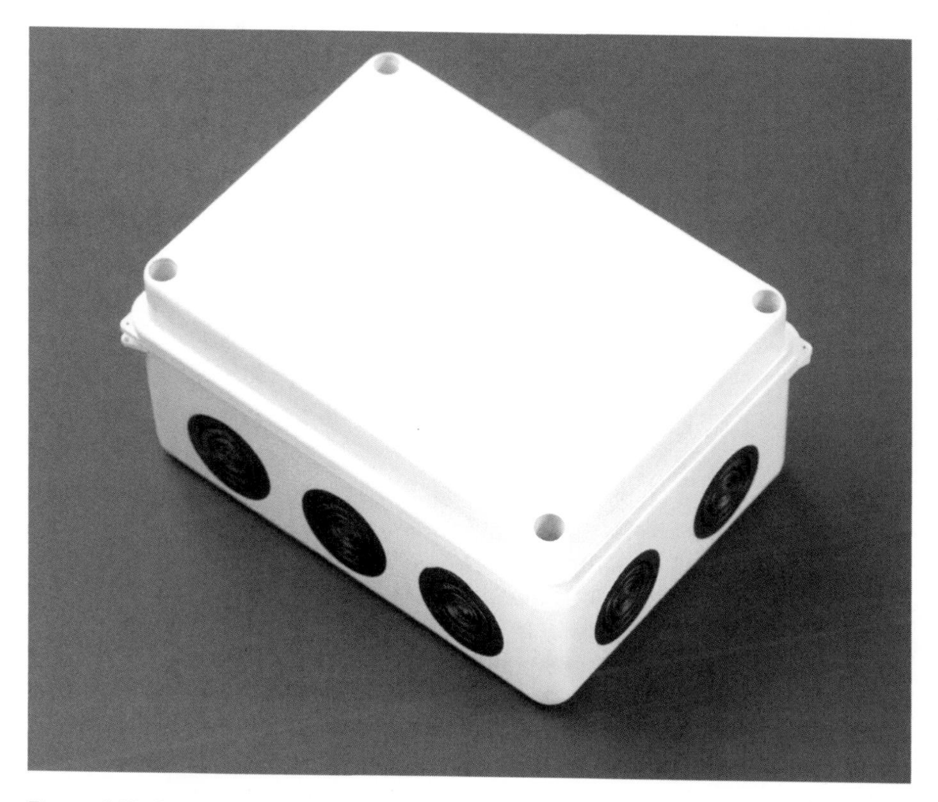

Figure 4-10. *General-purpose enclosure*

A hole for the smoke sensor is the only one you need to drill (see Figure 4-11). Others can be made by carefully applying a sharp knife to the rubber-covered openings. Actually, you could use those for the sensor also, but as they tend to look pretty rough we prefer to use them for the cables only. A 19 mm drill bit was a perfect match for the sensor and we didn't need any adhesives to keep it in place, as shown in Figure 4-12.

Figure 4-11. *Hole for the smoke sensor*

Figure 4-12. *Sensor in place*

To prevent the Raspberry Pi from moving around inside the box, we secured it with Velcro to the bottom, as shown in Figure 4-13. This way it's easy to remove it for debugging or to use in other projects.

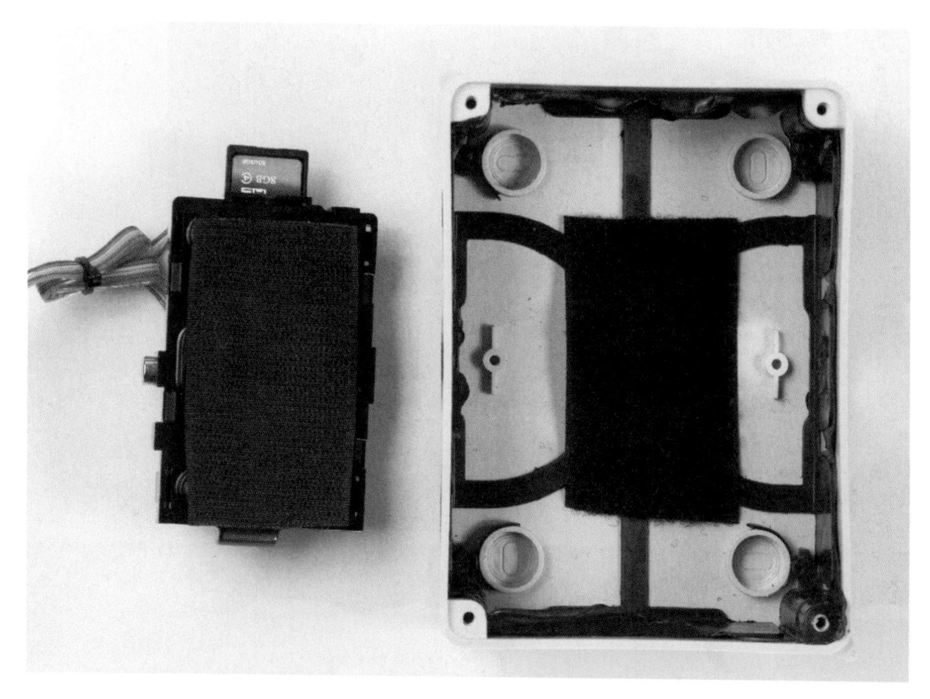

Figure 4-13. *Raspberry Pi's Velcro attachement*

We glued a mini breadboard to the bottom of the lid of the box (see Figure 4-14). Then, we used a zip tie to keep the jumper wires in a neat bundle.

We made one large hole (~2 cm) to the side for the LAN and power cables (see Figure 4-15). We wanted the smallest possible holes in the final gadget, so in our version we don't have inlets for the other cables (see Figure 4-16).

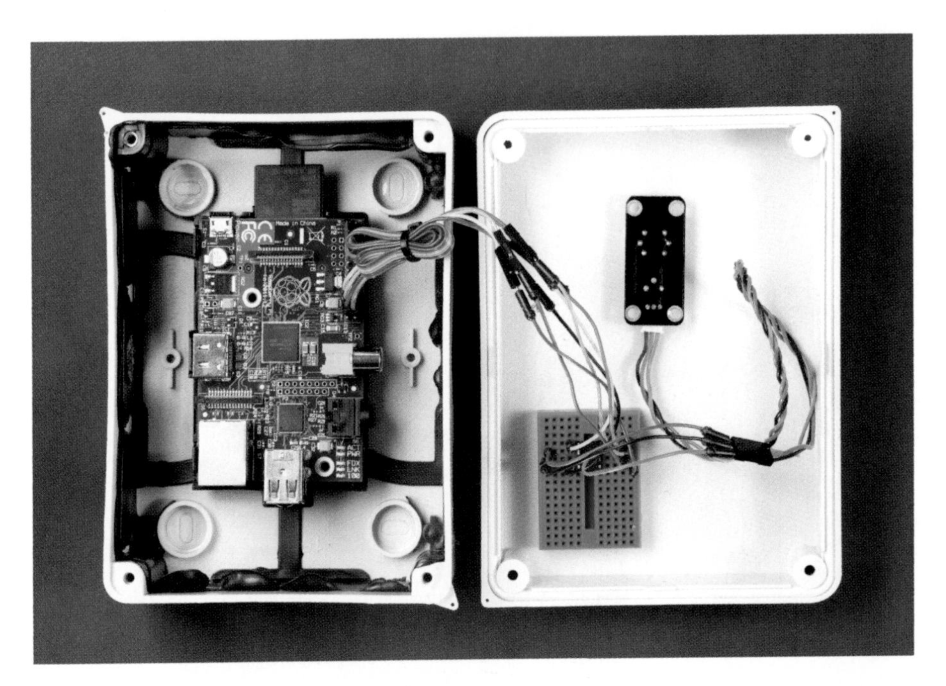

Figure 4-14. *Mini breadboard with zip-tied wires*

Figure 4-15. *Hole for the cables*

Figure 4-16. *Cables attached*

Now just close the lid and you're ready for some testing as shown earlier in Figure 4-9. And while placing your alarm, remember which way the smoke goes.

Your gadget now has an "electrical nose." Using the family of MQ sensors, you can detect gases that are poisonous (CO), flammable (hydrocarbons), explosive (hydrogen), or that reveal intoxication (ethanol in breath).

In the project, you wrote a Python program to react to changes in the environment. You can adapt this program to react to events from other sensors. Alternatively, the reaction could be things other than email, such as contacting a web service or activating an output.

Gas sensors detect things that are difficult or impossible for our human senses to see. In the next chapter, you'll sense a more visible phenomenon, touch.

Touch 5

Touch is the most common way to tell a device what to do. Flip a switch to turn on the lights, press a button on a remote control, or click a key on your keyboard.

In this chapter, you'll meet a button, a squeeze sensor, and a capacitive touch sensor. You'll learn to use pull-up resistors to avoid floating pins. Finally, you'll build a haunted bell to try your touch skills in a project.

A button is the most basic sensor, so it's also the most basic touch sensor. When a button is pressed, its leads are connected, closing a circuit. A microswitch is a type of small, durable button.

A FlexiForce sensor detects how much force is being put on it. You can use it to test the might of your fingers or to detect when someone is sitting on a chair.

A touch sensor senses touch without any moving parts. And the weird part of a touch sensor is that it doesn't even require touch! You'll learn to detect a hand placed on a table, through the table.

Even though touch sensors serve a similar purpose in your projects, their methods of measuring touch differ. A button simply connects two wires. The FlexiForce pressure sensor is based on thin layers of pressure-sensitive ink. A touch sensor measures capacitance—the ability to store electricity.

Experiment: Button

Light up an LED when a button is pressed.

A button (Figure 5-1) is the simplest sensor. Pressing the button down connects its leads, so that the button acts as a wire. Releasing the button breaks the circuit. All digital switch sensors (such as a reed switch or tilt switch) work like a button.

Figure 5-1. *Push button*

Buttons come in many sizes and forms. When using a breadboard, it's convenient to use a button with four leads. The leads are in pairs, so that two adjacent leads are always connected to each other. When you turn the button upside-down, you can see which leads are connected. There is a beveled line showing the connected leads (Figure 5-2).

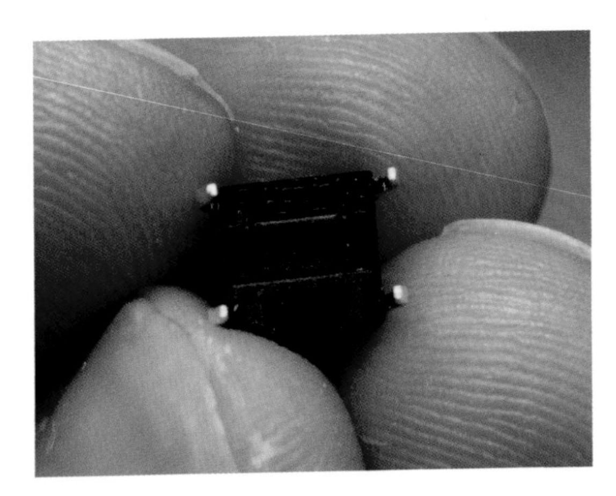

Figure 5-2. *Bottom of a push button*

Pull-Up Resistor

To get a reading from a data pin, it must be connected somewhere. You should not read an unconnected *floating pin*. A *pull-up resistor* pulls a pin HIGH when it's not connected to anything else, avoiding the floating state.

The state of a data pin can be read with commands such as `digitalRead()`, `analogRead()`, `botbook_gpio.read()`, `botbook_mcp2003.readAnalog()`.

A pin that's connected to ground or HIGH is an obvious case: when you read its state, it's clear which state it's in. If the pin is connected to ground (GND, 0 V), a digital read returns LOW. If the pin is connected to HIGH (+5 V or +3.3 V), the read returns HIGH.

If you read an unconnected, floating data pin, you would get an undefined value. It could be HIGH, it could be LOW, it could change every second or stay the same forever. Such an undefined value is completely useless, so there is no point in reading a floating pin.

Consider a circuit with a button between a data pin and ground (Figure 5-3). When the button is pressed, the data pin is connected to ground so it's LOW.

What about when the button is up? You must use a *pull-up* resistor to bring the data pin HIGH.

The pull-up resistor is big. We often use one in the tens of thousands ohms range. This way, when the button is pressed and the data pin is connected to both ground and the pull-up, there is no short-circuit between +5 V and ground because the path of least resistance is between the data pin and ground, rather than between +5 V and ground.

For convenience, the code below uses built-in pull-up resistors. For Arduino, you can enable the onboard pull-up resistors with a line of code. For Raspberry Pi, you'll use one of the pins that has an always-on pull-up resistor.

Code and Connection for Arduino

Build the circuit shown in Figure 5-3 and upload the code shown in Example 5-1. Press the button to light up the surface-mounted LED.

If pressing the button doesn't affect the LED at all, check that you have connected the button the right way.

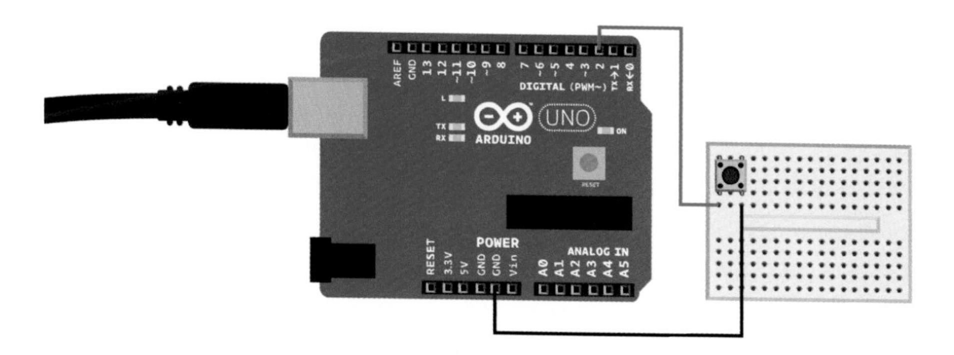

Figure 5-3. *Button circuit for Arduino*

Example 5-1. **button.ino**

```
// button.ino - light an LED by pressing a button
// (c) BotBook.com - Karvinen, Karvinen, Valtokari


int buttonPin=2;
int ledPin=13;
int buttonStatus=-1;

void setup() {
  pinMode(ledPin, OUTPUT);
  pinMode(buttonPin, INPUT); // ❶
  digitalWrite(buttonPin, HIGH); // internal pull-up // ❷
}

void loop() {
  buttonStatus=digitalRead(buttonPin); // ❸
  if (LOW==buttonStatus) { // ❹
    digitalWrite(ledPin, HIGH); // ❺
  } else {
    digitalWrite(ledPin, LOW);
  }
  delay(20);     // ❻
}
```

❶ Put the pin in INPUT mode to be able to check its voltage with `digitalRead()` later.

❷ Connect the D2 pin to GND with large (20 kilohm) internal resistor. This prevents the pin from floating when the button is open. The internal pull-up resistors save you the trouble of connecting an extra resistor to your breadboard. The latest versions of Arduino offer a new option for `pinMode()`, `INPUT_PULLUP`, which allows you to combine this line and the preceding line into a single command: `pinMode(buttonPin, INPUT_PULLUP);`. But the older method works fine and you will still see it in many example programs written before the new mode was introduced.

❸ Read the voltage of D2. HIGH means +5 V, LOW means 0 V, ground level.

❹ If the button level is LOW (meaning the button is pressed down)…

❺ …light up an LED. Data pin D13 is connected to surface mounted LED on the Arduino.

❻ Pause briefly to not overtax the CPU.

This code keeps the LED lit when the button is held down. If you want to instead **toggle** *the LED each time you press the button, you must* **debounce** *the input. For example, creating a device where one click turns the LED on and the next turns it off requires* **debouncing**. *You can find a debounce example program in the Arduino IDE under File→Examples→2.Digital: Debounce.*

Code and Connection for Raspberry Pi

Wire up the Raspberry Pi as shown in Figure 5-4 and run the Python program shown in Example 5-2.

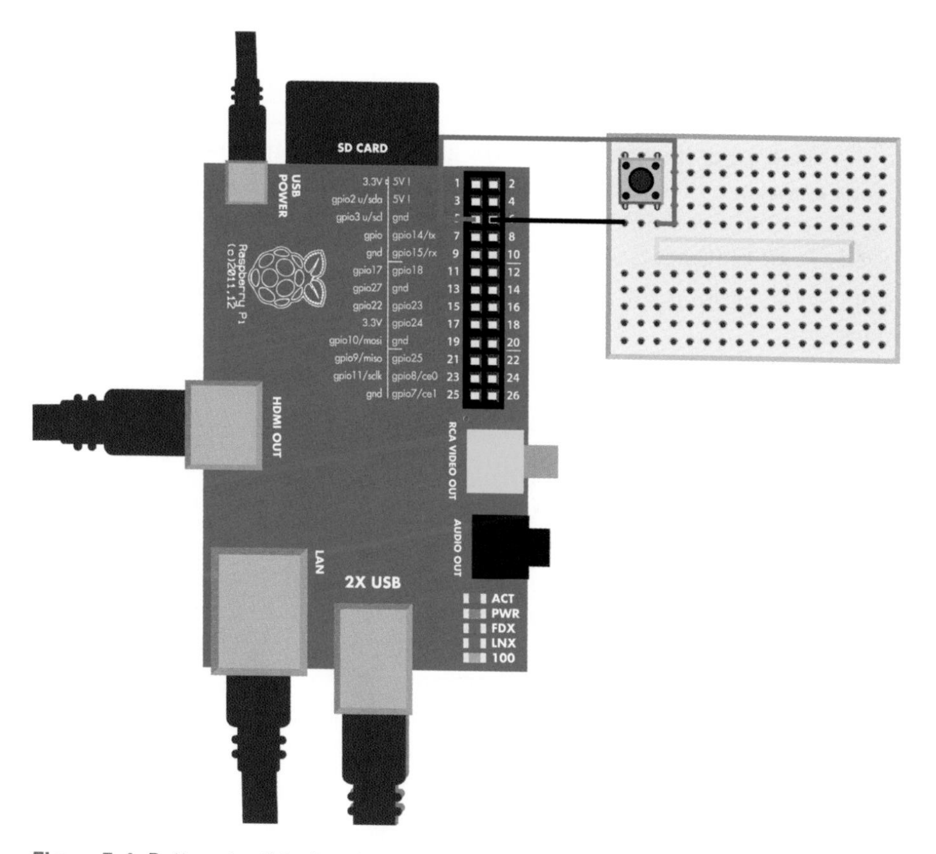

Figure 5-4. *Button circuit for Raspberry Pi*

Example 5-2. **button.py**

```
# button.py - write to screen if button is down or up
# (c) BotBook.com - Karvinen, Karvinen, Valtokari

import time
```

```
import botbook_gpio as gpio     # ❶

def main():
        buttonpin = 3   # has internal pull-up  # ❷
        gpio.mode(buttonpin, "in")      # ❸

        while (True):   # ❹
                buttonUp = gpio.read(buttonpin) # ❺
                if(buttonUp == gpio.HIGH):
                        print "Button is up"
                else:
                        print "Button is pressed"
                time.sleep(0.3) # seconds       # ❻

if __name__ == "__main__":
        main()
```

❶ Make sure there's a copy of the *botbook_gpio.py* library in the same directory as this program. You can download this library along with all the example code from *http://botbook.com*. See "GPIO Without Root" on page 19 for information on configuring your Raspberry Pi for GPIO access.

❷ The gpio3 pin has an internal pull-up resistor—it's connected to +3.3 V through an 1800 Ohm resistor.

❸ Set gpio3 to "in" mode, for reading its voltage with `gpio.read()` later.

❹ Keep running until the user kills the program with Control-C.

❺ Read the value of the pin. It will be either True or False.

❻ Sleep for a while to prevent the `while(True)` loop from taking 100% of the CPU. Also, the short wait makes it much easier to read the printed text.

Experiment: Microswitch

A microswitch is a button (see Figure 5-5). When you press the button, the leads are connected. This experiment writes "0" to the serial port when the microswitch is pressed.

You can use a microswitch just like a regular button (see "Experiment: Button" on page 89). Just as with a button, the specific model of microswitch doesn't matter, and you can use the same code and connection for any typical microswitch.

Figure 5-5. *Microswitch*

Microswitches are popular because of their low price, small size, and durability. A typical microswitch will last more than a million presses. The clicky sound and the feel of the click come from a tiny arm turning around a pivot.

This connection avoids floating pin by using a pull-up resistor ("Pull-Up Resistor" on page 90).

Microswitch Code and Connection for Arduino

Build the circuit shown in Figure 5-6 and upload the code shown in Example 5-3.

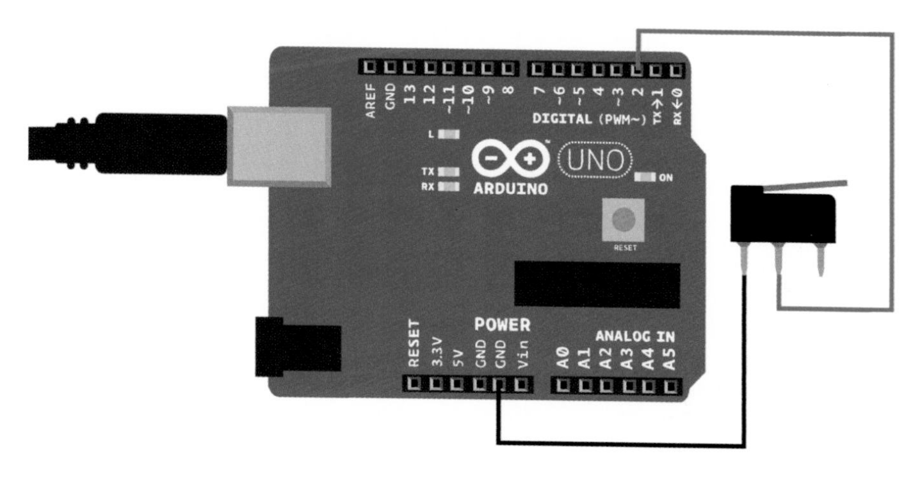

Figure 5-6. *Microswitch circuit for Arduino*

Example 5-3. **microswitch.ino**

```
// microswitch.ino - print to serial if microswitch is down or up
// (c) BotBook.com - Karvinen, Karvinen, Valtokari


const int switchPin = 2;
int switchState = -1;    // ❶

void setup() {
  Serial.begin(115200);
  pinMode(switchPin, INPUT);
  digitalWrite(switchPin, HIGH); // internal pull-up // ❷
}

void loop() {
  switchState = digitalRead(switchPin);
  Serial.println(switchState);  // ❸
  delay(10);
}
```

❶ Initialize to an impossible value.

❷ Writing HIGH to INPUT pin connects it to +5 V with a large (20 kilohm) resistor.

❸ If button is pressed, the D2 pin is directly connected to ground through the button, and switchState is LOW. When D2 is directly connected to ground, the 20 kOhm pull-up resistor's resistance is so large that the pull-up doesn't affect the state of D2. If the button is not pressed, the internal pull-up resistor pulls it to HIGH (+5 V).

You can make simple feeler antennas for your robot from a microswitch just by hot gluing a zip tie to it (Figure 5-7).

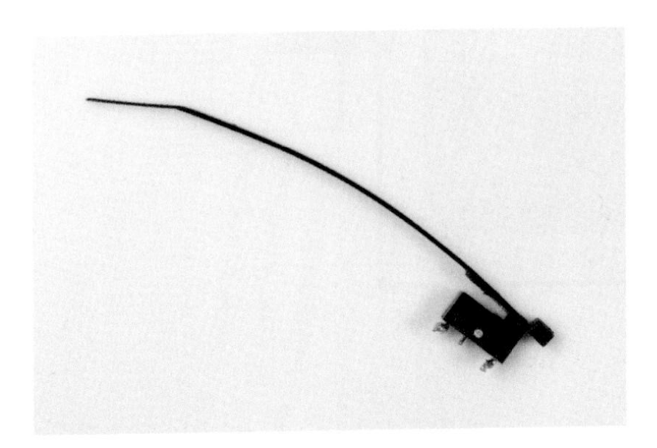

Figure 5-7. *Feeler antenna from microswitch*

Microswitch Code and Connection for Raspberry Pi

Wire up the Raspberry Pi as shown in Figure 5-8 and run the Python program shown in Example 5-4.

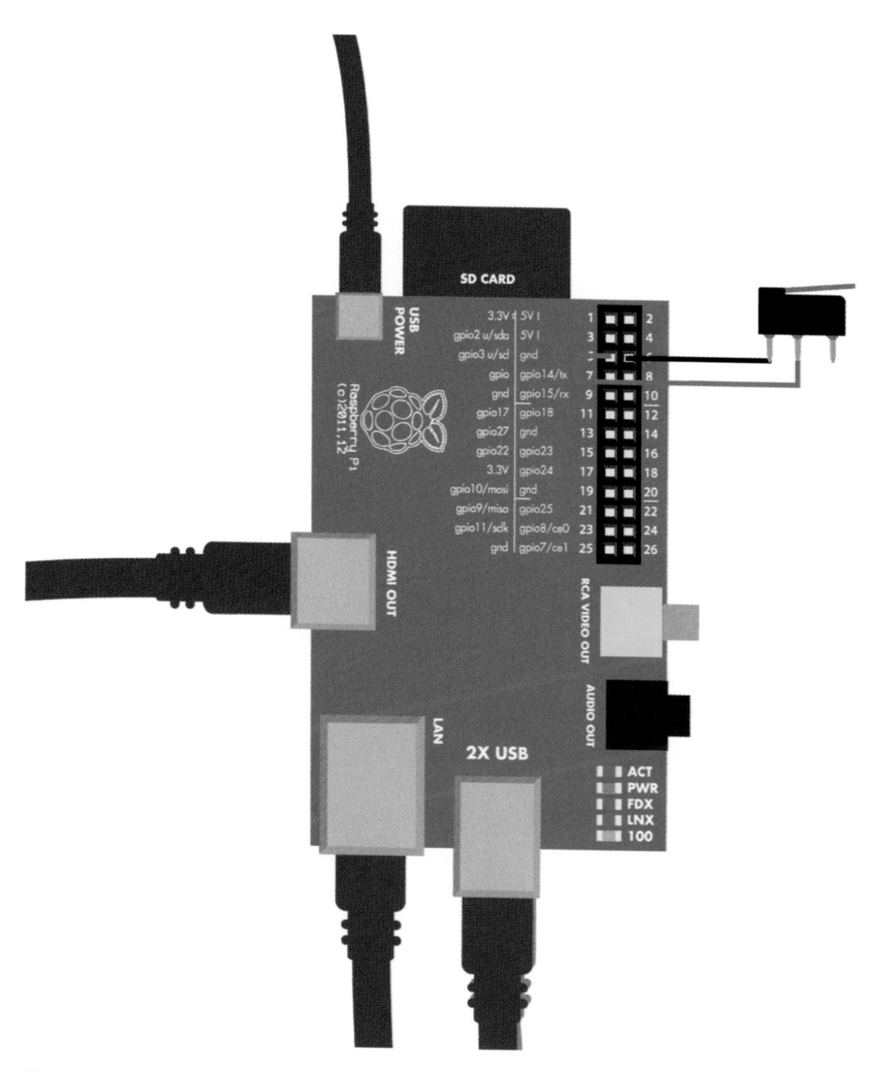

Figure 5-8. *Microswitch circuit for Raspberry Pi*

Example 5-4. microswitch.py

```
# microswitch.py - write to screen if switch is pressed or not
# (c) BotBook.com - Karvinen, Karvinen, Valtokari
import time
import botbook_gpio as gpio     # ❶
```

```
def main():
        switchpin = 3   # has internal pull-up  # ❷
        gpio.mode(switchpin, "in")

        while (True):
                switchState = gpio.read(switchpin)       # ❸
                if(switchState == gpio.LOW):
                        print "Switch is pressed"
                else:
                        print "Switch is up"
                time.sleep(0.3) # seconds

if __name__ == "__main__":
        main()
```

❶ Make sure there's a copy of the *botbook_gpio.py* library in the same directory as this program. You can download this library along with all the example code from *http://botbook.com*. See "GPIO Without Root" on page 19 for information on configuring your Raspberry Pi for GPIO access.

❷ gpio2 and gpio3 pins have always-on internal pull-up resistors.

❸ When the button is pressed down, gpio3 is connected to ground and thus LOW. The internal pull-up resistor's resistance is so large that the pull-up resistor doesn't affect gpio3 state when gpio3 is directly connected to ground. When the button is up, the pull-up resistor pulls it HIGH.

Experiment: Potentiometer (Variable Resistor, Pot)

In this experiment, you'll control how quickly an LED blinks by turning a potentiometer.

A potentiometer is a *variable resistor*. You turn its knob to change its resistance.

A normal, non-variable resistor has just two leads. If you want to use a potentiometer in place of a single resistor, use the middle lead and either of the side leads.

Why does a potentiometer (or *pot*) have three leads, then? When you read a data pin, the pin must always be connected somewhere.

The easiest way to use a pot is to connect one side to positive (+5 V) and another side to ground (GND). Then the middle lead of the pot is connected to data pin. The voltage of the pot's middle lead then changes between +5 V and GND as you turn it. This makes the pot a *voltage divider*, which is a circuit that lowers a voltage by dividing it between two resistors (one connected to positive voltage, the other to ground). In the case of a pot, the resistances on either side of the pot's *wiper* change as you turn the knob.

You can see a diagram of a potentiometer in Figure 5-9. The green middle lead is connected to a data pin that measures voltage. The side leads are connected to ground (black) and +5 V (red). There is an arc-shaped, round resistor (red-black) from positive to negative.

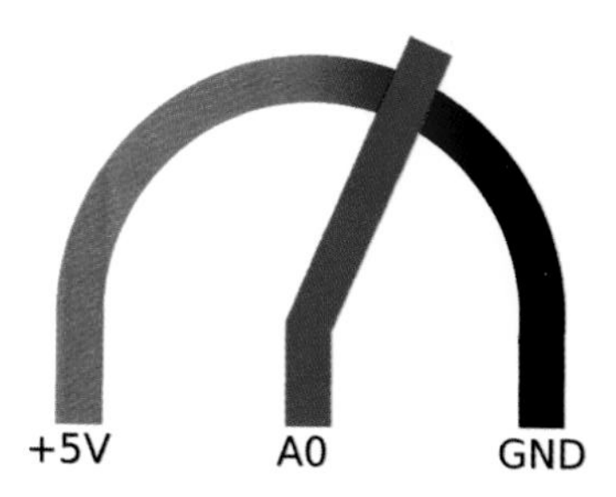

Figure 5-9. *Potentiometer diagram*

By turning the knob, you choose where the wiper touches the resistor somewhere on its arc. Positive is not short-circuited to ground, because there is always the long arc-shaped resistor between positive and ground.

If you turn the pot to the minimum, the green wiper is touching black ground (GND). The data pin connected to the wiper then reads 0 V. The entirety of the arc-shaped resistor separates the wiper from the positive terminal.

If you turn the pot to the maximum, the green wiper touches the red positive terminal (+5 V). The data pin connected to the wiper then reads +5 V.

As you turn the wiper on the arc-shaped resistor, you can select any voltage between 0 V and +5 V.

This example is using Arduino's +5 V. A pot works the same way in Raspberry Pi, but there you must use +3.3 V or you will damage your data pin.

Potentiometer Code and Connection for Arduino

Figure 5-10 shows the connection diagram for Arduino. Set up your circuit as shown, and then upload and run the sketch shown in Example 5-5.

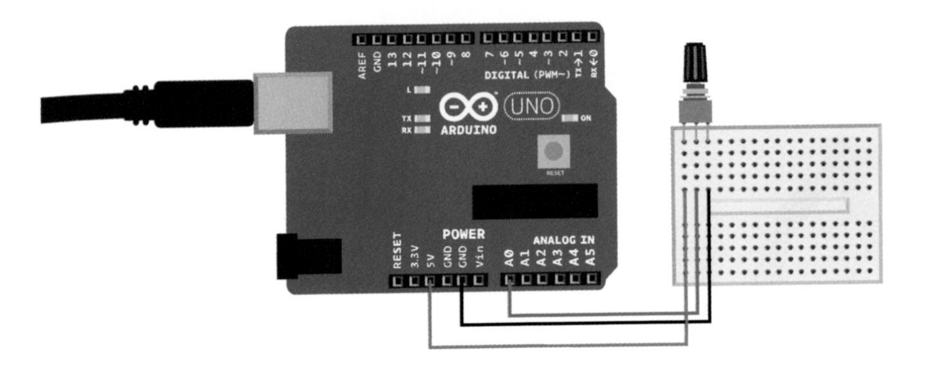

Figure 5-10. *Potentiometer circuit for Arduino*

Example 5-5. pot.ino

```
// pot.ino - control LED blinking speed with potentiometer
// (c) BotBook.com - Karvinen, Karvinen, Valtokari


int potPin=A0; // ❶
int ledPin=13;
int x=0; // 0..1023

void setup() {
  pinMode(ledPin, OUTPUT);
  pinMode(potPin, INPUT);
}

void loop() {
  x=analogRead(potPin); // ❷
  digitalWrite(ledPin, HIGH);
  delay(x/10); // ms    // ❸
  digitalWrite(ledPin, LOW);
  delay(x/10);
}
```

❶ You must use analog pins to read analog resistance sensors.

❷ Now x has a raw analogRead() value, from 0 (0 V) to 1023 (5 V).

❸ Potentiometer resistance controls the blinking delay. The delay varies from 0 ms to about 100 ms (1023/10).

Potentiometer Code and Connection for Raspberry Pi

Figure 5-11 shows the circuit diagram for using a pot with Raspberry Pi. Wire it up as shown, and run the program shown in Example 5-6.

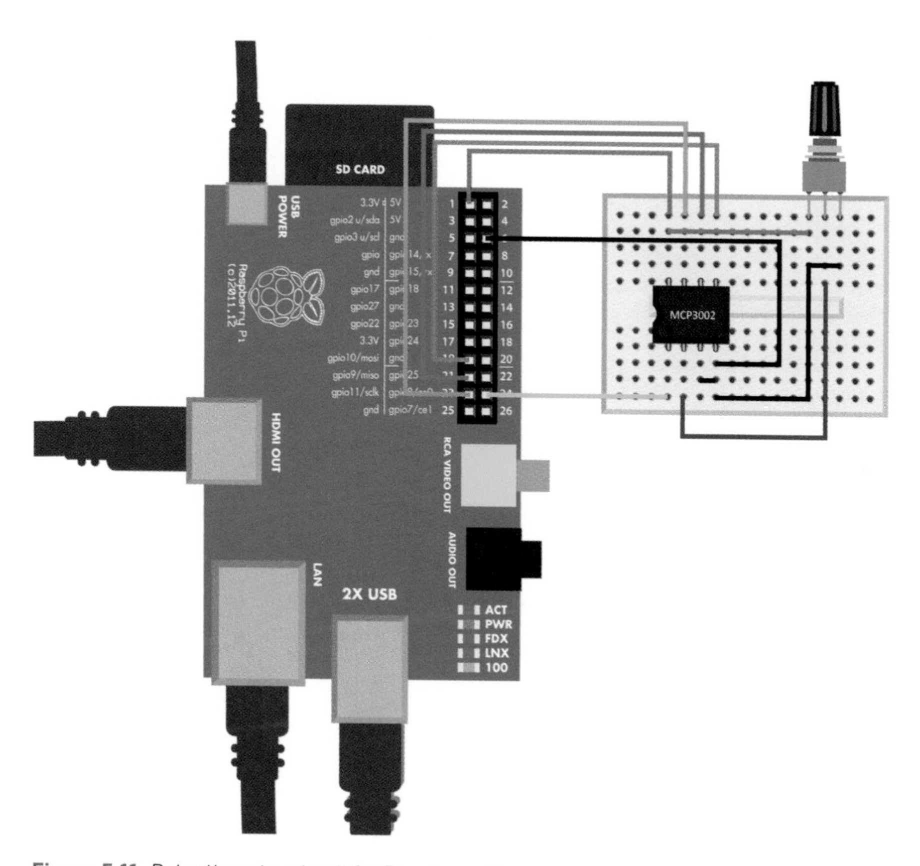

Figure 5-11. *Potentiometer circuit for Raspberry Pi*

Unlike Arduino, Raspberry Pi doesn't have a built-in analog-to-digital converter. This means that all circuits using analog resistance sensors are more complicated with Raspberry Pi than with Arduino.

A potentiometer is a very simple component, as you can see by looking at the Arduino code for it. It's taking that analog reading that's a bit harder for Raspberry Pi.

In an earlier chapter, you took the easy way and read analog values with our *botbook_mcp3002* library ("Compound Eye Code and Connections for Raspberry Pi" on page 54). But did it make you wonder how the *botbook_mcp3002* library itself works?

This Raspberry Pi code introduces the underlying code needed to talk to the MCP3002 analog-to-digital converter. So instead of just having you use the library, this code shows you how to read values directly from that chip with the SPI interface, thus making it easier to understand how that library works (should you decide to delve into its inner workings).

Other code examples use the library, making them much simpler.

Example 5-6. pot.py

```
# pot.py - potentiometer with mcp3002
# (c) BotBook.com - Karvinen, Karvinen, Valtokari
import spidev # installation help in book and botbook_mcp3002.py        # ❶
import time

def readPotentiometer():
        spi = spidev.SpiDev()   # ❷
        spi.open(0, 0)  # ❸
        command = [1, 128, 0]   # ❹
        reply = spi.xfer2(command)      # ❺
        #Parse reply 10 bits from 24 bit package
        data = reply[1] & 31    # ❻
        data = data << 6        # ❼
        data = data + (reply[2] >> 2)   # ❽
        spi.close()     # ❾
        return data

def main():
        while True:
                potentiometer = readPotentiometer()     # ❿
                print("Current potentiometer value is %i " % potentiometer)
                time.sleep(0.5)

if __name__ == "__main__":
        main()
```

❶ The *spidev* library makes it much easier to use SPI.

❷ Create a new `SpiDev` object and store it to the variable `spi`.

❸ Open the first channel of the first connected device.

❹ The first channel is 128; the second is 128+64.

❺ Transfer one byte and read the reply.

❻ Now you have to *parse* the reply, and find the interesting 10 bits out of 24 bits. We learned the format of the reply from the MCP3002 data sheet: take the second byte (1) from the reply, and perform a bitwise AND with 31 (which in binary is 00011111, which can also be written as `0b 0001 1111`). After the operation, the `data` variable contains the five rightmost bits of reply[1]. If you want to understand bitwise operations in detail, see "Bitwise Operations" on page 221.

❼ Shift `data` six bits left. For example, `0b 0001 1001` becomes `0b 0110 0100 0000` (we added spaces between digits to make it easier to read the long binary number). Now the rightmost six bits are zeroes.

❽ Fill the six rightmost bits with the third byte of `data`, `data[2]`.

❾ Release the SPI bus.

❿ The main program doesn't care about the implementation details we just described. Here, the main program's life is made simple with a call to `readPotentiometer()`.

Look demanding? The rest of the code in this book uses the *botbook_mcp3002* library, so you can concentrate on the bigger picture and let the library deal with all the bitwise operations.

Experiment: Sense Touch Without Touch (Capacitive Touch Sensor QT113)

The secret of a capacitive touch sensor is that it doesn't sense touch at all. Instead, it measures how long it takes to load a piece of wire electrically. If there is a human (a big sack of water) nearby, it takes longer to load the wire.

Figure 5-12. *QT113 capacitive touch sensor IC*

The QT113 is an IC (integrated chip) for capacitive sensing. The protocol is simple: when a touch is detected, the output pin goes LOW.

A good capacitive sensor needs some kind of surface. In this experiment, you'll use a piece of metal wire turned into a spiral. A piece of aluminum foil could also work. The circuit also uses a 10-500 nF capacitor.

There are many ways to do capacitive sensing:

- A piece of wire and a simple timer
- A piece of wire and CapSense library
- A specialized chip (like QT113)

You may use a capacitive sensor every day; if you have a smartphone, its touchscreen is probably capacitive.

Ground your sensor. This kind of capacitive sensing requires a real ground, and a battery is not enough. For example, using an Arduino powered by a laptop doesn't work reliably, until you connect your laptop to main power.

QT113 Code and Connection for Arduino

Figure 5-13 shows the circuit for using the QT113 with Arduino. Hook it up as shown, and then run the sketch shown in Example 5-7.

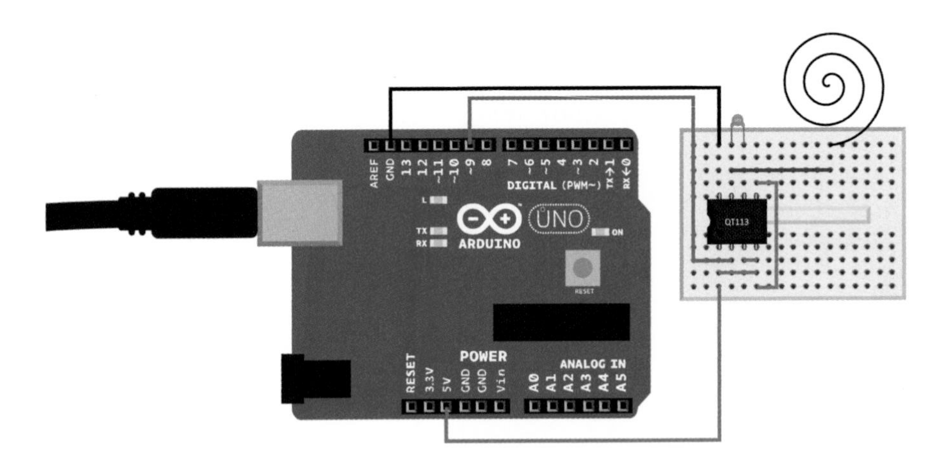

Figure 5-13. *QT113 circuit for Arduino*

Example 5-7. *qt113.ino*

```
// qt113.ino - qt113 touch sensor
// (c) BotBook.com - Karvinen, Karvinen, Valtokari

int sensorPin = 9;

void setup()
{
```

```
  pinMode(sensorPin, INPUT);
  Serial.begin(115200);
}

void loop()
{
  int touch = digitalRead(sensorPin);    // ❶
  if(touch == LOW) {
    Serial.println("Touch detected");
  } else {
    Serial.println("No Touch detected");
  }
  delay(100);
}
```

❶ It's a simple digital switch sensor.

QT113 Code and Connection for Raspberry Pi

Figure 5-14 shows the connections for using Raspberry Pi with the QT113. Wire it up and run the program shown in Example 5-8.

Example 5-8. qt113.py

```
# qt113.py - read touch information from QT113
# (c) BotBook.com - Karvinen, Karvinen, Valtokari

import time
import botbook_gpio as gpio

def main():
        limitPin = 23
        gpio.setMode(limitPin, "in")
        while True:
                if gpio.read(limitPin) == gpio.LOW:     # ❶
                        print("Touch detected!")
                time.sleep(0.5)

if __name__ == "__main__":
  main()
```

❶ It's a simple digital switch sensor.

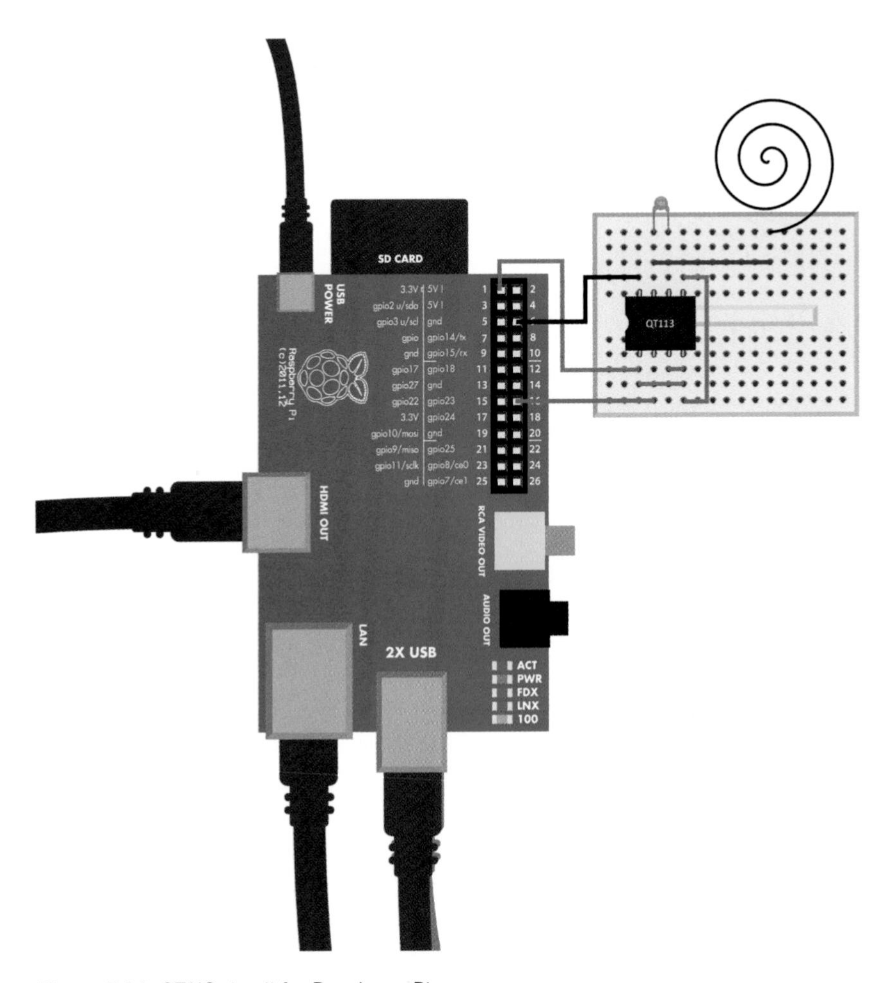

Figure 5-14. *QT113 circuit for Raspberry Pi*

Environment Experiment: Sensing Touch Through Wood

When properly adjusted, a capacitive touch sensor can detect a hand through solid objects. The result is very convincing, as it feels as though the object is the sensor (Figure 5-15). For example, we made a bookshelf that changed the color of RGB LED lights when touched.

In reality, the wood does not affect capacitive sensing. The sensor behind the wood works through wood almost as well as it does through air.

Figure 5-15. *For the user, the sensor looks like wood*

To try this out, use the code and the connection for QT113 as you learned earlier. This time you'll need to modify the touch wire to amplify its capability to detect changes in the electromagnetic field. Basically, you need a bigger chunk of metal on the other end the wire. Making a spiral or attaching some aluminum foil has worked well for us (Figure 5-16).

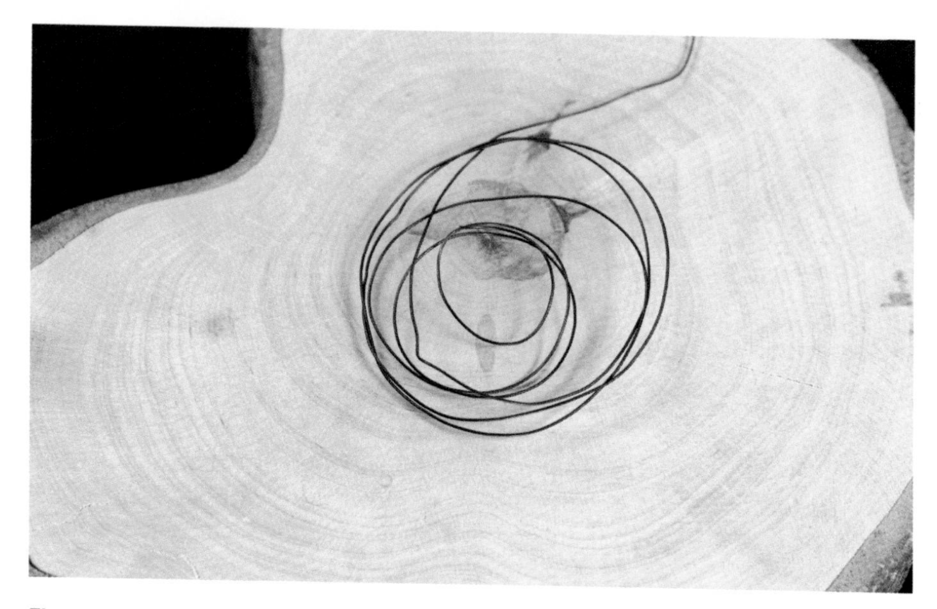

Figure 5-16. *Touch wire spiral, hidden from user*

Experiment: Feel the Pressure (FlexiForce)

The FlexiForce sensors measures squeeze (pressure) on its round head (Figure 5-17).

FlexiForce is a simple analog resistance sensor. The more you squeeze, the lower the resistance. It has just two leads. The third lead in the middle is not connected, but is there just to make connecting it easier. As it's just a resistor, it has no polarity, and it doesn't matter which way you connect it. You can even test it with a multimeter that's on the resistance setting.

Our students have used FlexiForce for an alarm that silences only when you get out of bed, a squeezing strong man competition, and for measuring squeezing precision. If you need a larger area for stepping over, you can put a piece of wood over the FlexiForce.

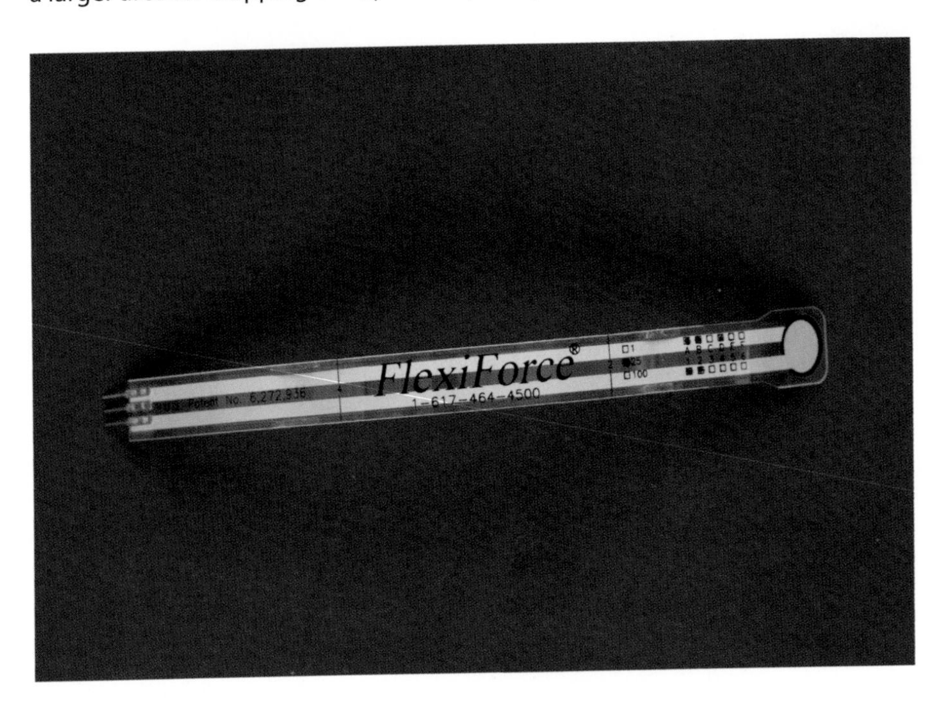

Figure 5-17. *FlexiForce*

FlexiForce Code and Connection for Arduino

FlexiForce is an analog resistance sensor. As Arduino has a built-in analog-to-digital converter, reading the value is a simple call to `analogRead()`.

A 1 megaohm pull-up resistor is used to avoid a floating pin. For an explanation of pull-up resistors, see "Pull-Up Resistor" on page 90. Figure 5-18 shows the wiring diagram, and Example 5-9 shows the sketch. After you wire up the Arduino, load the sketch and run it.

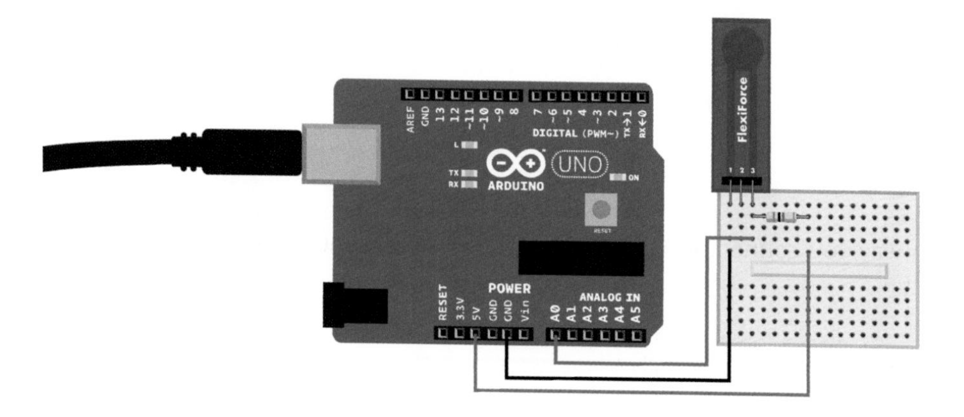

Figure 5-18. *FlexiForce connection for Arduino*

Example 5-9. flexiforce_25.ino

```
// flexiforce_25.ino - send flexiforce squeeze values to computer serial monitor
// (c) BotBook.com - Karvinen, Karvinen, Valtokari

int squeezePin=A0;      // ❶
int x=-1; // 0..1023

void setup() {
  pinMode(squeezePin, INPUT);
  Serial.begin(115200); // bit/s
}

void loop() {
  x=analogRead(squeezePin);  // ❷
  Serial.println(x);
  delay(500); // ms
}
```

❶ Arduino's predefined constants (`A0`, `A1`, `A2...`) are the best way to refer to analog pins in Arduino.

❷ As with any analog resistance sensor, `analogRead()` returns the voltage of the pin, from 0 (0 V) to 1023 (+5 V).

FlexiForce Code and Connection for Raspberry Pi

The Raspberry Pi connection is similar to other analog resistance sensors (Figure 5-19). Because Raspberry Pi doesn't have a built-in analog-to-digital converter, we use an external MCP3002 chip. Because the FlexiForce has just two leads, we use a pull-up resistor to avoid having a floating pin.

As the implementation details of MCP3002 analog converter are handled in the *botbook_mcp3002* library, the main program itself is quite simple. Wire up the Raspberry Pi as shown in Figure 5-19, and then run the code shown in Example 5-10.

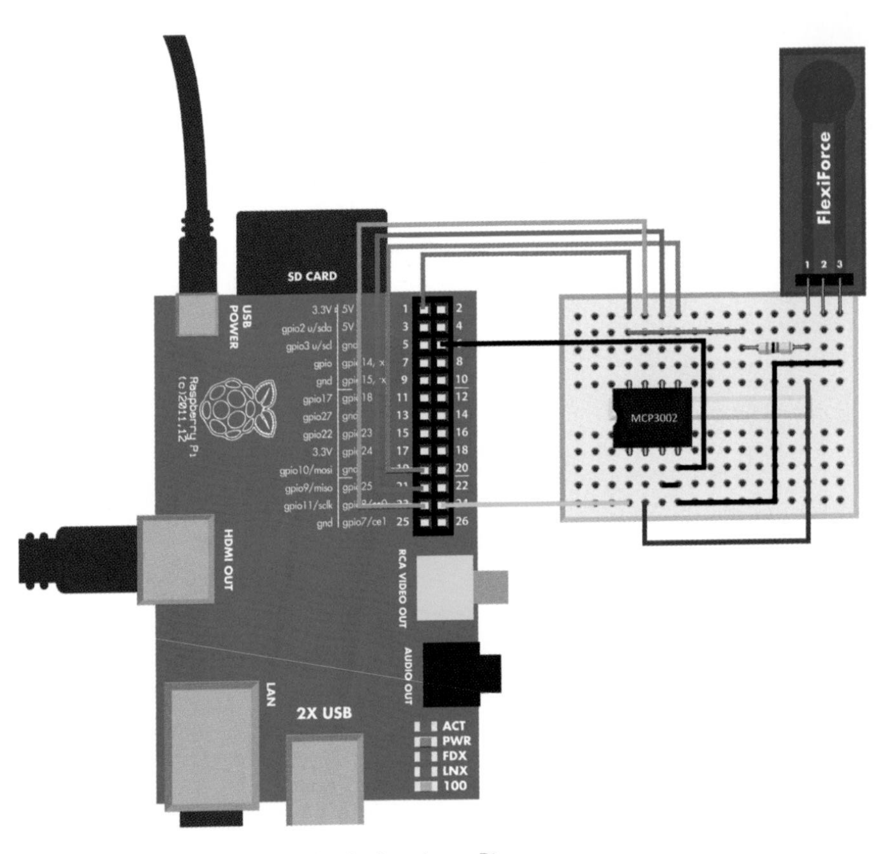

Figure 5-19. *FlexiForce connection for Raspberry Pi*

Example 5-10. *flexiforce.py*

```
# flexiforce.py - sense force and print value to screen.
# (c) BotBook.com - Karvinen, Karvinen, Valtokari
import time
import botbook_mcp3002 as mcp # ❶

def readFlexiForce():
        return mcp.readAnalog() # ❷

def main():
        while True:     # ❸
                f = readFlexiForce()    # ❹
                print("Current force is %i " % f)       # ❺
                time.sleep(0.5) # s     # ❻
```

```
if __name__ == "__main__":
        main()
```

❶ The *botbook_mcp3002.py* library file must be in the same directory as *flexiforce.py*. You must also install the *spidev* library, which is imported by *botbook_mcp3002*. See the comments in the beginning of *botbook_mcp3002/botbook_mcp3002.py* or "Installing SpiDev" on page 56.

❷ Read the first sensor connected to MCP3002. The device and channel parameters are not needed when both are 0.

❸ Embedded devices usually run as long as there is power. Use Control-C if you want to kill the program. When running a long or infinite loop, remember to add `delay()` to the end of the loop.

❹ You can easily use `readFlexiForce()` in bigger projects when it's in a separate function like this. The function name explains what it does, so no comment is needed.

❺ Print the force. The string printed is formatted with a *format string*, where the value of `f` replaces `%i`.

❻ Some delay is needed for the long loop (to avoid 100% CPU utilization) and also to allow the user to read the printed text.

Experiment: Build Your Own Touch Sensor

If capacitive sensing is just measuring the time it takes to load an electrical charge, could you build one yourself, and avoid using the QT113 chip? It's possible to sense touch with just aluminum foil, a resistor, and good ground for your Arduino board.

Capacitive sensing is a practical matter. Even though the principle is simple, implementation is precise business. In addition to good grounding, the measurements must use sliding averages to smooth out any random fluctuation.

For reliable ground, the power source of Arduino must be connected to wall socket. If you are using a laptop, connect the laptop power cord into a wall socket. If you're using a desktop, it's of course always connected. You can also use a USB charger connected to wall socket.

The measuring part, aluminum foil, is connected between two data pins. Connecting something between data pins like this is quite rare—you usually connect one wire to ground or +5 V and the other to a data pin.

The code charges one data pin, and waits for however long it takes for the other data pin to reach the same charge (same voltage). As capacitance is the ability to hold charge, a big object like a human nearby will affect capacitance and result in a different charging time.

The smaller 10 kOhm resistor (brown-black-orange) helps protect against static electricity.

The big 1 MOhm to 50 MOhm resistor selects the sensitivity. The bigger the resistor, the farther away it detects a human. The trade-off for bigger detection distance is a slower reading speed.

Figure 5-20 shows the connection diagram for the Arduino, and Example 5-11 shows the sketch. Wire up the Arduino as shown, and then upload and run the code.

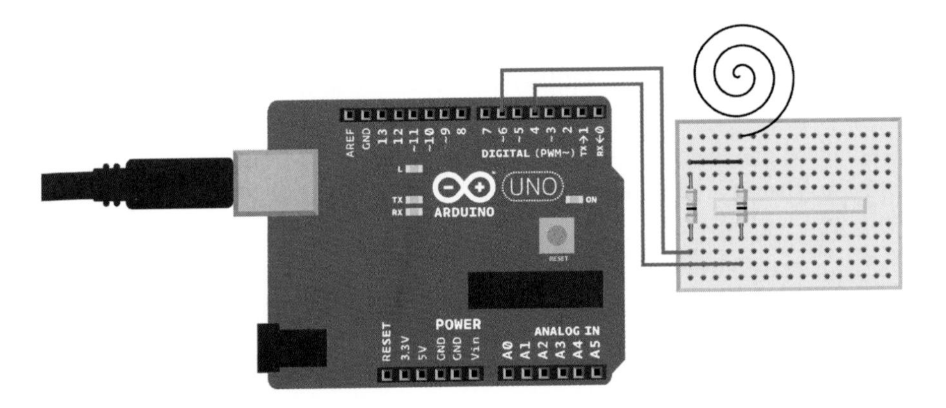

Figure 5-20. *Capacity sensor connection for Arduino*

Example 5-11. diy_capacitive_sensor.ino

```
// diy_capacitive_sensor.ino - measure touch
// (c) BotBook.com - Karvinen, Karvinen, Valtokari

const int sendPin = 4;
const int readPin = 6;

void setup() {
   Serial.begin(115200);
   pinMode(sendPin,OUTPUT);
   pinMode(readPin,INPUT);
   digitalWrite(readPin,LOW);
}

void loop() {
  int time = 0;

  digitalWrite(sendPin,HIGH);
  while(digitalRead(readPin) == LOW) time++;
  Serial.println(time);
  digitalWrite(sendPin,LOW);
  delay(100);
}
```

Capsense Code and Connection for Raspberry Pi

Figure 5-21 shows the wiring diagram for Raspberry Pi. Wire it up as shown, and then run the code shown in Example 5-12.

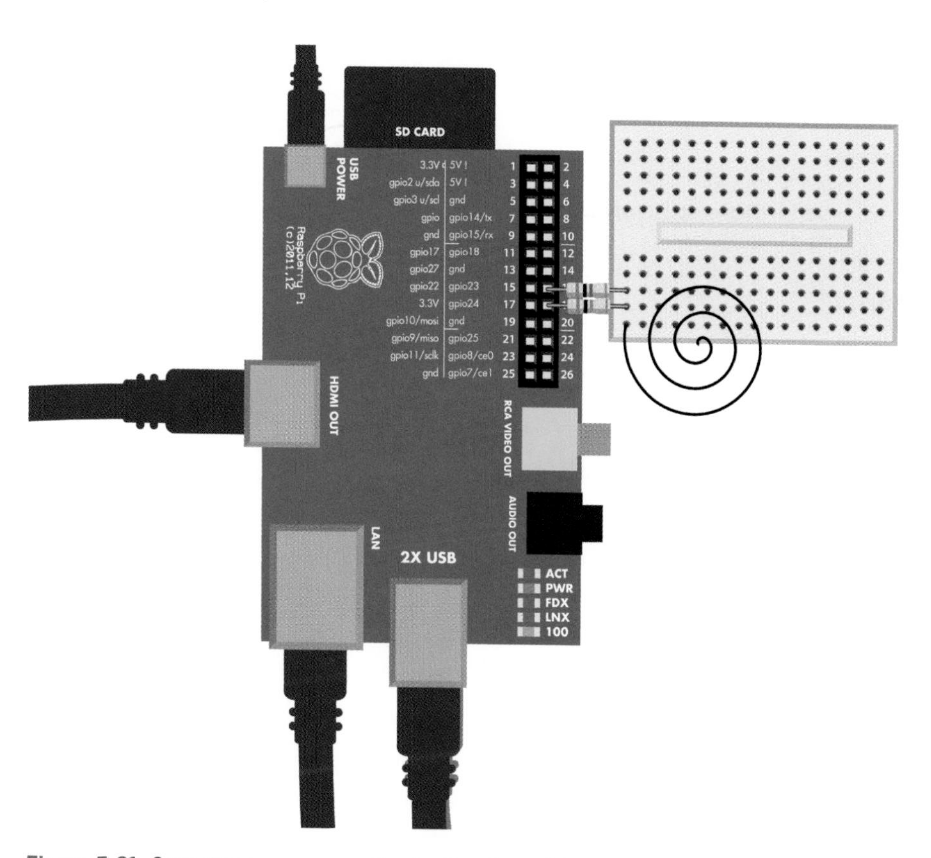

Figure 5-21. *Capacity sensor connection for Raspberry Pi*

Example 5-12. *diy_capacity_sensor_simple.py*

```
# diy_capacity_sensor_simple.py - read touch from diy capacity sensor.
# (c) BotBook.com - Karvinen, Karvinen, Valtokari
import time
import botbook_gpio as gpio

def sample(count):
        sendPin = 23
        recievePin = 24
        gpio.mode(sendPin,"out")
        gpio.mode(recievePin,"in")
        gpio.write(sendPin,0)
        total = 0
        # set low
        for x in xrange(1,count):
```

```
                time.sleep(0.01)
                gpio.write(sendPin,gpio.HIGH)
                while(gpio.read(recievePin) == False):
                        total += 1
                gpio.write(sendPin,gpio.LOW)
        return total


def main():
        while True:
                touch = sample(30)
                print("Touch: %d" % touch)
                time.sleep(0.5)

if __name__ == "__main__":
        main()
```

Test Project: Haunted Ringing Bell

You enter a silent-looking reception desk. There is nobody in sight to check you in. You reach for the bell on the desk (Figure 5-22), but before you touch it… it rings!

Figure 5-22. *Haunted bell*

In this project we combine a capacitive touch sensor with an old-school ringing bell. The result is a bell that makes a sound just before you touch it. Even the bell knob moves by itself, giving this gadget a nice ghostly charm. You'll also learn how to use a versatile new component, a servo motor.

What You'll Learn

In the *Haunted Ringing Bell* project, you'll learn how to:

- Build a gadget that reacts to your hand before you touch it.
- Make things move with servo motors.
- Control servo motors.
- Package a project to look like an innocent, everyday object.

Servo Motors

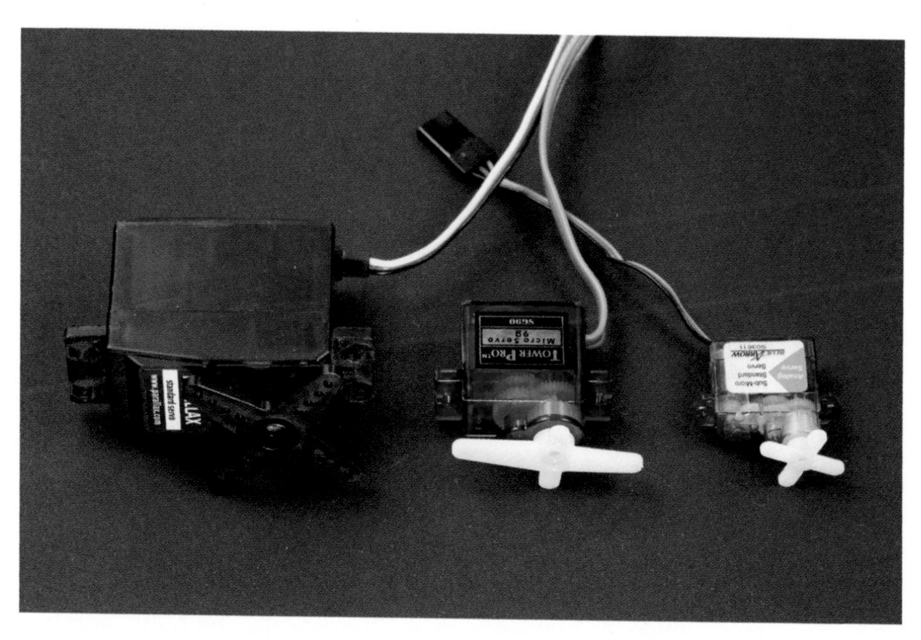

Figure 5-23. *Different servos*

A servo is a motor you can precisely control (Figure 5-23). You can tell a standard servo to turn to a specific angle, such as 90 degrees. There are also full rotation servos where you control just the direction and rotation speed.

If you ever think you need a motor for your project, consider a servo first. The servo is in contrast to common DC motors, which are difficult to control and require extra components just for changing direction. Most moving things in Arduino and Raspberry Pi prototyping projects are done with servos.

A servo has three wires: black ground (0 V), red positive (+5 V), and control (yellow or white). Arduino sends a continuous stream of pulses to control wire. The length of these pulses tell the servo which angle to move to. The servo turns to this angle and then maintains it as long as pulses keep coming from Arduino.

The square wave (the pulse) is easy to create by rapidly changing a digital output pin between LOW and HIGH. To control the servo, Arduino must send 50 pulses a second, in other words, at the rate of 50 Hz.

```
50/s = 50 * 1/s = 50 * Hz = 50 Hz
```

The pulses are about 1 ms to 2 ms long. The longer the pulse, the bigger the angle. The pulse length between minimum and maximum angle centers the servo. Often, the center is about 1.5 ms.

For most servos, it's not easy to find data sheets. But the update frequency (pulses per second) is usually the same, and it does not have to be exact.

The actual pulse length varies between servos, but it's easy to find it out experimentally. To get an idea of pulse length versus angle, see the example values in Table 5-1.

Table 5-1. Pulse length controls servo angle, example values

Pulse length ms	Pulse length µs	Angle	Comment
0.5 ms	500 µs	< -90 deg	Trying to turn over range, ugly sound from gears
1 ms	1000 µs	-90 deg	Extreme left
1.5 ms	1500 µs	0%	Centered
2 ms	2000 µs	90%	Extreme right
2.5 ms	2500 µs	> 90 deg	Over range, ugly sound

Finding Servo Range

Even though you already know that servos use pulses from 1 ms to 2 ms, how does this help with the specific servo you have? You can find out the servo range experimentally. This code is essentially the "Hello World" for servos.

We always test a servo's range when we buy new servos. Theoretically, the pulse length information should be available on the manufacturer's website somewhere. In practice, many servos don't have data sheets, or the correct data sheet is not easy to find.

We test every sensible pulse length from too small to too large (Example 5-13). As the code prints the pulse length to screen, we notice the key points and write them down (Table 5-2). The wiring diagram is shown in Figure 5-24.

Run the program. Open the serial monitor on Arduino IDE to see the pulse lengths printed (Tools→Serial Monitor). Set the serial monitor to the same speed ("baud", bit/s) that the code uses.

Table 5-2. Key pulse lengths for servo calibration

Pulse length µs	Angle	Comment
	-90 deg	Extreme left (test with code)
	0 deg	Middle, mean of extreme left and right
	+90 deg	Extreme right test with code)

First, the pulse is way too short; it's asking the servo to turn past its range. The servo doesn't turn, but it might shake a bit and the gears could make a tiny but ugly noise. A short burst of this ugly noise is not harmful to the servo. When the servo starts turning, notice the point on the serial monitor. This is the -90 deg pulse length for extreme left.

When the servo stops turning, notice the pulse length. This is the pulse length for +90 degrees, extreme right.

The pulse length for the middle is the mean of maximum left and right. For example, if the extremes are 1 ms and 2 ms, the middle is 1.5 ms.

Some servos have a built-in potentiometer to set the center. If you have a servo like this, you can send the pulse for middle and experiment with the pot.

Example 5-13. servo_range.ino

```
// servo_range.ino - turn servo and print values to serial
// (c) BotBook.com - Karvinen, Karvinen, Valtokari
int servoPin=2;

void pulseServo(int servoPin, int pulseLenUs)   // ❶
{
        digitalWrite(servoPin, HIGH);    // ❷
        delayMicroseconds(pulseLenUs);   // ❸
        digitalWrite(servoPin, LOW);     // ❹
        delay(15);      // ❺
}

void setup()
{
        pinMode(servoPin, OUTPUT);
        Serial.begin(115200);
}

void loop()
{
        for (int i=500; i<=3000; i=i+2) {       // ❻
                pulseServo(servoPin, i);        // ❼
                Serial.println(i);
        }
}
```

❶ To send one short pulse, call `pulseServo()`. The function is not run here where it's defined, only when it's called later. As the servo pin is given as a parameter, multiple calls to this function can control multiple servos.

❷ The pin is expected to be LOW first, meaning no pulse is being sent. Then you turn it HIGH, and the pulse starts.

❸ Wait for a very short time. As the variable name implies, the unit is microseconds (µs), millionths of a second. Typical values are in Table 5-1. The purpose of this sketch is to find the exact values for your servo.

❹ Turn the pin LOW, ending the pulse.

❺ Wait for a while. The 15 ms time is short for humans but about 10 times longer than the length of the pulse. The pulses must be sent at least 50 times a second, so a pulse must be sent every 1/50 s = 0.02 s = 20 ms. We chose to wait a bit less than that.

❻ Run the loop body with increasing pulse lengths. Start from the way-too-small 500 µs (0.5 ms), and end up at the way-too-high 3000 µs (3 ms). If you've forgotten how `for` loops work, see "What For?".

❼ Send a short pulse to the servo. As a delay is included in `pulseServo()`, you don't need to wait in the main loop.

What For?

The for loop is an eloquent way to say "run this code for a number of iterations." Here is the syntax:

```
for(initialization; condition;
afterthought)
{
  body
}
```

For example:

```
for(int i=0; i<3; i++) {
  Serial.print(i);
}
```

Only once, when the loop starts, is the initialization run. That's also where you declare the loop variable.

When loop is about to start, the condition is checked. If the condition is false, the for loop exits. If true, then the body of the loop is run.

Finally, the afterthought is run. Typically, the loop counter is increased there. Then the condition is checked again.

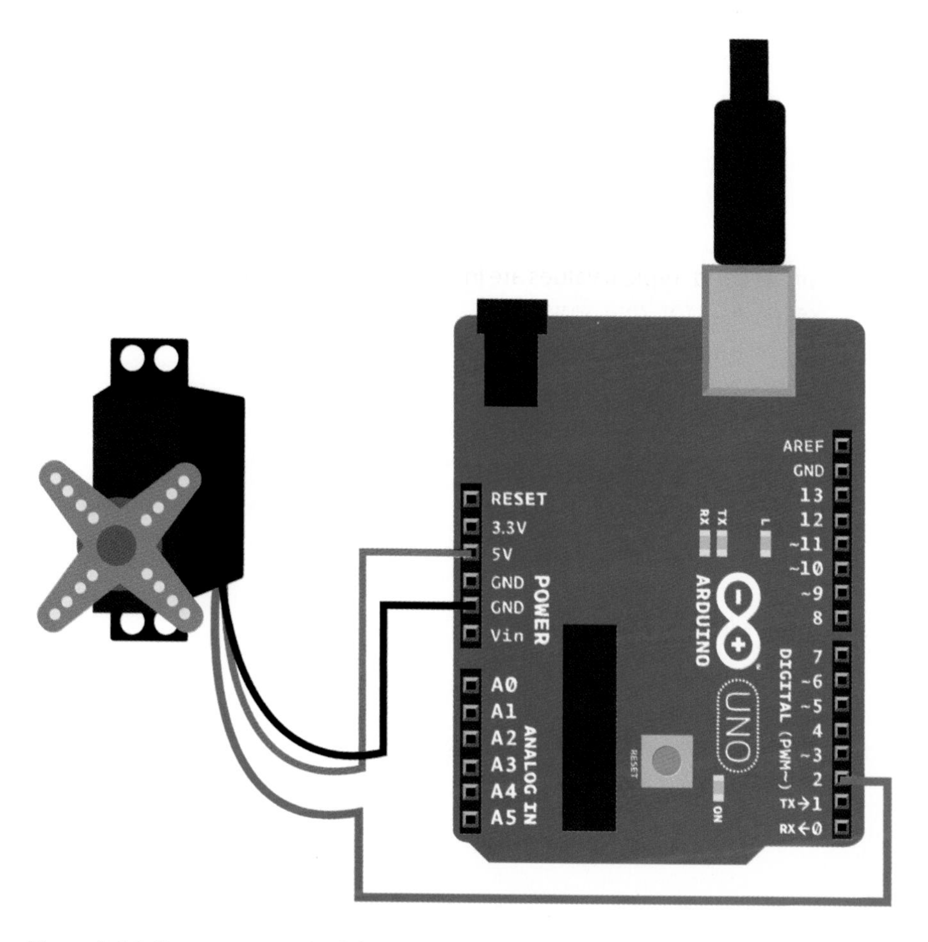

Figure 5-24. *Servo connected to D2 for servo_range.ino*

It's also possible to control servos with the built-in Arduino library, ***Servo.h***. ***Servo.h*** *comes with the Arduino IDE and provides an object-oriented interface to servos. It lets you control the servos by specifying degrees, like* `myservo.write(180)`*. You could find it handy when you need to control many standard servos. However, using the Arduino Servo library disables the use of* `analogWrite()` *on pins 9 and 10, and we prefer to keep that functionality enabled, so we don't use the library. See http://arduino.cc/en/reference/servo for more information.*

Haunted Bell Code and Connection for Arduino

Connect the servo motor and capacitive touch sensor as in the earlier examples (see Figure 5-25). Again you'll need to use an 10-500 nF capacitor with QT113. The best value for us was 300 nF.

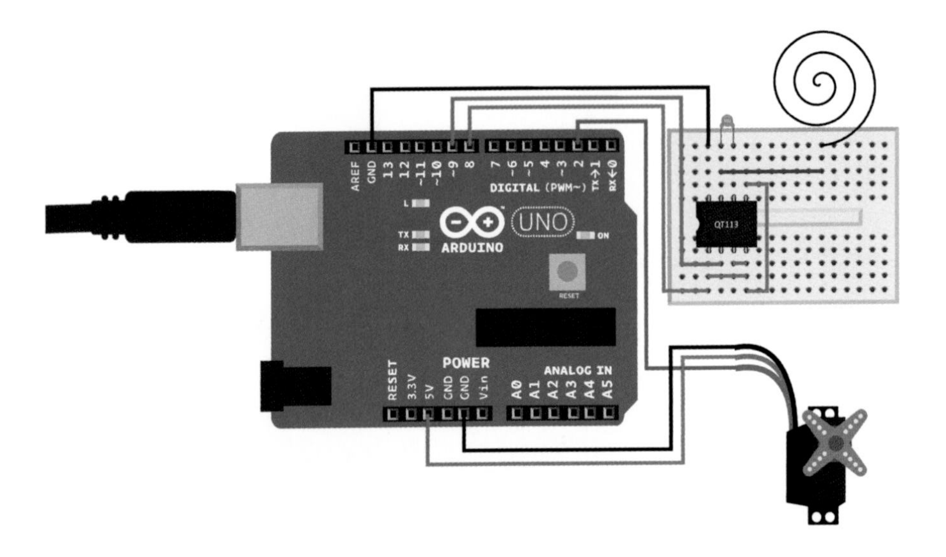

Figure 5-25. *Haunted bell connections*

Example 5-14. *haunted_bell.ino*

```
// haunted_bell.ino - bell rings just before you touch it
// (c) BotBook.com - Karvinen, Karvinen, Valtokari


int servoPin=2;
int sensorPin = 9;
int sensorPowerPin = 8;
int hasRang = 0;          // ❶

void pulseServo(int servoPin, int pulseLenUs)    // ❷
{
  digitalWrite(servoPin, HIGH);
  delayMicroseconds(pulseLenUs);
  digitalWrite(servoPin, LOW);
  delay(20);
}

void cling()    // ❸
{
    for (int i=0; i<=3; i++) {  // ❹
      pulseServo(servoPin, 2000);
    }
    for (int i=0; i<=100; i++) {
      pulseServo(servoPin, 1000);
    }

}
```

```
void setup()
{
  pinMode(servoPin, OUTPUT);
  pinMode(sensorPowerPin, OUTPUT);
  digitalWrite(sensorPowerPin,HIGH);
  pinMode(sensorPin,INPUT);
}

void loop()
{
  int touch = digitalRead(sensorPin);   // ❺
  if(touch == HIGH) {   // ❻
    hasRang = 0;
  }
  if(touch == LOW && hasRang == 0) {    // ❼
    cling();    // ❽
    hasRang = 1;         // ❾
    digitalWrite(sensorPowerPin,LOW);
    delay(1);
    digitalWrite(sensorPowerPin,HIGH);
  }
  delay(100);
}
```

❶ The variable `hasRang` will be 1 if the bell has rung without the touch being removed. This will help make sure the bell rings only once. (We used integer 1 for true and 0 for false to avoid introducing a lot of Boolean logic, but feel free to use Bool in your programs if you prefer.)

❷ Servo control is explained in Example 5-13.

❸ A function to ring the bell once. It's in a separate function to make main `loop()` easier to read, and because it's one simple thing to do.

❹ To give the servo some time to move and hit the bell, you must send multiple pulses. As one `pulseServo()` takes about 20 ms, 100 iterations is 2 seconds. So one `ring()` runs about two seconds.

❺ `SensorPin` is LOW when there is touch.

❻ If there is no touch (HIGH), reset `hasRang`. Now the bell will ring again if a hand comes near.

❼ If a hand is near *and* (`&&`) it has not rung for this touch yet…

❽ …ring the bell once.

❾ Remember that the bell has rung, to avoid ringing twice for the same near-touch.

Attaching Servo to Ringing Bell

Use hot glue to attach the servo inside the ringing bell (Figure 5-26). Before gluing, make sure that the servo arm pushes the moving part inside the bell so that it gives a solid sound. Servo movement should also allow the bell button to move naturally. It might take a few tries to get the servo attached exactly the right way. Now your haunted ringing bell is ready to spook unsuspecting victims.

Figure 5-26. *Servo inside the bell*

You have now seen many kinds of touch and near-touch. You've played with buttons, microswitches, touch switches, and even used touch sensors without touch. After you've let the bell haunt your friends for a while, you can start applying this knowledge to your own projects. Think of any electronic devices around you: most sense touch in one way or another.

Movement

As you walk to your yard, a light comes on automatically. It's time to relax a bit with your favorite gaming console (we prefer Ouya). To save your ears, you turn down the volume.

Automatic garden lights detect the moving heat radiated by humans. Joysticks convert motion to a change in resistance. A volume knob is a potentiometer, another kind of variable resistor.

Experiment: Which Way Is Up? (Tilt Ball Switch)

If you decide to build a pinball machine, you might want to detect excessive nudging and end the player's turn with a TILT alarm. Also, a burglar alarm could use a tilt sensor.

Figure 6-1. *Tilt ball switch*

Tilt Sensor Code and Connection for Arduino

Figure 6-2 shows the circuit diagram for an Arduino-based tilt sensor. Build it, and then run the code shown in Example 6-1.

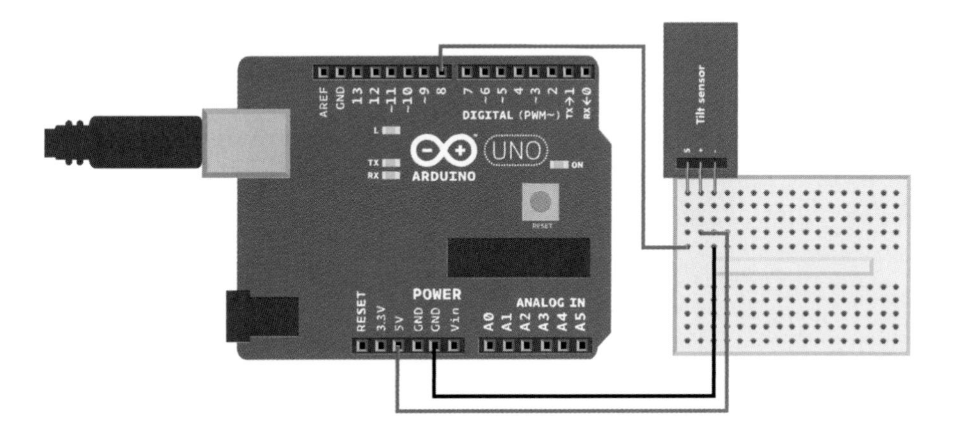

Figure 6-2. *Tilt sensor circuit for Arduino*

Example 6-1. tilt_sensor.ino

```
// tilt_sensor.ino - detect tilting and print to serial
// (c) BotBook.com - Karvinen, Karvinen, Valtokari

const int tiltPin = 8;
int tilted = -1;

void setup() {
  Serial.begin(115200);
  pinMode(tiltPin, INPUT);
  digitalWrite(tiltPin, HIGH);
}

void loop() {
  tilted = digitalRead(tiltPin);         // ❶
  if(tilted == 0) {
    Serial.println("Sensor is tilted");
  } else {
    Serial.println("Sensor is not tilted");
  }

  delay(100);
}
```

❶ It's a simple digital switch sensor.

Tilt Sensor Code and Connection for Raspberry Pi

Figure 6-3 shows the connections for Raspberry Pi. Wire it up and run the program shown in Example 6-2.

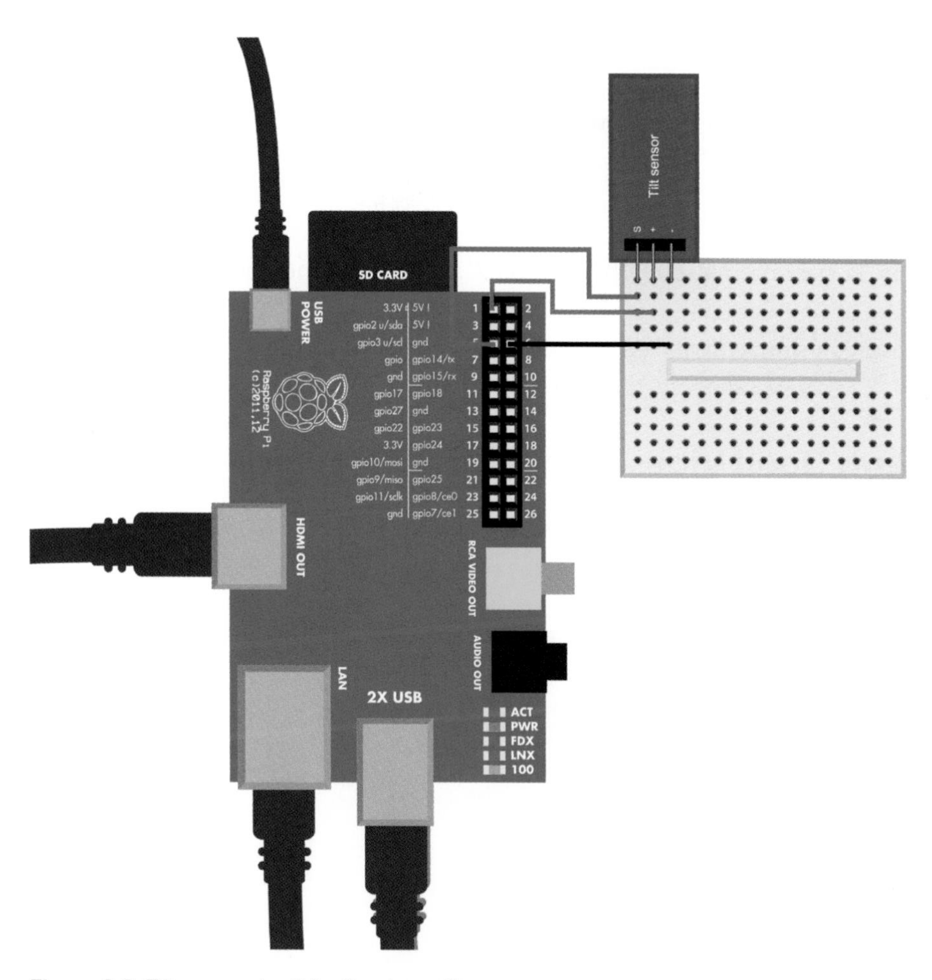

Figure 6-3. *Tilt sensor circuit for Raspberry Pi*

Example 6-2. ***tilt_sensor.py***

```
# tilt_sensor.py - print if sensor was tilted
# (c) BotBook.com - Karvinen, Karvinen, Valtokari

import time
import botbook_gpio as gpio

def main():
        tiltpin = 3     # has internal pull-up  # ❶
        gpio.mode(tiltpin, "in")
```

```
        while (True):
                isNotTilted = gpio.read(tiltpin)        # ❷
                if(isNotTilted == gpio.LOW):
                        print "Sensor is tilted"
                else:
                        print "Sensor is not tilted"
                time.sleep(0.3) # seconds

if __name__ == "__main__":
        main()
```

❶ Raspberry Pi has internal pull-ups on gpio2 and gpio3. They are permanently enabled.

❷ A tilt sensor is a simple digital switch sensor.

Experiment: Good Vibes with Interrupt (Digital Vibration Sensor)

A vibration sensor can detect tiny vibrations, like the ground shaking (Figure 6-4).

Figure 6-4. *Vibration sensor*

Vibration Code and Connection for Arduino

The signal sent by a vibration sensor is very short. You must use an *interrupt* to catch it.

For most sensors, you could just *poll* them. Polling means that you check the state of the digital pin, wait a while, and then check again.

With an interrupt, you can tell Arduino to run a function whenever some event happens. Interrupts make it harder to follow the flow of the code you write, but they allow you to catch very short-lived events, like a pin going up and then down quickly.

Figure 6-5 shows the circuit diagram for the vibration sensor. Wire it up as shown, and then run the code shown in Example 6-3.

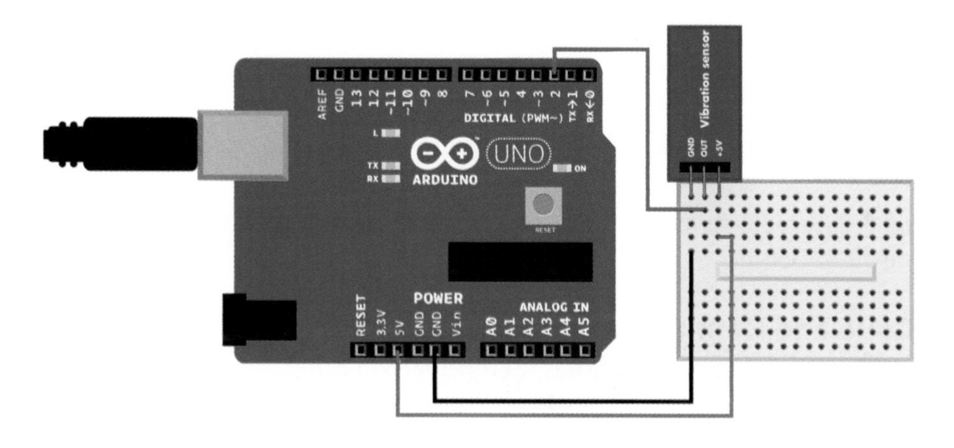

Figure 6-5. *Vibration sensor circuit for Arduino*

Example 6-3. vibration_sensor.ino

```
// vibration_sensor.ino - detect vibration using interrupt
// (c) BotBook.com - Karvinen, Karvinen, Valtokari

const int sensorPin = 0; //UNO,Mega pin 2, Leonardo pin 3
volatile int sensorState = -1;

void setup() {
  Serial.begin(115200);
  attachInterrupt(sensorPin, sensorTriggered, RISING);  // ❶
}

void loop() {
  if(sensorState == 1) {        // ❷
    Serial.println("Vibrations detected");
    delay(1);   // ms
    sensorState = 0;
  }
```

```
  delay(10);
}

void sensorTriggered() {        // ❸
  sensorState = 1;  // ❹
}
```

❶ This tells Arduino to call `sensorTriggered()` when `sensorPin` goes from LOW to HIGH. This is known as a *callback*. Notice that the name of the function in the callback does not have parentheses after it. Without the parentheses, a function name is a *pointer* to the function (whereas including the parentheses would cause the function to be called once and return its value). After all, `attachInterrupt()` needs to know about the function, and isn't ready to run it yet.

❷ The main loop can go on its merry way as slowly as it wants. It checks the value of a global variable.

❸ When the interrupt is triggered, `sensorTriggered()` is called. The interrupt reacts very quickly.

❹ The `sensorTriggered()` function simply sets a global variable, which can be read at the main loop's leisure.

Vibration Code and Connection for Raspberry Pi

Figure 6-6 shows the wiring diagram for the Raspberry Pi version of this sensor circuit. Wire it up as shown and then run the program shown in Example 6-4.

Example 6-4. vibration_sensor.py

```
# vibration_sensor.py - detect vibration
# (c) BotBook.com - Karvinen, Karvinen, Valtokari
import time
import botbook_gpio as gpio     # ❶

def main():
        vibPin = 3
        gpio.mode(vibPin, "in")
        while (True):
                vibrationInput = gpio.read(vibPin)      # ❷
                if(vibrationInput == gpio.LOW):
                        print "Vibration"
                        time.sleep(5)   # ❸
                time.sleep(0.01) # seconds      # ❹

if __name__ == "__main__":
        main()
```

❶ As with previous examples, this library must be in the same directory along with *vibration_sensor.py*.

❷ The vibration sensor is used like a digital switch sensor. You can read it like a button.

❸ If vibration was detected, report it only once instead of filling the screen with "Vibration" texts.

❹ We use a very short 10 ms delay, so `vibPin` is polled 100 times a second (100 Hz).

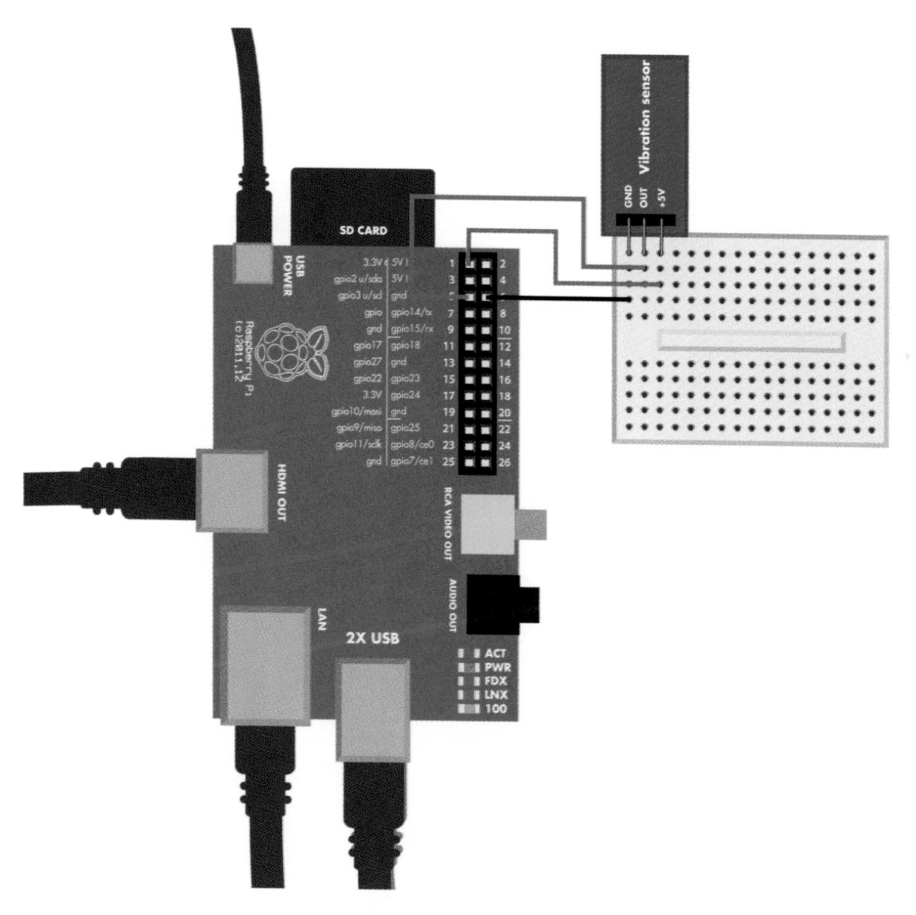

Figure 6-6. *Vibration sensor circuit for Raspberry Pi*

Raspberry Pi code can't detect as short vibrations as Arduino can. The code doesn't use an interrupt, because it would complicate the code, would still need fast looping, and still couldn't match Arduino. If you need very precise vibration detection in Raspberry Pi, you can connect Arduino to Raspberry Pi ("Talking to Arduino from Raspberry Pi" on page 337).

Experiment: Turn the Knob

A rotary encoder measures turning (see Figure 6-7). An absolute rotary encoder tells the position (position is 83 deg), and relative rotary encoders tell the change (turned 23 deg from previous position).

The rotary encoder we use here is a relative one. When you turn the knob, the encoder sends clock pulses on the clock pin. On the *rising edge* (the clock goes from LOW to HIGH), one data pulse comes on the data pin; another comes on the *falling edge* (HIGH to LOW transition).

If the data pulse is HIGH, you're turning right (clockwise, negative direction). If the data pulse is LOW, you're turning left.

Figure 6-7. *Rotary encoder*

Rotary Encoder Code and Connection for Arduino

This code uses interrupts, so it might look different from the Arduino codes you've worked with before. Wire up the circuit as shown in Figure 6-8, and run the sketch shown in Example 6-5.

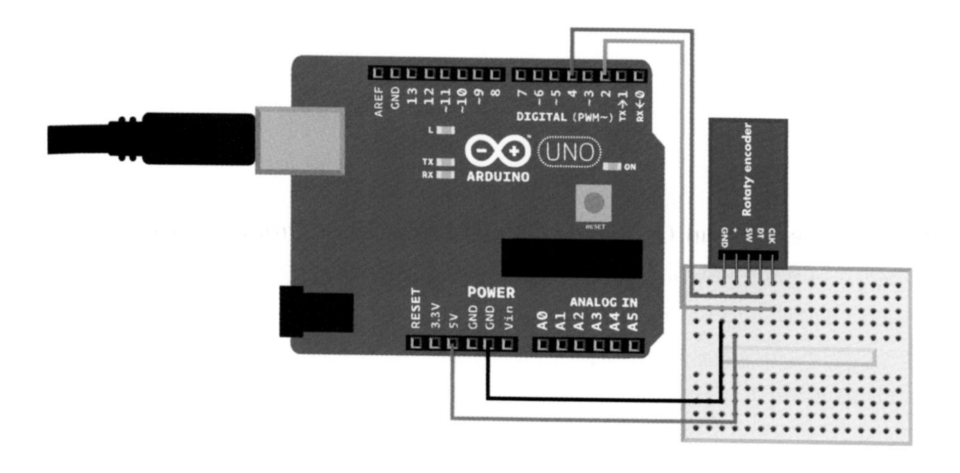

Figure 6-8. *Rotary encoder circuit for Arduino*

Example 6-5. **rotary_encoder.ino**

```
// rotary_encoder.ino - print encoder position
// (c) BotBook.com - Karvinen, Karvinen, Valtokari

const int clkPin = 2;
const int dtPin = 4;

volatile unsigned int encoderPos = 0;    // ❶

void setup()
{
  Serial.begin(115200);
  pinMode(clkPin, INPUT);
  digitalWrite(clkPin, HIGH); // pull up        // ❷
  pinMode(dtPin, INPUT);
  digitalWrite(dtPin, HIGH); // pull up

  attachInterrupt(0, processEncoder, CHANGE);   // ❸
}

void loop()
{
  Serial.println(encoderPos);
  delay(100);
}

void processEncoder()    // ❹
{
  if(digitalRead(clkPin) == digitalRead(dtPin)) // ❺
  {
    encoderPos++;        // turning right
  } else {
```

```
        encoderPos--;
    }
}
```

❶ `EncoderPos` is marked as *volatile*, because it's modified in the interrupt function. The volatile keyword lets the Arduino compiler know that the value could change "behind the back" of the main loop of the Arduino sketch.

❷ Writing `HIGH` to an `INPUT` pin connects it to +5 V through a 20 kOhm pull-up resistor. This is done to avoid a floating pin.

❸ Anytime there is `CHANGE` (rising or falling edge), call the function `processEncoder`. In Arduino Uno, interrupt 0 monitors digital pin 2. The callback function `processEncoder` doesn't have parentheses after it, because it's not needed to run yet. It will run later, when called by the interrupt (this happens behind the scenes).

❹ Anything else is put on hold while the interrupt function runs.

❺ When `dtPin` is in the same state as the clock pin, it indicates a right turn. Otherwise, it's a left turn.

Rotary Encoder Code and Connection for Raspberry Pi

Figure 6-9 shows the wiring diagram for using a rotary encoder with Raspberry Pi. Wire it up as shown and then run the program shown in Example 6-6.

Example 6-6. rotary_encoder.py

```
# rotary_encoder.py - read rotary encoder
# (c) BotBook.com - Karvinen, Karvinen, Valtokari
import time
import botbook_gpio as gpio     # ❶

def main():
        encoderClk = 3
        encoderDt = 2
        gpio.mode(encoderClk, "in")
        gpio.mode(encoderDt, "in")
        encoderLast = gpio.LOW
        encoderPos = 0
        lastEncoderPos = 0
        while True:
                clk = gpio.read(encoderClk)
                if encoderLast == gpio.LOW and clk == gpio.HIGH:        # ❷
                        if(gpio.read(encoderDt) == gpio.LOW):   # ❸
                                encoderPos += 1
                        else:
                                encoderPos -= 1
                encoderLast = clk
                if encoderPos != lastEncoderPos:
                        print("Encoder position %d" % encoderPos)
```

```
                lastEncoderPos = encoderPos
                time.sleep(0.001) # s   # ❹

if __name__ == "__main__":
        main()
```

❶ Make sure there's a copy of the *botbook_gpio.py* library in the same directory as this program. You can download this library along with all the example code from *http://botbook.com*. See "GPIO Without Root" on page 19 for information on configuring your Raspberry Pi for GPIO access.

❷ If a rising edge (LOW to HIGH) is detected…

❸ …when data is LOW on the clock edge, the encoder is turning left (counterclockwise, positive direction).

❹ Sleep only 1 millisecond, so that we don't miss any clicks.

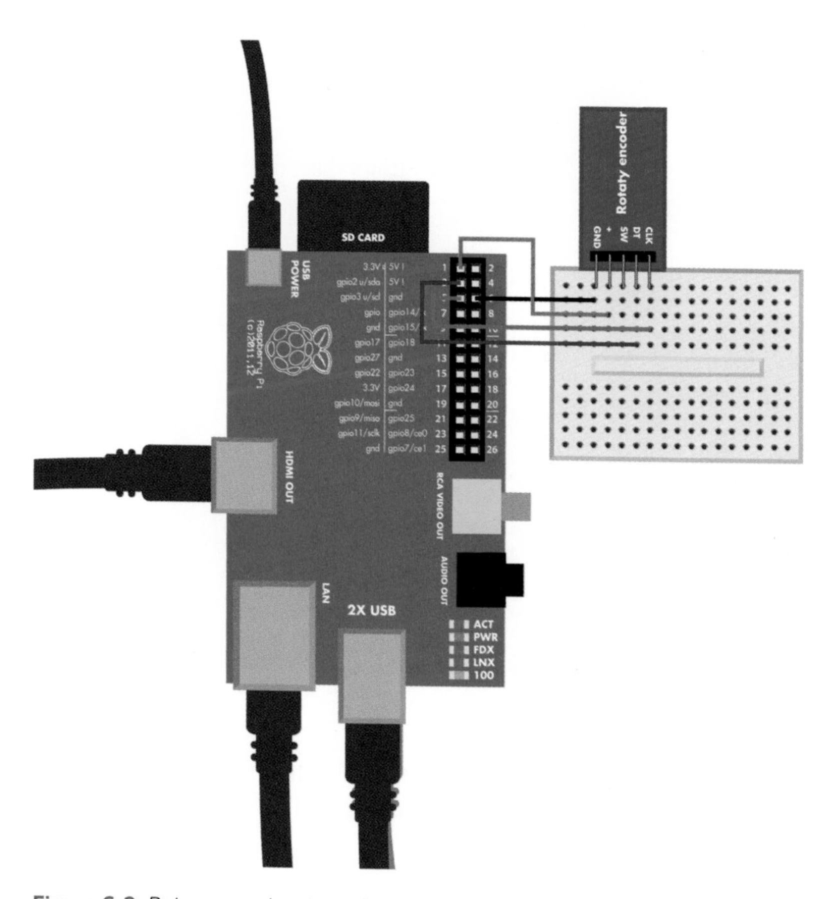

Figure 6-9. *Rotary encoder circuit for Raspberry Pi*

Even though the Raspberry Pi has more processing power, Arduino is much faster (as in more real time). If you feel that Raspberry Pi is missing too many pulses when you turn fast and run a lot of other software on your Raspberry, consider reading the pulses through an Arduino. See "Talking to Arduino from Raspberry Pi" on page 337.

Experiment: Thumb Joystick (Analog Two-Axis Thumb Joystick)

If you've played on any videogame consoles, you have used a joystick. In the old days, you grabbed a big joystick with your strong hand. Modern gaming consoles like Xbox, PlayStation, or Ouya use multiple thumb joysticks.

Figure 6-10. *Two-axis thumb joystick*

Typically, a joystick has two potentiometers (pots), which are variable resistors. Tilting the joystick along the vertical y-axis changes the resistance of one pot. Tilting along the horizontal x-axis changes the resistance of another pot.

In many joysticks, the potentiometers are in three-lead configuration. They have one lead on the ground, another on +5 V, and one in the middle, giving a varying voltage between 0 V and +5 V. In this three-lead configuration, no pull-up or pull-down resistors are needed.

Mobile phones and Wii consoles can use accelerometers instead of joysticks. The device attitude is measured against gravity and is used for controlling games. (See "Experiment: Hacking Wii Nunchuk (with I2C)" on page 225.) Gravity has the same effect as acceleration. To get precise attitude measurements, see Chapter 8.

Joystick Code and Connection for Arduino

Figure 6-11 shows the circuit design for using Arduino with a joystick. Wire it up and run the sketch shown in Example 6-7.

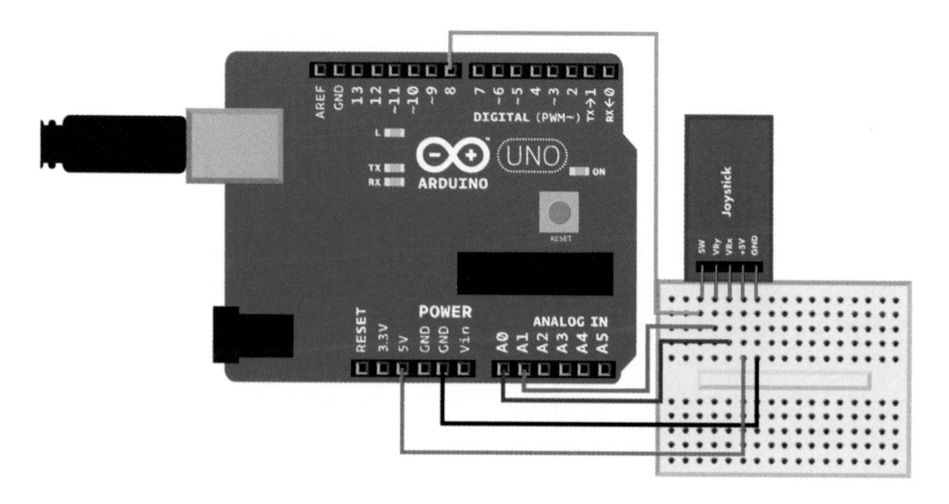

Figure 6-11. *Joystick circuit for Arduino*

Example 6-7. ky_023_xyjoystick.ino

```
// ky_023_xyjoystick.ino - print joystick position to serial
// (c) BotBook.com - Karvinen, Karvinen, Valtokari


const int VRxPin = 0;   // ❶
const int VRyPin = 1;
const int SwButtonPin = 8;

int button = -1; // LOW or HIGH // ❷
int x = -1; // 0..1023
int y = -1; // 0..1023

void readJoystick() { // ❸
```

```
  button = digitalRead(SwButtonPin);    // ❹
  x = analogRead(VRxPin);
  y = analogRead(VRyPin);
}

void setup()  {
  pinMode(SwButtonPin, INPUT);
  digitalWrite(SwButtonPin, HIGH); // pull-up resistor  // ❺
  Serial.begin(115200);
}

void loop()  {
  readJoystick();        // ❻
  Serial.print("X: ");
  Serial.print(x);
  Serial.print(" Y: ");
  Serial.print(y);
  Serial.print(" Button: ");
  Serial.println(button);
  delay(10);
}
```

❶ Store the pin numbers into global constants: a variable resistor for x, a variable resistor for y, and a button.

❷ These global variables store the state of the button and the tilt of the joystick along y- and x-axes. We initialize these variables to impossible values (values that won't get generated by calls to `analogRead` or `digitalRead`) to help debugging. The intended range of values is indicated in the comments.

❸ `readJoystick()` doesn't return a value (it's said to be of a *void type*). In C++, which Arduino is based on, functions can't conveniently return multiple values. Instead, `readJoystick()` updates global variables.

❹ The button state is stored to the global variable `button`.

❺ Enable the internal pull-up resistor to avoid having a floating pin.

❻ Update the global variables that store the joystick state.

Joystick Code and Connection for Raspberry Pi

Figure 6-12 shows the circuit layout for Raspberry Pi and a joystick. Hook everything up as shown, and then run the program shown in Example 6-8.

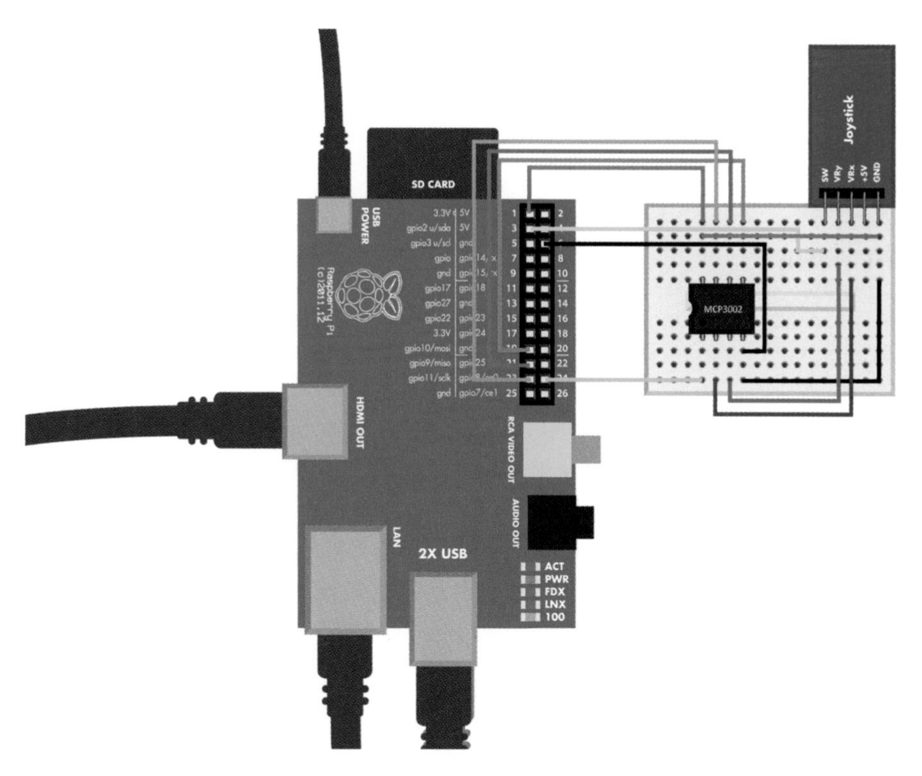

Figure 6-12. *Joystick circuit for Raspberry Pi*

Example 6-8. *xy_joystick.py*

```
# xy_joystick.py - print KY 023 joystick tilt and button status
# (c) BotBook.com - Karvinen, Karvinen, Valtokari
import time
import botbook_mcp3002  as mcp  # ❶
import botbook_gpio as gpio     # ❷

def readX():     # ❸
        return mcp.readAnalog(0, 0)

def readY():
        return mcp.readAnalog(0, 1)      # ❹

def readButton():
        buttonPin = 25
        gpio.mode(buttonPin, "in")
        return gpio.read(buttonPin)

def main():
        while True:      # ❺
                xAxel = readX() # ❻
                yAxel = readY()
                button = readButton()
```

```
                print("X: %i, Y: %i, Button: %r" % (xAxel, yAxel, button))      # ❼
                time.sleep(0.5)

if __name__ == "__main__":
        main()
```

❶ Import the library for the MCP3002 analog-to-digital converter chip. The library *botbook_mcp3002.py* must be in the same directory as *xy_joystick.py*. You must also install the *spidev* library, which is imported by *botbook_mcp3002*. See comments in the beginning of *botbook_mcp3002/botbook_mcp3002.py* or "Installing SpiDev" on page 56.

❷ Use more convenient namespace to keep the code from getting too verbose. The `as` keyword allows you to write `gpio.read()` instead of `botbook_gpio.read()`.

❸ Write functions according to their purpose. The purpose of `readX()` (read the state of the x-axis) will look obvious in the main program.

❹ Measure the voltage on the second channel (number 1).

❺ Keep running until you hit Control-C to kill the program (or shut down the Raspberry Pi).

❻ Read the tilt along the x-axis and store it to new variable `xAxel`.

❼ When using multiple variables in a format string, the variables must be in a tuple (a group of values separated by commas). You can think of a tuple `(1, 2, 3)` as a list `[1, 2, 3]` that you can't modify.

Environment Experiment: Salvage Parts from an Xbox Controller

If you have an old game console (Xbox, PlayStation, etc.) control pad lying around, you can salvage two sensors from it and use them with your Arduino or Raspberry Pi (Figure 6-13). Opening the controller is quite easy, but you'll need to use a soldering iron to detach the components from the circuit board.

For detaching components, you may also find these tools/supplies helpful: a desoldering pump, solder wick (also known as a braid), or even a flux pen.

Many controllers also have a simple force feedback system. For example, if a player takes damage in a game, force feedback makes the controller shake. This is done with vibration motors—eccentric DC motors that spin and shake when powered. Take these out, too, and give them a new home in your own force feedback gadget (Figure 6-14).

Figure 6-13. *Thumb joysticks inside Xbox controller*

Figure 6-14. *Salvaged vibration motors*

Experiment: Burglar Alarm! (Passive Infrared Sensor)

A passive infrared (PIR) sensor is probably the most common burglar alarm. All warm objects radiate invisible infrared light. A PIR sensor reacts to changing infrared light. This change is typically caused by a warm human moving.

Be sure to let your PIR sensor adapt to its environment first. Because a PIR only detects change, it has to first know the heat pattern in the room when there is no burglar. After you turn on the power, the PIR sensor needs 30 seconds to adapt to the environment. There must be no movement or people in the watched area during the adaptation period.

You can use a box to limit the area watched by a PIR sensor. When you want to quickly test a PIR sensor, put it in an upside-down box, so that it can only see upward. While the PIR sensor learns what the environment looks like, you can keep writing code without needing to be absolutely still. When you want to send an alarm to test your code, wave your hand above the box.

The Parallax PIR has two modes of operation:

- H for stay High as long as there is movement
- L for return to Low after an alarm

So in L mode, the PIR just sends one pulse even if movement continues after it's first detected.

The PIR has a tiny jumper wire to choose the mode of operation. The jumper is a tiny rectangular part covered in black plastic. In this project, set the jumper to H (stay High) mode.

Burglar Alarm Code and Connection for Arduino

Figure 6-15 shows the circuit diagram for Arduino. Wire it up and run the sketch shown in Example 6-9.

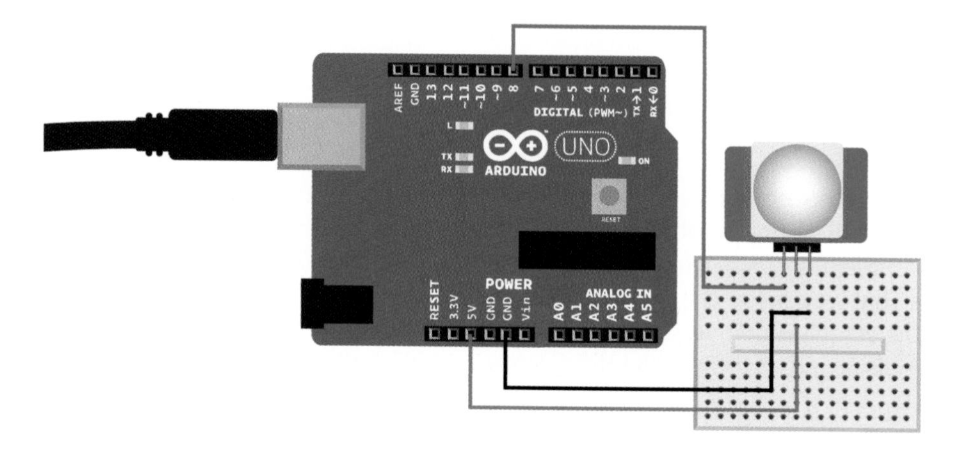

Figure 6-15. *Parallax PIR sensor Rev A circuit for Arduino*

Example 6-9. ***parallax_pir_reva.ino***

```
// parallax_pir_reva.ino - print movement detection
// (c) BotBook.com - Karvinen, Karvinen, Valtokari


const int sensorPin = 8;
const int ledPin = 13;
const int learningPeriod = 30*1000; // ms       // ❶
int learningCompleted = 0;

void setup() {
  Serial.begin(115200);
  pinMode(sensorPin, INPUT);
  Serial.println("Starting to learn unmoving environment..."); // ❷
  pinMode(ledPin, OUTPUT);
}

void loop() {
   if(millis() < learningPeriod) {       // ❸
     delay(20); // ms   // ❹
     return;    // ❺
   }
   if(learningCompleted == 0) { // ❻
     learningCompleted = 1;
     Serial.println("Learning completed.");
   }

   if(digitalRead(sensorPin) == HIGH) { // ❼
     Serial.println("Movement detected");
     digitalWrite(ledPin,HIGH);
   } else {
     Serial.println("No movement detected");
```

```
      digitalWrite(ledPin,LOW);
    }
    delay(100);
}
```

❶ The PIR needs movement-free time to adapt. Here, we use 30,000 ms (30 seconds). The value is stored into a global constant. Avoid the temptation to repeat the value in the comment. Instead, use your comment to tell the unit (milliseconds) of the variable. That way, if you change a value, you don't need to change the comment. Also, if the standard unit version of the value (seconds) is a result of a calculation (number of seconds * 1000), write the calculation into code (`30*1000`).

❷ Print instructions to serial. To open the serial monitor, use Tools→Serial Monitor. Remember to choose the same speed ("baud", bit/s) in both the serial monitor and your code.

❸ This is a common pattern to measure time in Arduino. The `millis()` function returns milliseconds since last boot.

❹ A short delay prevents repeated calls to `loop()` from consuming 100% of CPU time.

❺ A return statement finishes the `loop()` function. As always, after `loop()` finishes, it's automatically called again.

❻ The `learningCompleted` variable ensures that you print "Learning completed" only once. Here, 1 and 0 are used like true and false.

❼ When PIR detects warm movement, `sensorPin` goes HIGH.

Burglar Alarm Code and Connection for Raspberry Pi

Figure 6-16 shows the circuit for Raspberry Pi. Wire it up as shown, and then run the code in Example 6-10.

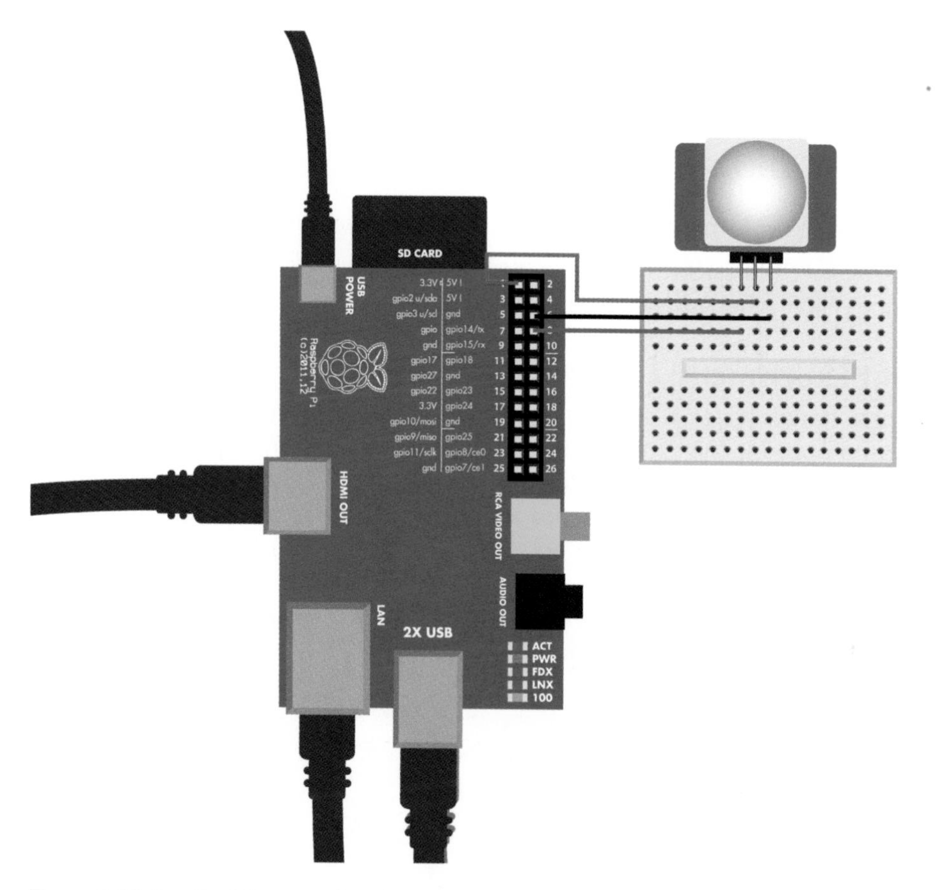

Figure 6-16. *Parallax PIR sensor Rev A circuit for Raspberry Pi*

Example 6-10. ***parallax_pir_reva.py***

```
# parallax_pir_reva.py - write to screen when movement detected
# (c) BotBook.com - Karvinen, Karvinen, Valtokari
import time
import botbook_gpio as gpio

learningPeriod = 60

def main():
        pirPin = 14
        gpio.mode(pirPin, "in")
        #Learning period
        time.sleep(learningPeriod) # ❶
        while (True):
                movement = gpio.read(pirPin) # ❷
                if(movement == gpio.HIGH):
                        print "Movement detected"
                else:
                        print "No movement detected"
```

```
        time.sleep(0.3)

if __name__ == "__main__":
    main()
```

❶ The sensors need movement-free time to adapt. Here we use 60 seconds, but you can experiment with different times.

❷ The pin goes HIGH when movement is detected.

Environment Experiment: Cheating an Alarm

Sometimes a penetration tester (a security expert who probes for vulnerabilities) must move by alarms unnoticed. In this experiment, you can practice in the privacy of your home. This way, you don't have to worry about customers watching (if you are a professional pentester) or a lifetime of free room and board with no way to check out (if you dream of a criminal career).

Figure 6-17. *Sometimes a penetration tester must bypass alarms*

You can try cheating your PIR sensor, but as you will soon notice it's not quite as easy as in action films. We'll use the same code as before but add a piezo beeper. This way, it's easier to know when movement is detected. Connect the piezo according to the circuit diagram in Figure 6-18 and upload the code in Example 6-11.

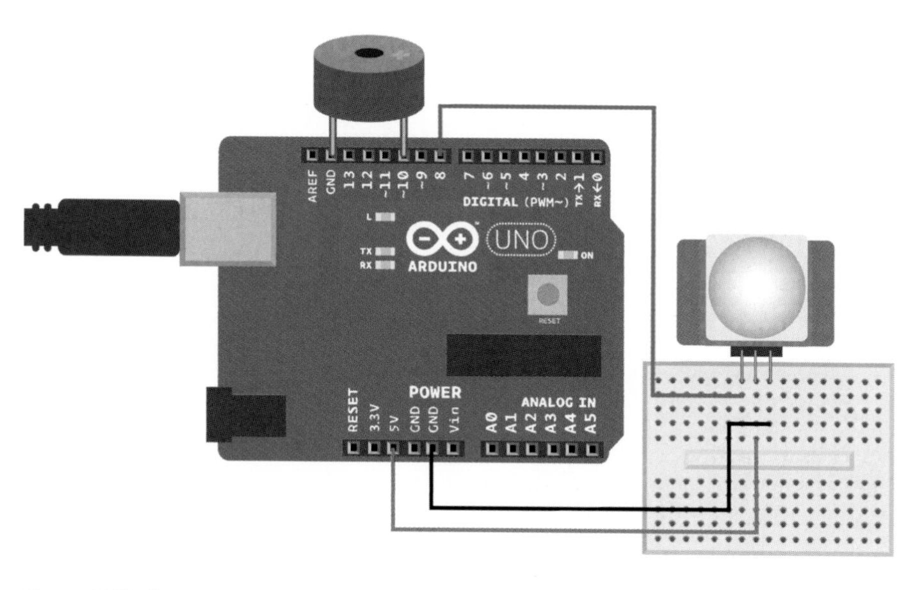

Figure 6-18. *Parallax PIR sensor Rev A circuit for Arduino with LED and speaker*

Example 6-11. parallax_PIR_revA_cheating_pir.ino

```
// parallax_PIR_revA_cheating_pir.ino - light an LED when movement detected
// (c) BotBook.com - Karvinen, Karvinen, Valtokari



const int sensorPin = 8;
const int ledPin = 13;
int speakerPin = 10;
const int learningPeriod = 30*1000; // 30 seconds for learning period.
int learningCompleted = 0;

void setup() {
  Serial.begin(115200);
   pinMode(speakerPin, OUTPUT);
  pinMode(sensorPin, INPUT);
  Serial.println("Start learning for next 30 seconds.");
  pinMode(ledPin, OUTPUT);
}

void alarm()
{
  wave(speakerPin, 440, 40);
  delay(25);
  wave(speakerPin, 300, 20);
  wave(speakerPin, 540, 40);
  delay(25);
}

void wave(int pin, float frequency, int duration)
```

```
{
  float period=1/frequency*1000*1000; // microseconds (us)
  long int startTime=millis();
  while(millis()-startTime < duration) {
    digitalWrite(pin, HIGH);
    delayMicroseconds(period/2);
    digitalWrite(pin, LOW);
    delayMicroseconds(period/2);
  }
}

void loop() {
   if(millis() < learningPeriod) {
     return; // Sensor has not yet learned its environment.
   }
   if(learningCompleted == 0) {
     learningCompleted = 1;
     Serial.println("Learning completed.");
   }
   if(digitalRead(sensorPin) == HIGH) {
     Serial.println("Movement detected");
     alarm();
     digitalWrite(ledPin, HIGH);
   } else {
     Serial.println("No movement detected");
     digitalWrite(ledPin, LOW);
   }
   delay(100);
}
```

Now you should hear an alarm when you move your hand before the sensor. Try to approach the PIR from a few meters away without triggering it. You'll have to move really slowly, and even then, it's hard. Using a bed sheet or a big towel makes it much easier. Fully cover yourself, cartoon ghost style (no eye holes, though!) and slowly start moving toward the sensor. Covering yourself limits the sensor's ability to detect your body heat radiation. This way, you could almost touch the sensor before it notices you.

In real-life physical pentesting, you must work with movement alarms that combine ultrasonic distance sensing to passive infrared. And of course, security cameras automatically detect changes in the picture.

You already learned how to confuse an ultrasonic distance sensor in "Environment Experiment: Invisible Objects" on page 43. Even though cameras can be difficult to cheat, you can sneak in between cameras. Many cameras use active infrared illumination to see in the dark. You learned how to see infrared light in "Environment Experiment: How to See Infrared" on page 48.

Test Project: Pong

Detecting movement is much more interesting when you can actually present that information to the user. In this project you'll learn how to use sensor data to move objects on the screen. To keep things simple, we use a joystick as an input for a Pong game in this example. What you learn here can be easily adapted to other projects. Any sensor could be the input device, and only your imagination limits what is shown on the screen.

Figure 6-19. *Game on!*

Pong, originally manufactured by Atari in 1972, is the classic game where you move a paddle up and down. Your goal is to keep the ball out of your goal. See <<[pong-game>>.

This project introduces you to pyGame, one of the easiest libraries for programming games. You'll build your own game console and learn to use sensor input for moving things in the big screen. For added effect, use a video projector for output.

This project is easy to do with Raspberry Pi, as you can connect your normal television or video projector to Raspberry Pi's HDMI connector. Doing the same in Arduino would not be as straightforward.

That's not to say it would be impossible. Pong comes from a time when CPUs were slower and had less RAM than the Arduino. They would draw scan lines on a screen and perform their computations during the horizontal and vertical blank. See *http://bit.ly/1f0GgHt* for a simple Arduino Pong project. If you want to make all kinds of old-fashioned video games with Arduino, check out the Video Game Shield from Wayne and Layne (*http://bit.ly/QD3zeL*).

Figure 6-20. *Pong playfield*

What You'll Learn

In the *Pong* project, you'll learn how to:

- Use data from a sensor to move objects on the screen.
- Display full high-def graphics with Raspberry Pi.
- Make Raspberry Pi react faster by drawing directly to screen, without going through the desktop environment or the X Window System.
- Program a simple game with Python's pyGame.
- Automatically start your program when Raspberry Pi boots.

Figure 6-21 shows the wiring diagram for the Pong project. Wire it up as shown, and then run the program shown in Example 6-12. Be sure that the *botbook_gpio.py* library is in the same directory as the *pong.py* program.

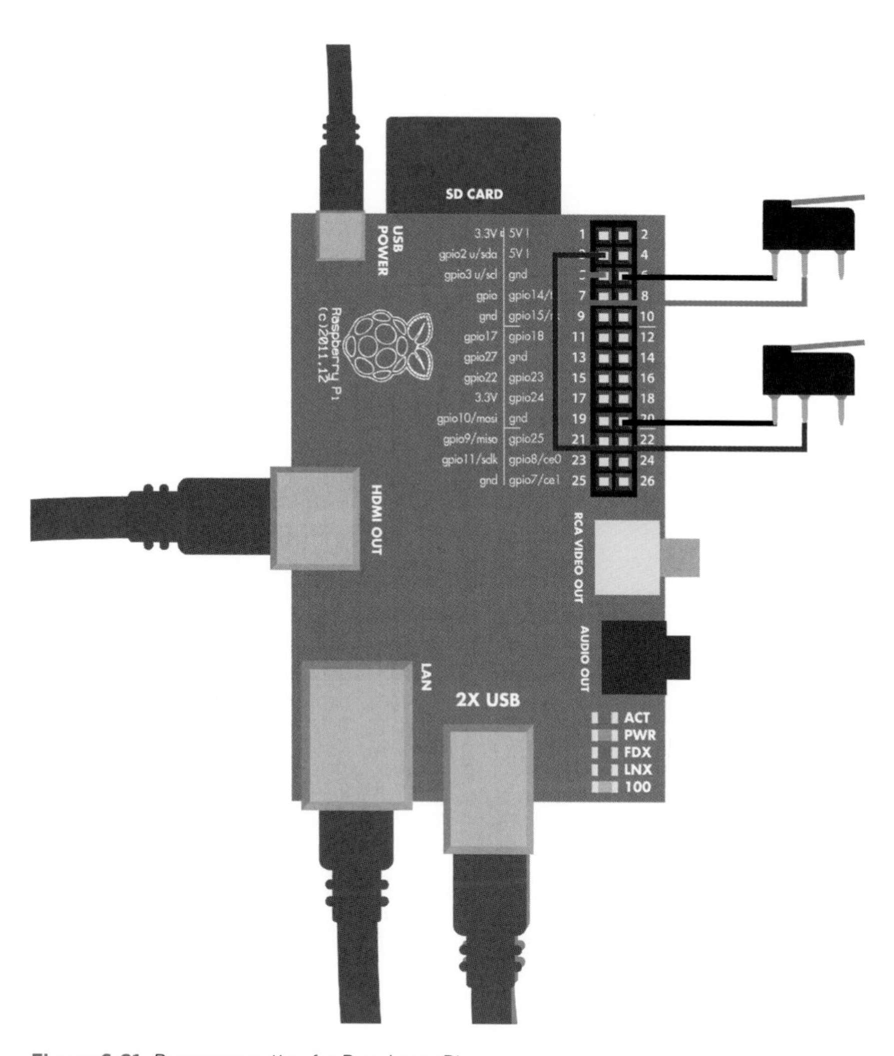

Figure 6-21. *Pong connection for Raspberry Pi*

Example 6-12. **pong.py**

```
# pong.py - play ball game classic with joystick and big screen
# (c) BotBook.com - Karvinen, Karvinen, Valtokari

import time
import sys
import pygame
import botbook_gpio as gpio
from pygame.locals import *

print "Loading BotBook.com Pong..."
pygame.init()    # ❶
```

```
width = pygame.display.Info().current_w    # ❷
height = pygame.display.Info().current_h

size = width, height    # ❸
background = 0, 0, 0    # ❹
screen = pygame.display.set_mode(size,pygame.FULLSCREEN)    # ❺
normalSpeed = 512
ballrect = Rect(width/2, height/2, 16, 16)    # ❻
computerrect = Rect(width-20, 0, 20, 120)    # ❼
playerrect = Rect(0, 0, 20, 120)    # ❽
#movement is diff in x and y. ball can only move in 45 degree angles.
speed = [normalSpeed, normalSpeed]    # ❾
clock = pygame.time.Clock()    # ❿
pygame.mouse.set_visible(False)
mainloop = True

uppin = 2
downpin = 3
gpio.mode(uppin, "in")
gpio.mode(downpin, "in")

while mainloop:    # ⓫
    seconds = clock.tick(30) / 1000.0 # seconds since last frame    # ⓬

    # User input

    for event in pygame.event.get():    # ⓭
        if event.type == pygame.QUIT: mainloop = False    # ⓮
        if (event.type == KEYUP) or (event.type == KEYDOWN):
            if event.key == K_ESCAPE: mainloop = False

    # Movement and collisions
    playerspeed = 0
    if gpio.read(uppin) == gpio.LOW:
        playerspeed = -normalSpeed
    if gpio.read(downpin) == gpio.LOW:
        playerspeed = normalSpeed
    ballrect.x += speed[0] * seconds    # ⓯
    ballrect.y += speed[1] * seconds
    if ballrect.left < 0 or ballrect.right > width:    # ⓰
        ballrect.x = width/2;
    if ballrect.top < 0 or ballrect.bottom > height:
        speed[1] = -speed[1]

    computerrect.y = round(ballrect.y)    # ⓱
    playerrect.y += playerspeed * seconds    # ⓲
    if playerrect.top < 0: playerrect.top = 0    # ⓳
    if playerrect.bottom > height: playerrect.bottom = height    # ⓴
    if computerrect.colliderect(ballrect):    # ㉑
        speed[0] = -normalSpeed

    if playerrect.colliderect(ballrect):
        speed[0] = normalSpeed
```

```
        # Draw frame
        screen.fill(background)
        pygame.draw.circle(screen, (255, 255, 255), (int(round(ballrect.x+8)),
                                            int(round(ballrect.y+8))), 10)    # ㉒
        pygame.draw.rect(screen, (255, 255, 255), computerrect)    # ㉓
        pygame.draw.rect(screen, (255, 255, 255), playerrect)
        pygame.display.update() # ㉔
```

❶ For pyGame to work, you must first initialize it.

❷ PyGame uses predefined dimensions for canvas. This allows you to work with actual pixels onscreen, instead of some intermediate units. These two lines retrieve the width and height.

❸ The canvas size is specified as a tuple: `(width, height)`.

❹ Colors in pyGame are red, green, and blue (RGB). You've probably worked with RGB colors if you have ever specified colors for web pages. `(0, 0, 0)` is black.

❺ `screen` is the object where the actual drawing will happen. It will be used near the end of the main loop.

❻ Create bounding rectangles for onscreen objects in the game: `Rect(x, y, width, height)`. The ball starts from the top-left corner (y==0, x==0). The ball's bounding rectangle size is 16 × 16 pixels.

❼ This places the computer player's paddle near the right edge of the canvas. The paddle is 20 pixels (px) thick and 120 px tall.

❽ The human player's paddle starts from top left (0,0). Its dimensions are 20 × 120 pixels, like the computer's paddle.

❾ The *speed vector* for the ball. Each second, the ball will move 64 pixels on the x-axis and 64 pixels on the y-axis.

❿ Create a new `Clock` object, and store it into the newly declared variable `clock`. This will be used for keeping the speed consistent even if you've overclocked your Raspberry Pi.

⓫ PyGame uses a main loop programming style. Main loop is a very common pattern in games. Just like the typical Arduino `loop()` or the `while(True)` you've seen in Python examples, the main loop starts over each time it completes. Typical game main loop tasks include getting user input, moving objects on the screen, checking for collisions, and finally, drawing one frame on the screen.

⓬ `clock.tick()` returns the time since its last call. When you call it once in a frame (once every main loop run), you get the elapsed time since the last frame. The parameter 60 specifies a maximum frame rate of 60 Hz, that is, 60 frames per second. If the game is going faster, `tick()` will wait before returning. For more human-friendly units, we convert the milliseconds (1/1000 s) returned by `tick()` to seconds.

⓭ Keyboard input is handled through the `pygame.event` object. The `pygame.event.get()` returns an object of events that contains a collection of items that you can *iterate* over. The "for ITEM in LIST" loop goes through the LIST one at a time. It sets ITEM to be the first item in LIST in the first iteration, then the second in the list in the second iteration, and so forth until it is finished.

⓮ Compare each individual event to predefined pyGame constants, and react if it matches. For example, if the user performs a user interface action that would quit the game, the QUIT event comes through, and it's time to exit.

⓯ Move the ball. Because we account for elapsed time each frame, it will move at the same 64 pixels per second despite changes in frame rate.

⓰ If the ball hits the limits of playing field, bounce!

⓱ Move the computer paddle to exactly the same height as the ball. You have one tough opponent here!

⓲ Move the player paddle vertically according to acceleration from the input keys. The shorthand `a+=2` means the same as `a=a+2` .

⓳ If a player's movement would take the paddle through the top of the canvas, just stay on the top.

⓴ If a player's movement would take the paddle through the bottom of the canvas, just stay put.

㉑ Check to see if the ball collided with either paddle. If so, set the ball's direction to a direction away from the paddle it collided with.

㉒ Draw the ball at its calculated position.

㉓ The computer paddle is drawn at the location calculated earlier. As its shape is a rectangle, the bounding box is exactly the same as the object itself.

㉔ Show everything in this frame, all at the same time.

Pong Packaging Tips

We used a drop-forged aluminum box to make a super robust casing with street credibility for our game console (Figure 6-22). This construction makes an attractive alternative to the flimsy plastic gadgets we usually see and use.

In the back of the box, we made a hole for power and HDMI cables (Figure 6-23). To make a wide hole like ours, drill two large holes separately and then remove the metal in between with a jigsaw blade.

Figure 6-22. *Pong case*

Figure 6-23. *Hole for power and HDMI*

A traditional arcade joystick makes a perfect match for the indestructible box (Figure 6-24). To attach the joystick, you need three holes: two 5 mm to mount the frame and one 10 mm to get the actual stick through the cover (Figure 6-25).

Figure 6-24. *Arcade joystick*

Figure 6-25. *Arcade joystick taken apart*

As we are going to control only up and down movement, we need to solder wires to two of the joystick's microswitches, as shown in Figure 6-26.

Figure 6-26. *Soldering wires*

Raspberry Pi was attached simply by hot gluing the Raspberry Pi's cover box bottom to the bottom of our aluminum box (see Figure 6-27).

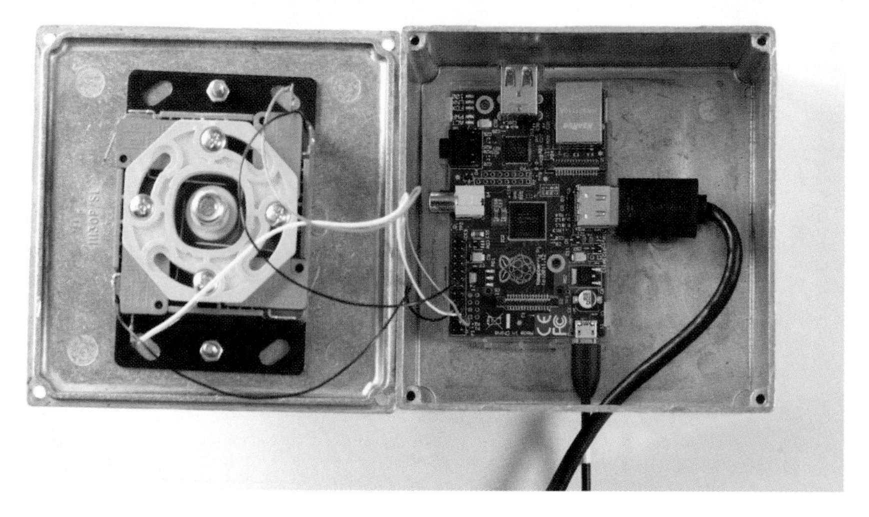

Figure 6-27. *Everything inside*

Automatically Start Your Game When Raspberry Pi Boots

Now that you have a beautifully packaged game console of your own, it's time to have another look at user experience. Wouldn't it be nice if the game started on boot? Without a keyboard, it's difficult to type `python pong.py` each time you want to start the game.

To start the game as a normal user, you'll set the system to log in as the user *pi* automatically. Then you'll configure things to start the game in the user's login script. This way, the game starts immediately when you boot Raspberry Pi.

Run Game on Login

When you log in, `bash` opens. Bash is your shell: it interprets the commands that you type at the command prompt ($).

If your Raspberry Pi automatically boots into the graphical desktop (the default), you should change it to start up in a text mode shell, because you need to run it from text mode in order for it to run automatically on login. Open LXTerminal and run `sudo raspi-config`*. There you can choose Enable Boot To Desktop and choose Console Text console.*

The first thing that bash does is to run scripts, such as *.profile*, *.login*, *.bash_profile*, and *.bash_login*. Just like all per-user configuration files, they are hidden files in the user home directory. To see them, you must use the `-a` flag with `ls`. If you're not already in your home directory, you can quickly change to it (*/home/pi/*) with *cd* without arguments. Here's a set of commands (don't type the $, since that's the shell prompt) that change to your home directory and list all the files there:

```
$ cd
$ ls -a
```

The *.bash_login* file is a shell script: it has commands to run, one after another. Before you add this command to the file, try it out on the command line:

```
$ python /home/pi/makesensors/pong/pong.py
```

Replace *makesensors* with the path where you have put the *pong.py* program. Press Esc to quit the game. If everything worked OK, you can now open the *.bash_login* file with the nano editor:

```
$ nano .bash_login
```

Add the line `python /home/pi/makesensors/pong/pong.py` to *.bash_login*. If there are any other lines already in the file, delete them. Example 6-13 shows what the *.bash_login* file should now look like. Save the file by typing Control-X, then press y when prompted to Save, and finally type Enter/Return to confirm the filename.

Example 6-13. *bash_login*

```
# /home/pi/.bash_login - automatically start pong game on login

python /home/pi/makesensors/pong/pong.py
```

Log out:

```
$ exit
```

Next, log back in. The game starts automatically.

Automatic Login

Now it's time to make your user "pi" log in automatically on boot. Because the game runs immediately after "pi" logs in, this will start the game after you power up the Pi. If you're still in the game, press Esc to escape from it.

The `init` program controls system boot, so you will need to modify its settings. All system-wide settings are in */etc/*, and the configuration file usually starts with the name of the thing: */etc/init** (in this case, */etc/inittab*). Because logging users in is a process that requires full privileges, you need to edit the file as root with the `sudoedit` command:

```
$ sudoedit /etc/inittab
```

To edit text files as root, you use ***sudoedit*** *instead of using* ***nano*** *with* ***sudo****. This way, you won't get errors about nano's history file ownership when using nano as a normal user later.*

Modify the line that governs the first "virtual terminal" so that it automatically logs you in (don't add this line, modify the one that starts with `1:2345:respawn:/sbin/getty`):

```
1:2345:respawn:/sbin/getty --noclear 38400 tty1 --autologin pi
```

Save with Control-X, answer y when asked about saving, and then press Enter or Return to confirm the filename. You can see a complete, modified */etc/inittab* in Example 6-14.

Shut down the Raspberry Pi:

```
$ sudo shutdown -P now
```

Then disconnect and reconnect USB power.

Relax as you are automatically logged in. The game starts, and your very own game console is ready. Time to play Pong!

Example 6-14. inittab

```
# /etc/inittab - automatically log in user pi on boot (also disable serial)

id:2:initdefault:
si::sysinit:/etc/init.d/rcS
~~:S:wait:/sbin/sulogin
l0:0:wait:/etc/init.d/rc 0
l1:1:wait:/etc/init.d/rc 1
l2:2:wait:/etc/init.d/rc 2
l3:3:wait:/etc/init.d/rc 3
l4:4:wait:/etc/init.d/rc 4
l5:5:wait:/etc/init.d/rc 5
l6:6:wait:/etc/init.d/rc 6
z6:6:respawn:/sbin/sulogin
ca:12345:ctrlaltdel:/sbin/shutdown -t1 -a -r now
pf::powerwait:/etc/init.d/powerfail start
pn::powerfailnow:/etc/init.d/powerfail now
po::powerokwait:/etc/init.d/powerfail stop
# modified:
1:2345:respawn:/sbin/getty --noclear 38400 tty1 --autologin pi
2:23:respawn:/sbin/getty 38400 tty2
3:23:respawn:/sbin/getty 38400 tty3
4:23:respawn:/sbin/getty 38400 tty4
5:23:respawn:/sbin/getty 38400 tty5
6:23:respawn:/sbin/getty 38400 tty6
# removed serial console, so that serial port can be used with sensors:
# T0:23:respawn:/sbin/getty -L ttyAMA0 115200 vt100
```

Run as normal user whenever possible. Only perform system administration (e.g., apt-get, raspi-config) as root. It's especially important that games run as a normal user because they often need direct access to hardware, and programs that reach deeper into your system can do more damage if they have a serious bug. That's why you should use autologin and login script to start the game (as described here), instead of simply putting the game into a startup script such as rc.local (which always runs as root).

To stop playing and return to the command line, just press Esc.

You have now tested movement in many ways, starting from basic buttons and potentiometers. These sensors are the archetypes of resistance sensors. The button is the simplest digital (on/off or zero/infinite resistance) resistance sensor, and the potentiometer is the simplest analog resistance sensor. You'll be able to use similar circuits and code with many other sensors.

For more specialized sensors, you've detected touch—even through wood. And you've measured pressure, which could be useful in your projects to see if a bed or a seat is occupied, or just to have a finger strength competition.

Next, you'll measure a less tangible form of energy: light.

7 Light

A robot follows a complicated path, seemingly without difficulty. A closer look shows a black line on the floor for the bot to follow. Later in the evening, backyard lights light up automatically when darkness falls.

Because color is reflected light, sensors can detect the color of a surface. With some creatively applied tubing, the direction of light can be detected, too. And if fire is the thing for your bot, there is a sensor for flame.

Do you want to measure human movement with infrared (IR) light? See "Experiment: Burglar Alarm! (Passive Infrared Sensor)" on page 140. Need to know if an object is nearer than a given distance, using IR? See "Experiment: Detect Obstacles With Infrared (IR Distance Sensor)" on page 44.

Experiment: Detecting Flame (Flame Sensor)

Flames emit a range of infrared light not very common in ambient light. The KY-026 flame sensor reports the level of infrared light with a change of resistance (Figure 7-1).

The code you'll write for flame detection is the same code you'd use for an analog resistance sensor: you use `analogRead()` to read the voltage of a pin.

The KY-026 flame sensor provides two ways to measure flame: `digitalRead()` and `analogRead()`. Even though the code in this experiment implements both, you can just use whichever one you need in your own code.

Using digital mode only is especially convenient with Raspberry Pi, because Raspberry Pi doesn't have an analog-to-digital converter.

Figure 7-1. *Flame sensor*

Figure 7-2. *Flame-following robot prototype (robot workshop in Austria)*

Flame Sensor Code and Connection for Arduino

Figure 7-3 shows the wiring diagram for the flame sensor with Arduino. Wire it up as shown, and then run the sketch shown in Example 7-1.

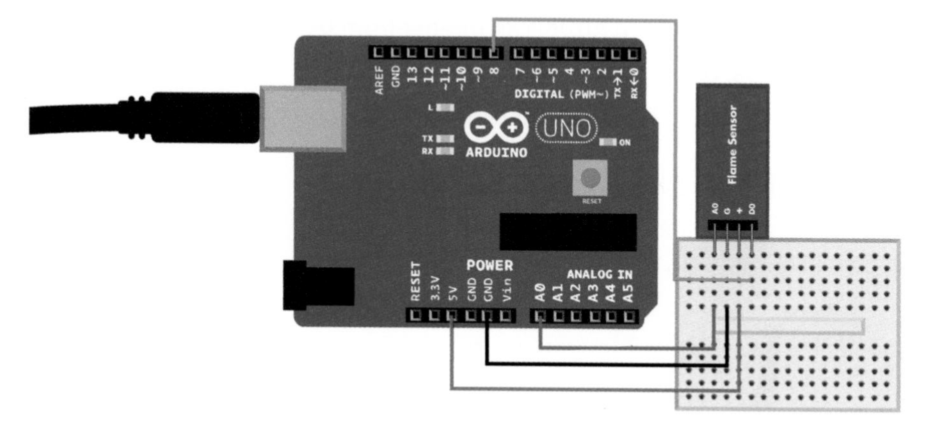

Figure 7-3. *Flame sensor circuit for Arduino*

Example 7-1. ***ky_026_flame.ino***

```
// ky_026_flame.ino - report level IR light from flame to serial
// (c) BotBook.com - Karvinen, Karvinen, Valtokari

const int analogPin = A0;
const int digitalPin = 8;
const int ledPin = 13;

void setup() {
  Serial.begin(115200);
  pinMode(digitalPin,INPUT);
  pinMode(ledPin,OUTPUT);
}

void loop()
{
  int threshold = -1; // HIGH or LOW
  int value = -1; // 0..1023
  value = analogRead(analogPin);          // ❶
  threshold = digitalRead(digitalPin);    // ❷
  Serial.print("Raw: ");
  Serial.print(value);
  Serial.print(" Over threshold: ");
  Serial.println(threshold);
  delay(10);
  if(threshold==HIGH) { // ❸
    digitalWrite(ledPin, HIGH);
  } else {
    digitalWrite(ledPin, LOW);
  }
}
```

❶ Get the raw voltage reading from A0. It is an integer from 0 (0 V) to 1023 (+5 V).

❷ Read the voltage of D8: LOW (0 V) or HIGH (+5 V).

❸ Turn on the built-in LED (D13) if you've detected a flame.

Flame Sensor Code and Connection for Raspberry Pi

Figure 7-4 shows the circuit for connecting the sensor to a Raspberry Pi. Wire it up and run the program shown in Example 7-2.

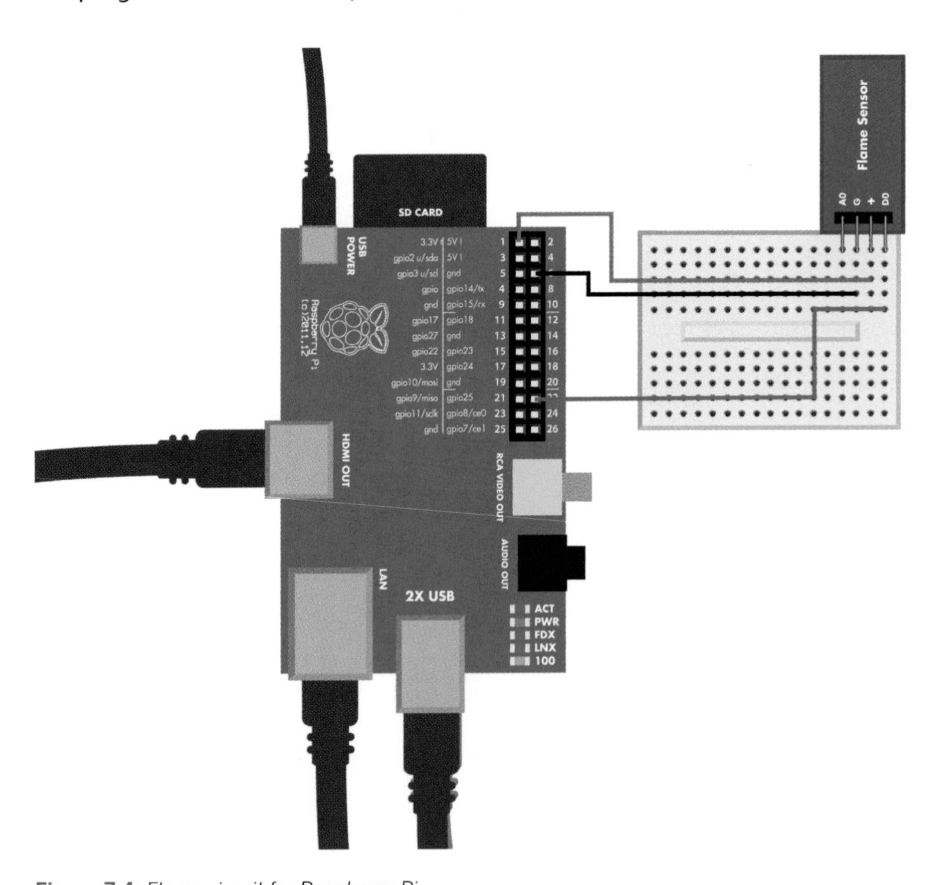

Figure 7-4. *Flame circuit for Raspberry Pi*

Example 7-2. *ky_026_flame.py*

```
# ky_026_flame.py - report presence of IR light from flame to serial
# (c) BotBook.com - Karvinen, Karvinen, Valtokari

import time
import botbook_gpio as gpio     # ❶

def main():
        triggerPin = 25
```

```
        gpio.mode(triggerPin, "in")     # ❷
        flame = gpio.read(triggerPin)   # ❸
        if(flame == gpio.HIGH): # ❹
                print "Flame detected"
        else:
                print "No flame detected"
        time.sleep(0.5)

if __name__ == "__main__":
        main()
```

❶ Make sure there's a copy of the *botbook_gpio.py* library in the same directory as this program. You can download this library along with all the example code from *http://botbook.com*. See "GPIO Without Root" on page 19 for information on configuring your Raspberry Pi for GPIO access.

❷ Pin is set to *in* mode to read its voltage in the next line of code.

❸ Read the status of pin 23. The value is True for HIGH (+3.3 V) and False for LOW (0 V).

❹ Note that a Boolean (true or false) value can be used in an `if` without a comparison operator. That is, you can say `if(b)` instead of `if(b==True)`.

Environment Experiment: Flame Precision

The flame sensor's built-in sensitivity resistor is very useful. It is especially important to adjust it so that ambient light won't trigger the sensor constantly. It can also be used to set the flame sensor to react to very specific level of flame.

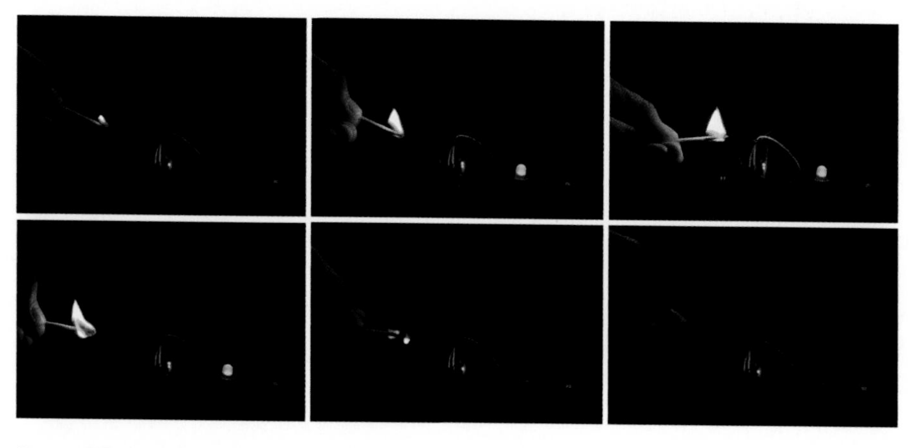

Figure 7-5. *Adjusting and testing flame sensor sensitivity*

First, put an extra LED between GND and pin 13 (see Figure 7-6). It's easier to see a full-sized LED than the onboard LED, so you'll be more likely to notice when the sensor is activated. Upload the flame sensor code (Example 7-1) to your Arduino. Turn the potentiometer all the way right, then turn it left until the LED goes out. It's probably a good idea to close the curtains, as strong sunlight can overpower other light sources.

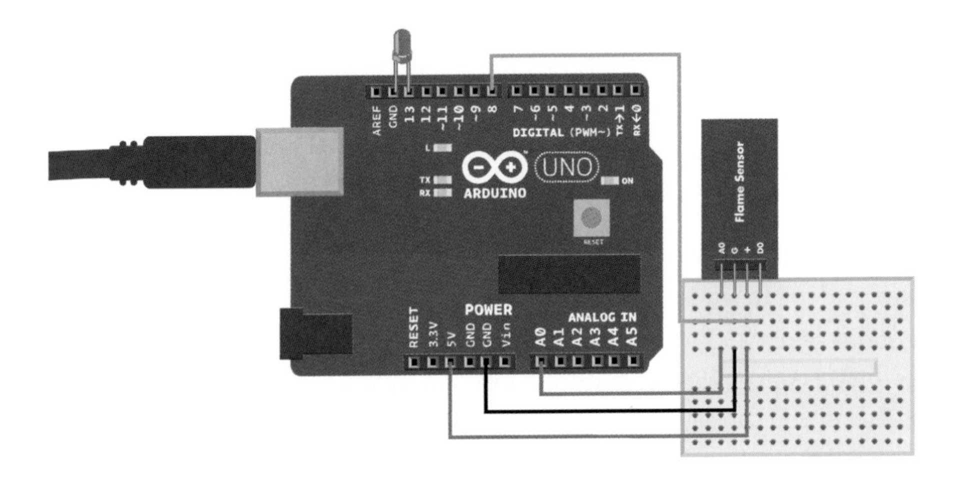

Figure 7-6. *Flame sensor with LED*

Now, light a match. The LED should light again. Try adjusting the sensitivity potentiometer so that the sensor will react only to a full-size flame. Don't get distracted watching the LED, or you might burn your fingers as the match burns down!

Experiment: See the Light (Photoresistor, LDR)

A light-dependent resistor (LDR) changes its resistance according to the level of visible light. Its resistance is lower in bright light. LDR is also known as a photoresistor (see Figure 7-7).

Photoresistors can turn on the lights when it's dark, detect if a dark box is opened in a lit room, and help create robots that love light. With some creative use of heat shrink tubing, an LDR can also detect the direction of light. When testing, you can simply put your finger on an LDR to make darkness fall. If you want bright light, you can point a flashlight at the LDR.

To test a light-seeking robot in your lab, try covering it with a blanket (see Figure 7-8). This way, you don't have to shut down the lights in the lab or run between your lab and a dark room.

Figure 7-7. *Photoresistor*

Figure 7-8. *Ambient light hideout (robot workshop in Austria)*

LDR Code and Connection for Arduino

A photoresistor is just a two-legged variable resistor. With Arduino, it's just a matter of configuring the sensor with another resistor as a voltage divider (see "Experiment: Potentiometer (Variable Resistor, Pot)" on page 98), then using `analogRead()`. In this way, a photoresistor is similar to many other analog resistance sensors.

Wire up the LDR as shown in Figure 7-9, and then run the sketch shown in Example 7-3.

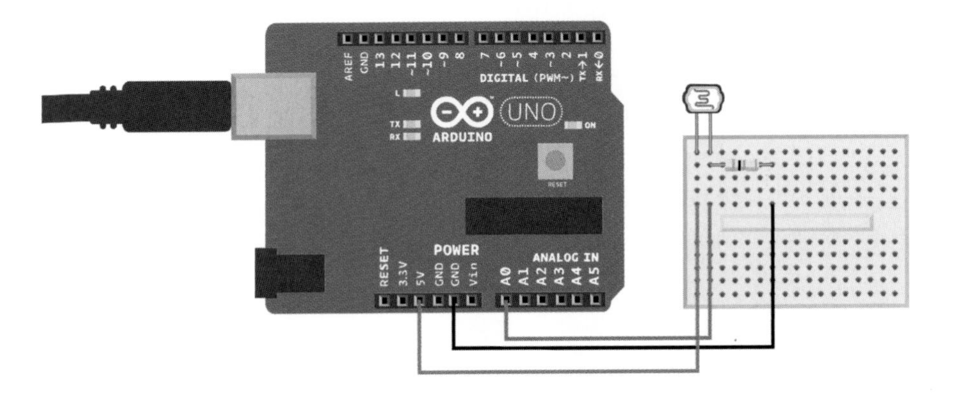

Figure 7-9. *Photoresistor circuit for Arduino*

Example 7-3. ldr_light_sensor.ino

```
// ldr_light_sensor.ino - report high level of light with built-in LED
// (c) BotBook.com - Karvinen, Karvinen, Valtokari

const int lightPin = A0;
const int ledPin = 13;
int lightLevel = -1;

void setup() {
  Serial.begin(115200);
  pinMode(ledPin, OUTPUT);
}

void loop() {
  lightLevel = analogRead(lightPin);     // ❶
  Serial.println(lightLevel);
  if(lightLevel < 400) {         // ❷
    digitalWrite(ledPin, HIGH);
  } else {
    digitalWrite(ledPin, LOW);
  }
  delay(10);
}
```

❶ To measure the voltage at A0, just use `analogRead()`.

❷ If printing the value to serial is not enough, you can also light an LED when the level of light exceeds an experimentally chosen value.

LDR Code and Connection for Raspberry Pi

Connect the components as shown in Figure 7-10, and then run the code shown in Example 7-4.

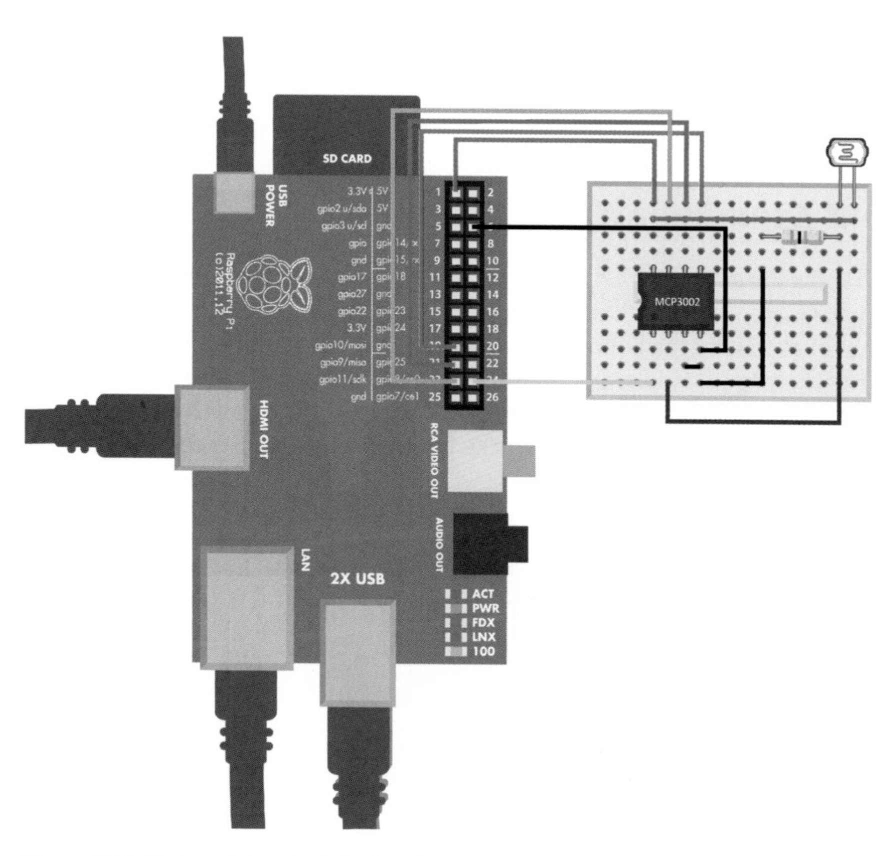

Figure 7-10. *Photoresistor circuit for Raspberry Pi*

Example 7-4. ldr.py

```
# ldr.py - sense light level and print to screen
# (c) BotBook.com - Karvinen, Karvinen, Valtokari
import time
import botbook_mcp3002 as mcp

def main():
        while True:
                lightLevel = mcp.readAnalog()
                print("Current light level is %i " % lightLevel)
```

```
                time.sleep(0.5)

if __name__ == "__main__":
        main()
```

The library botbook_mcp3002.py must be in the same directory as this program. You must also install the spidev library, which is imported by botbook_mcp3002. See comments in the beginning of botbook_mcp3002/botbook_mcp3002.py or "Installing SpiDev" on page 56.

Environment Experiment: One Direction

Would you like to know the direction that light is coming from, rather than just how bright it is? A naked photoresistor reacts to light coming from around it so you can't use it, for example, to turn a robot toward the light source. There's a very easy solution for this. Take a piece of heat-shrink tubing and make a hood for the sensor, as shown in Figure 7-11. This prevents it from uncontrollably seeing light from every direction. You can use a material other than heat-shrink tubing as long as it blocks light coming in from the side (see Figure 7-12).

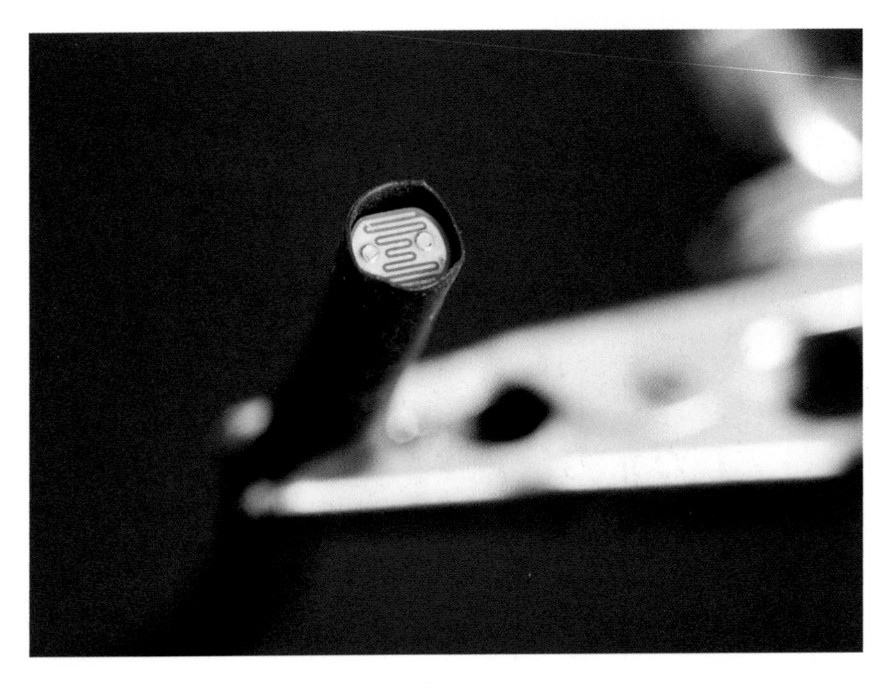

Figure 7-11. *Preventing light from reaching the sensor*

Figure 7-12. *Photoresistors inside plastic cup blinkers (robot workshop in Austria)*

When you have many photoresistors, it's a good idea to put all three wires and the resistor inside heat shrink tubing, as shown in Figure 7-13.

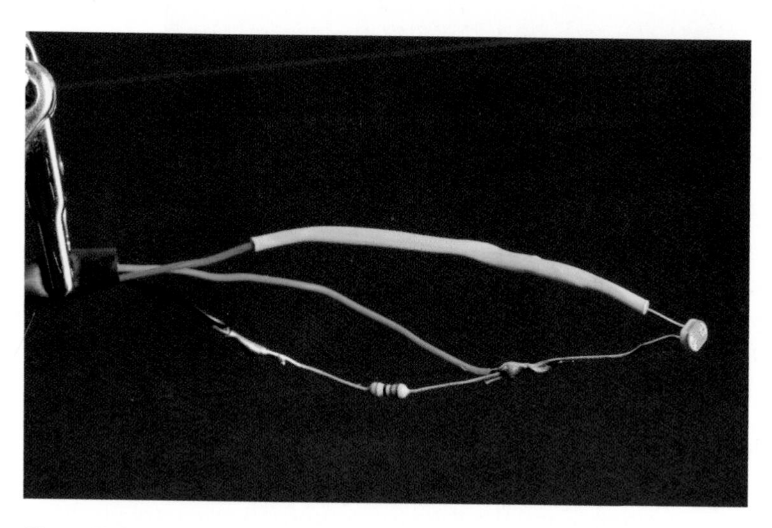

Figure 7-13. *With heat shrink tubing, you can neatly package the wires*

Experiment: Follow the Line

Line following is an easy way to move a robot along a predefined path. The most common use is creating "rails" with black tape. We have used line avoidance to keep our mind (EEG) controlled robot inside its playground. Figure 7-14 shows a line detector.

Figure 7-14. *Line tracking sensor*

Line detectors light the surface below with light, usually infrared. The surface is considered "white" if enough light is reflected back; anything else is considered a line. To know if your bot is going off the line from left or right, you can use two or three line detectors side by side (for example, if the center detector sees a line, but the other two see white, you know you're following the line). There are ready-made line detection sensors available that combine multiple sensors into one.

Line Sensor Code and Connection for Arduino

Because you are using a line detector with three leads here, no pull-up resistor is needed. One lead is for positive, another for signal, and the third for ground. The circuitry on the board includes any resistors or other components needed.

Wire up the circuit as shown in Figure 7-15, and run the sketch shown in Example 7-5.

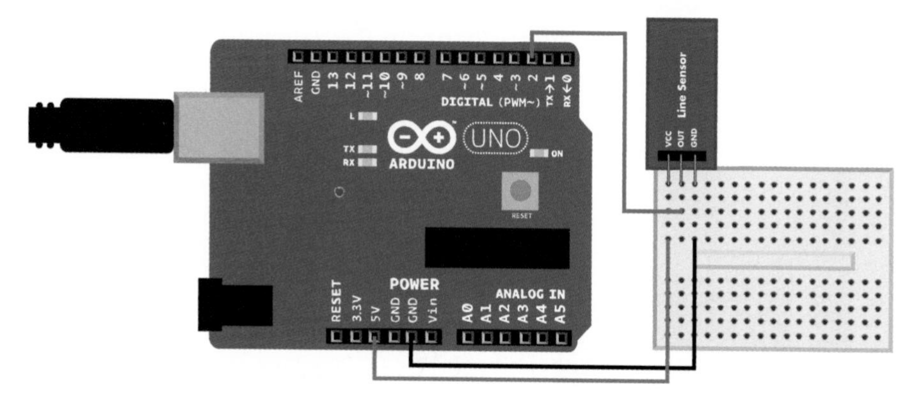

Figure 7-15. *Line sensor circuit for Arduino*

Example 7-5. line_sensor.ino

```
// line_sensor.ino - print to serial if we are on a line
// (c) BotBook.com - Karvinen, Karvinen, Valtokari

const int sensorPin = 2;
const int ledPin = 13;
int lineFound = -1;

void setup() {
  Serial.begin(115200);
  pinMode(sensorPin, INPUT);
  pinMode(ledPin, OUTPUT);
}

void loop() {
  lineFound = digitalRead(sensorPin);    // ❶
  if(lineFound == HIGH) {
    Serial.println("Sensor is on the line");
    digitalWrite(ledPin, HIGH);
  } else {
    Serial.println("Sensor is off the line");
    digitalWrite(ledPin, LOW);
  }
  delay(10);
}
```

❶ `digitalRead(8)` returns HIGH if the sensor has gone over the line.

Line Sensor Code and Connection for Raspberry Pi

As a line sensor is a digital sensor, its connection to Raspberry Pi is a simple as with Arduino. Figure 7-16 shows the circuit diagram. Wire it up as shown, and then run the code shown in Example 7-6.

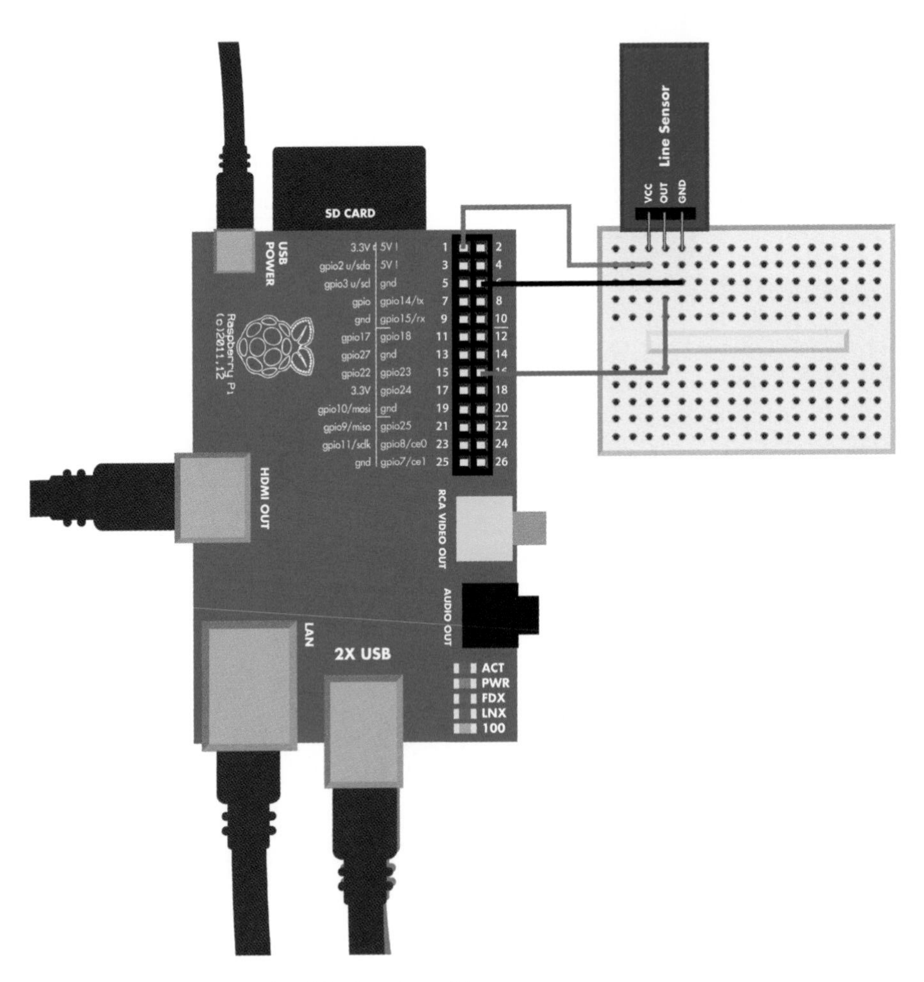

Figure 7-16. *Line sensor circuit for Raspberry Pi*

Example 7-6. line_sensor.py

```
# line_sensor.py - print to serial if we are on a line
# (c) BotBook.com - Karvinen, Karvinen, Valtokari
import time
import os
import botbook_gpio as gpio      # ❶

def main():
        linePin = 23
```

```
        gpio.mode(linePin, "in")
        while True:
                lineState = gpio.read(linePin)  # ❷
                if( lineState == gpio.HIGH ):
                        print "Sensor is on the line"
                else:
                        print "Sensor is off the line"
                time.sleep(0.5)

if __name__ == "__main__":
        main()
```

❶ Make sure there's a copy of the *botbook_gpio.py* library in the same directory as this program. You can download this library along with all the example code from *http://botbook.com*. See "GPIO Without Root" on page 19 for information on configuring your Raspberry Pi for GPIO access.

❷ The digital value is read just like any other sensor.

Environment Experiment: Black is White

Figure 7-17. *Believe it or not, everything you see here is white*

As you already know, an infrared sensor sees the world differently than we do. Different materials reflect light differently, and sometimes objects that appear dark can be so reflective that the sensor thinks they are white. If your line-following robot acts strange, it might be the surface texture—not the code—that's playing tricks on you.

Usually black is black and white is white, but the total opposite has happened to us. When we presented our mind-controlled robot at Maker Faire, we brought tape and cardboard in order to make a platform with black borders and white center. The idea was that our robot would turn back when it saw black, keeping it on the platform. Surprisingly, the line detector saw tape and cardboard in inverted colors, as one was very reflective and the other very matte.

You can adjust the line follower sensitivity by adjusting its onboard potentiometer as shown in Figure 7-18.

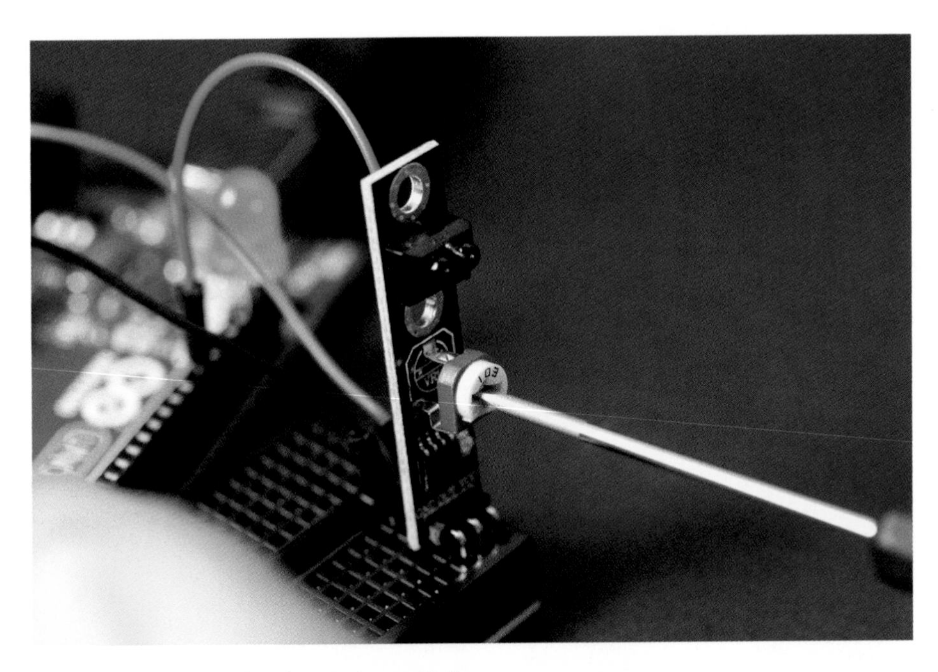

Figure 7-18. *Adjusting line detector's sensitivity*

Upload the code shown in Example 7-7 to Arduino. We have changed the previous example slightly so that now the serial monitor says BLACK or WHITE depending on what it sees.

Example 7-7. line_sensor_black_or_white.ino

```
// line_sensor_black_or_white.ino - line follow sensor. Signal low when over black line.
// (c) BotBook.com - Karvinen, Karvinen, Valtokari


const int sensorPin = 2;
const int ledPin = 13;
int lineFound = -1;
```

```
void setup() {
  Serial.begin(115200);
  pinMode(sensorPin, INPUT);
  // No need for pull-up as sensor has this already.
  pinMode(ledPin, OUTPUT);
}

void loop() {
  lineFound = digitalRead(sensorPin);
  if(lineFound == 1) {
    Serial.println("BLACK");
    digitalWrite(ledPin, HIGH);
  } else {
    Serial.println("WHITE");
    digitalWrite(ledPin, LOW);
  }
  delay(50);
}
```

This code combines serial printing with the code shown in "Line Sensor Code and Connection for Arduino" on page 172.

Try out different materials. Can you find one that's black for humans but white for the sensor?

Experiment: All the Colors of the 'Bow

A color sensor measures the color of a surface and returns values for red, green, and blue (Figure 7-19). For each basic color (red, green, blue), a color sensor has a color filter on top of a photodiode. The sensor for each color is read like any analog resistance sensor.

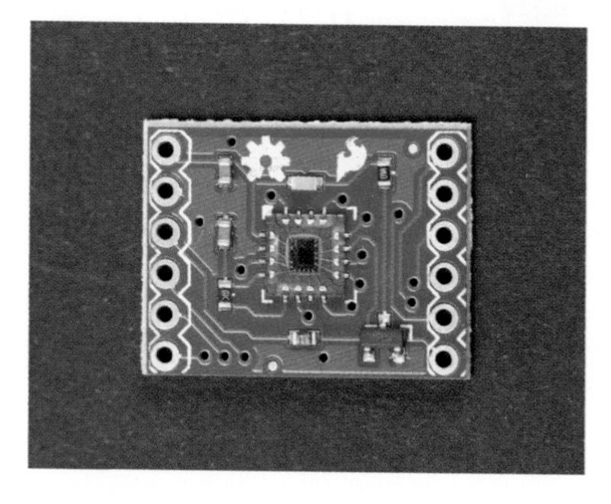

Figure 7-19. *Color sensor*

The experiment prints RGB (red, green, blue) values of the light it sees, one value for each color. You can use this sensor to build the Chameleon Dome ("Test Project: Chameleon Dome" on page 182).

Colors are simply names for specific wavelengths of light. For example, 555 nanometers (nm, which describes the light's wavelength) or 540 terahertz (THz, its frequency) is green. Some animals can see colors humans don't see, like ones in the infrared or the ultraviolet range.

The basic colors really are fundamental to humans. The only colors that the human eye can see are red, green, and blue. There are three types of cone cells in the retina, one type for each color.

Color Sensor Code and Connection for Arduino

Figure 7-20 shows the circuit diagram for the color sensor and Arduino. Wire it up as shown, and then run the sketch shown in Example 7-8.

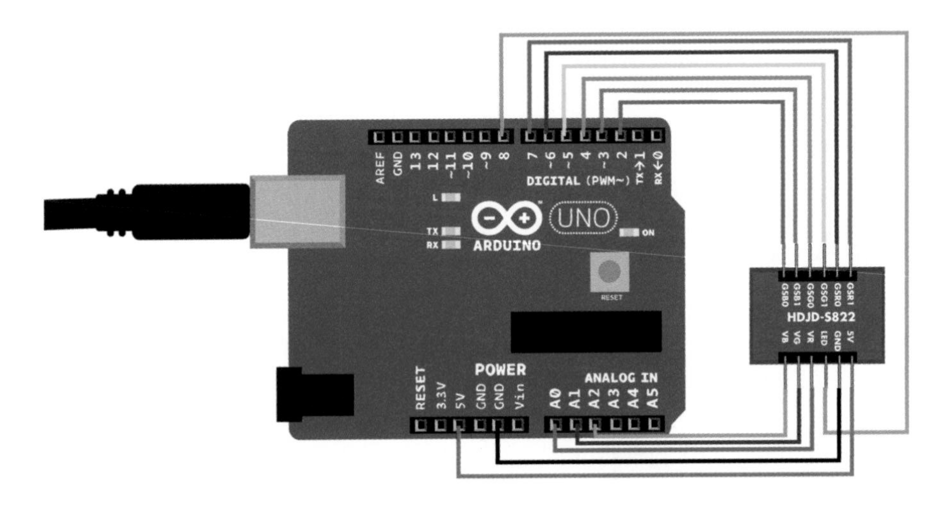

Figure 7-20. *Color sensor circuit for Arduino*

Example 7-8. color_sensor.ino

```
// color_sensor.ino - sense color with HDJD-S822-QR999 and print RGB value
// (c) BotBook.com - Karvinen, Karvinen, Valtokari


const int gsr1Pin = 7;  // ❶
const int gsr0Pin = 6;
const int gsg1Pin = 5;
const int gsg0Pin = 4;
const int gsb1Pin = 3;
```

```
const int gsb0Pin = 2;

const int ledPin = 8;   // ❷

const int redPin = A0;  // ❸
const int greenPin = A1;
const int bluePin = A2;

int red = -1;   // ❹
int green = -1;
int blue = -1;

void setup() {
  Serial.begin(115200);
  pinMode(gsr1Pin, OUTPUT);
  pinMode(gsr0Pin, OUTPUT);
  pinMode(gsg1Pin, OUTPUT);
  pinMode(gsg0Pin, OUTPUT);
  pinMode(gsb1Pin, OUTPUT);
  pinMode(gsb0Pin, OUTPUT);
  pinMode(ledPin, OUTPUT);

  digitalWrite(ledPin, HIGH);   // ❺

  digitalWrite(gsr1Pin, LOW);   // ❻
  digitalWrite(gsr0Pin, LOW);
  digitalWrite(gsg1Pin, LOW);
  digitalWrite(gsg0Pin, LOW);
  digitalWrite(gsb1Pin, LOW);
  digitalWrite(gsb0Pin, LOW);
}

void loop() {
  int redValue = analogRead(redPin);    // ❼
  int greenValue = analogRead(greenPin);
  int blueValue = analogRead(bluePin);

  redValue = redValue * 10 / 1.0;  // ❽
  greenValue = greenValue * 10 / 0.75;
  blueValue = blueValue * 10 / 0.55;

  Serial.print(redValue); Serial.print(" ");    // ❾
  Serial.print(greenValue); Serial.print(" ");
  Serial.print(blueValue); Serial.println(" ");
  delay(100);
}
```

❶ Specify the *gain selection pins*. To calibrate individual colors, you can later add gain to red (`gsr1Pin`, `gsr0Pin`), green (`gsg1Pin`, `gsg0Pin`), or blue (`gsb1Pin`, `gsb0Pin`) if needed. This way, you can increase sensitivity to individual colors as needed.

❷ This LED illuminates the surface and is critical for the operation of this sensor. The illuminating LED is built into the sensor.

❸ Pins for reading red, green, and blue levels.

❹ The variables for the values read from the sensor. The variables are initialized to impossible values to help debugging (if you see those values later when the code is running, then something's wrong).

❺ Turn on the sensor's built-in LED for illuminating the surface.

❻ Turn off gain selection for all colors. Each color has two bits (0 and 1), allowing for four levels (including off) of gain selection.

❼ Reading the values is a simple call to `analogRead()`.

❽ The elements for different colors have different sensitivity. On the other hand, you need all of RGB in the same scale if you want to recreate the color. The conversion factors (10 red, 14 green, 17 blue) were deduced from "HDJD-S822-QR999 RGB Color Sensor Datasheet."

❾ Print each of the RGB values to the serial monitor (Tools→Serial Monitor).

Color Sensor Code and Connection for Raspberry Pi

Figure 7-21 shows the circuit for Raspberry Pi. Hook it up as shown, and then run the code shown in Example 7-9.

A Tangled Mess

Does the Raspberry Pi circuit for color sensor make you want to scream? The number of wires makes off-by-one mistakes and loose wires more likely.

There are two basic solutions to this problem: either simplify the circuit here and there, or combine Arduino and Raspberry Pi. Our favorite solution is combining the two platforms.

You can read the sensor with Arduino, and then send the RGB values to Raspberry Pi using USB-Serial. With Arduino's built-in `analogRead()`, the circuit is simple (Figure 7-20). Then use USB to connect Arduino to Raspberry Pi.

In Raspberry Pi, you can read serial-over-USB using pySerial. *Make: Arduino Bots and Gadgets* has a tutorial project with pySerial (Chapter 7: "Remote for a Smart Home").

A third method that could slightly reduce wiring is to use an ADC (analog-to-digital converter) with more channels, such as MCP3008. To use such a chip, you would have to modify the *botbook_mcp3002* library.

If you never plan to use gain pins, you can try connecting them to ground, thus keeping them LOW and disabling all gains. This could also simplify the connections.

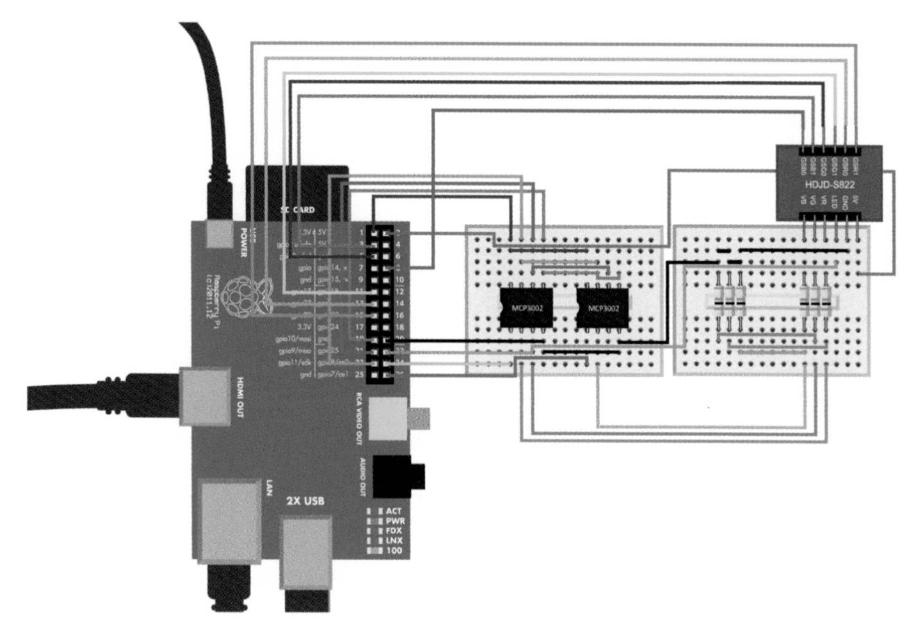

Figure 7-21. *Color sensor circuit for Raspberry Pi makes your head spin*

Example 7-9. *color_sensor.py*

```
# color_sensor.py - sense color and print RGB value to serial
# (c) BotBook.com - Karvinen, Karvinen, Valtokari

import time
import botbook_mcp3002 as mcp    # ❶
import botbook_gpio as gpio

def initializeColorSensor():
        ledPin = 25
        gpio.mode(2,"out")       # ❷
        gpio.mode(3,"out")
        gpio.mode(14,"out")
        gpio.mode(17,"out")
        gpio.mode(22,"out")
        gpio.mode(27,"out")

        gpio.write(2,gpio.LOW)
        gpio.write(3,gpio.LOW)
        gpio.write(14,gpio.LOW)
        gpio.write(17,gpio.LOW)
        gpio.write(22,gpio.LOW)
        gpio.write(27,gpio.LOW)

        gpio.mode(ledPin,"out")
        gpio.write(ledPin, gpio.HIGH)    # ❸

def main():
```

```
        initializeColorSensor()
        while True:      # ❹
                redValue = mcp.readAnalog(0, 0)
                greenValue = mcp.readAnalog(0, 1)
                blueValue = mcp.readAnalog(1, 0)          # ❺

                redValue = redValue * 10 / 1.0; # ❻
                greenValue = greenValue * 10 / 0.75;
                blueValue = blueValue * 10 / 0.55;

                print("R: %d, G: %d, B: %d" % (redValue,greenValue,blueValue))  # ❼

                time.sleep(0.1) # s

if __name__ == "__main__":
        main()
```

❶ Both libraries, *botbook_mcp3002.py* (analog) and *botbook_gpio.py* (digital), must be in the same directory as this program (*color_sensor.py*). You must also install the *spidev* library, which is imported by *botbook_mcp3002*. See the comments in the beginning of *botbook_mcp3002/botbook_mcp3002.py* or "Installing SpiDev" on page 56. You can download both libraries, along with all the example code, from *http://botbook.com*. For configuring access to GPIO without needing to be root, see "GPIO Without Root" on page 19.

❷ Turn off all gains by putting the gain select (`gs*`) pins to 0 (LOW).

❸ Light up the illuminating LED in the sensor. This light is needed for the sensor to see the colors of the surface it's measuring.

❹ The program will keep running until you press Control-C.

❺ Read the second (1, because we start counting at 0) MCP3002 chip on the first (0) channel: `readAnalog(device=1, channel=0)`. The preceding commands use a different combination of chip and channel.

❻ Equalize the color values according to the "HDJD-S822-QR999 RGB Color Sensor Datasheet." After equalization, all colors use the same scale.

❼ Create the printable string with a format string. The format string takes only one parameter, so multiple values are put inside a tuple, `(a, b)`.

Test Project: Chameleon Dome

Our final project for this chapter is a dome that changes color to match the surface it's sitting on. We'll use the color sensor code and display the color it senses with an RGB LED. When the whole thing is built in a solid package, the result is very impressive.

What You'll Learn

In the *Chameleon Dome* project, you'll learn how to:

- Build a device that changes color like a chameleon.
- Show any color with an RGB LED.
- Use a *moving average* to filter out random noise.
- Use an *easing function* to map input values to output values.

RGB LED

An RGB LED (Figure 7-22) packages three LEDs into one package. It looks just like a single LED. By mixing red, green, and blue, you can show any color.

The human eye has receptor cells for three colors: red, green, and blue. That's why those colors are used in televisions and other displays. Human perception has a strange feature (or a bug) that it sees combinations of frequencies as another frequency. This means that you can mix red light with green to get yellow.

An RGB LED typically has four leads: one lead for each color (red, green, blue) and one common lead.

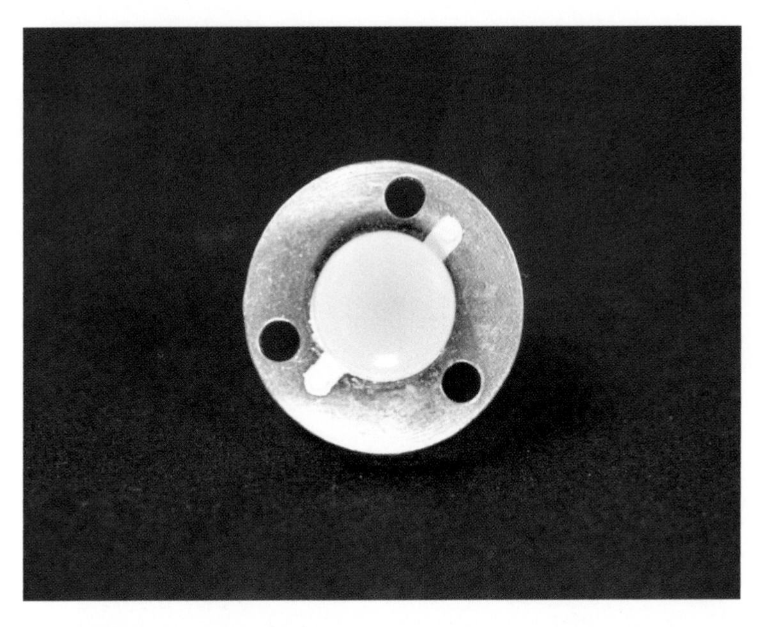

Figure 7-22. *RGB LED*

Contrary to what you might expect, in many RGB LEDs, the common lead is often positive. A common positive lead is also called *common anode*. Because the common lead is positive, you'll have to take each color lead (red, green, blue) LOW (0 V or GND) to make each color shine.

What About Common Cathode?

If your RGB LED is *common cathode* (common lead negative), you'll need to change the circuit *and* your code. For the circuit, you'll need to connect the common lead to ground instead of +5 V. For the code, instead of reversing the light intensity value with `255-color` (for example `255-red`) with `analogWrite()`, you must remove the `255-` part (for example `analogWrite(redPin, red)`).

But how do you know which lead is the common anode, and which one is each color?

Finding the Leads of an RGB LED

You can find the leads experimentally.

Turn on Arduino by connecting it to USB. As you'll just use +5 V and GND, it doesn't matter which code is running on the Arduino.

> *We are using a 5 V LED. If you only have a LED for lower voltage, you can use +3.3 V when testing for colors, and a resistor when connecting to Arduino.*

On Arduino, connect a red wire to +5 V and a black wire to GND.

Connect the two wires to any adjacent leads on the RGB LED. Keep trying adjacent leads around the LED until it lights up. Try a couple of other leads until you have lit the red, green, and blue elements of the RGB LED separately. Note that in order to do this, one lead (for a common anode LED, this would be the positive lead) had to stay connected to the red +5 V wire, while you had to connect each of the other three leads in turn to the black GND wire. Mark the common anode (positive) lead.

If you notice that the common lead is longer than the others, remember this fact so you can find it later. Otherwise, mark the common lead with a small piece of tape and write something on it (A for anode, + for positive, whatever helps you remember).

If instead, you kept a common lead connected to the black GND wire and had to connect each of the other three leads in turn to the red +5 V wire, then you have a common cathode LED. If that's the case, see "What About Common Cathode?" on page 184.

Mark each of the remaining leads with their corresponding color (R, G, or B).

RGB Code and Connection for Arduino

Example 7-10. *hellorgb.ino*

```
// hellorgb.ino - mix colors with RGB LED
// (c) BotBook.com - Karvinen, Karvinen, Valtokari

const int redPin=11;    // ❶
const int greenPin=10;
const int bluePin=9;

void setup()
{
        pinMode(redPin, OUTPUT);
        pinMode(greenPin, OUTPUT);
        pinMode(bluePin, OUTPUT);
}

void loop()
{
        setColor(255, 0, 0);    // ❷
        delay(1000);

        setColor(255, 255, 255);
        delay(1000);
}

void setColor(int red, int green, int blue)
{
```

```
        analogWrite(redPin, 255-red);    // ❸
        analogWrite(greenPin, 255-green);
        analogWrite(bluePin, 255-blue);
}
```

❶ Like many RGB LEDs, this one has common positive lead. This means that all the data pins will need to be taken negative (LOW, 0 V) to light up.

❷ To light the RGB LED, just call `setColor()` with the color you want. The parameters are R, G, and B, where each color is from (nothing) 0 to (maximum) 255.

❸ Because you need to take a pin low to light a color, things are reversed from what you're used to. For example, when `redPin` is `HIGH` (255), the red LED is off. When `redPin` is `LOW` (0), the red LED is at maximum brightness.

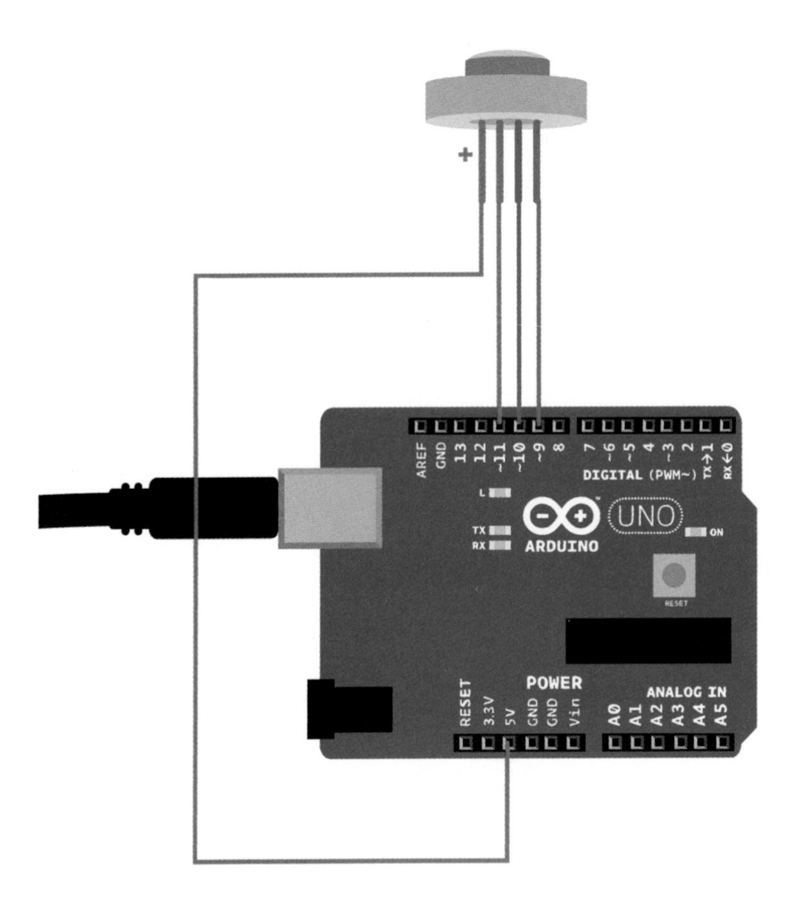

Figure 7-23. *RGB LED connected to Arduino*

Figure 7-24. *Color-changing robot prototype made by students during two-day robot workshop*

Do you just want to make your Chameleon Dome? Skip the math and jump to "Combining Codes" on page 190.

Moving Average

The Chameleon Dome looks nice when it smoothly changes values. It would look ugly if it erratically jumped from one color to another. But that's one of the perils of working with sensors: sometimes, for a blink of an eye, a red will look green for the sensor.

Random noise is a common problem in any measurement, and sensors are no exception.

Consider how I might measure the height of a tiny tree. I made multiple measurements to reduce the chance of error. I got these values:

```
102 cm, 100 cm, 180 cm, 103 cm, 105 cm
```

Most of us would likely drop the 180 cm measurement as a typo. Luckily, a computer can do that for you.

Initially, you might think of storing the values in an array, and then calculating the average for the array on every round. Even though this would work, it would require a lot of code for such a small thing: an array, a pointer, an average, and some calculation on each iteration.

Moving average to the rescue! You can calculate the average of current and previous value to get some smoothing. To get more data points without using a table, you can use a *weighted moving average*.

You could give 70% weight to new input, leaving 30% (100% minus 70%) for old values.

```
input = 0.7*input + 0.3*old
old = input
```

Because the old value is affected by the older data points, you don't need to use an array at all.

RGB LED Shows Any Color

To get a variety of colors, you'll need to mix red, green, and blue at varying levels. If you set all three of them to full brightness, you'll get something pretty close to white light. If you set all three of them to minimum brightness, you'll get nothing.

Arduino isn't capable of truly dimming an LED, because an LED can't be dimmed effectively. If you lower the voltage, the brightness will go down, but lower the voltage enough, and it simply turns off (usually well before you get all the way down to 0 V).

To work around this, Arduino uses *pulse width modulation* (PWM). To make an LED look like it's at 10% brightness, PWM will use a *duty cycle* of 10%. This means that Arduino will keep the LED on for 10% of the time, and off 90% of the time. But Arduino switches the LED on and off so fast (each cycle takes 2 milliseconds or 2000 microseconds) that you don't notice the flicker. At a 10% duty cycle, Arduino will turn the LED on for 200 microseconds, leave it off for 1800 microseconds, then start the on/off cycle again.

But in your code, you don't have to worry about these details. You can pretend that you're just sending a range of voltage (from 0 to 255) that corresponds to different brightness levels. Because the common lead is HIGH and each color lead is brightest when you take the lead LOW, you'll get the brightest red color with `analogWrite(redPin, 0)`.

For any values between the minimum and the maximum, we use the value of the color (such as red, `r`) subtracted from 255:

```
analogWrite(redPin, 255-r);
```

To get a feel for the basic colors, see Table 7-1.

Table 7-1. RGB LED colors and pin values

Name	RGB color	Data pin values	Comment
Black	(0,0,0)	(255,255,255)	All colors off
Red	(255,0,0)	(0,255,255)	
Green	(0,255,0)	(255,0,255)	
Blue	(0,0,255)	(255,255,0)	
White	(255,255,255)	(0,0,0)	Maximum brightness of all LEDs
(formula)	(r,g,b)	(255-r, 255-g, 255-b)	

You can get all the colors of the rainbow. Just experiment with mixing red, green and blue light!

Easing Input to Output

Inputs and outputs can have different kinds of values. Your code must convert between these values.

In the simplest case, output is just the input multiplied by a number. In that case, conversion is just a matter of multiplication.

For example, Arduino's `analogRead()` has range of 0 to 1023, but `analogWrite()` has a range of 0 to 255. To convert between the ranges, you must first calculate the percentage, `p`, of the maximum input value, that a given input value (`in`) represents:

```
p = in/1023
```

Then, the mapped output is `p` percent of the maximum output value:

```
out = p*255
```

Arduino's library even has a convenience function for this:

```
out = map(in, 0, 1023, 0, 255)
```

You can see some examples of linear conversion in Table 7-2.

Table 7-2. Linearly mapping input to output, using map()

Input analogRead()	Percent	Output analogWrite()
0	0.0 %	0.0
234	23 %	58
511	50 %	127
1023	100 %	255

However, some outputs don't work well with an output that increases linearly, so you need to use *easing*. The RGB LED in this project is a good example of an output that works better with easing.

Most RGB LED color mixing needs to happen with low values. Near the upper range of output values, everything becomes white, and individual colors are hard to discern.

An easing function maps inputs to outputs in non-linear fashion. For an RGB LED, an exponential function is good. Otherwise, most values would just result in bright white light instead of colors like orange or violet.

First, calculate the percentage for the input:

```
p = in/1023
```

Then create the output non-linearly. Both ends, bottom 0 and top 255, are still represented along the range of possible values. The Chameleon Dome uses an exponent function for easing:

$$out = 255 * p^4$$

Because percentage p has a maximum of 1.0 (100%), the exponent p^4 values are always between 0.0 (0%) and 1.0 (100%). The exponent function creates the classic hockey-stick figure. You can see sample values mapped in Table 7-3.

Table 7-3. Easing with exponent function

p	p**4	analogWrite()	Comment
0%	0.0%	0	min
	20%	0.2%	0
	40%	2.6%	6
	50%	6.2%	15
half	60%	13.0%	33
	80%	41.0%	104
	90%	65.6%	167
	100%	100.0%	255

Easing functions are also used in animation. When objects slide and then stop into place, the speed accelerates and decelerates according to an easing function.

Combining Codes

The Chameleon Dome combines an RGB LED with the color sensor you saw earlier in "Color Sensor Code and Connection for Arduino" on page 178.

Connect the color sensor and RGB LED to Arduino as shown in Figure 7-27. DuPont connector cables (see Figure 7-25) combined with a ScrewShield are a life-saver when you need to use a lot of pins (see Figure 7-26).

Figure 7-25. *DuPont connector cable*

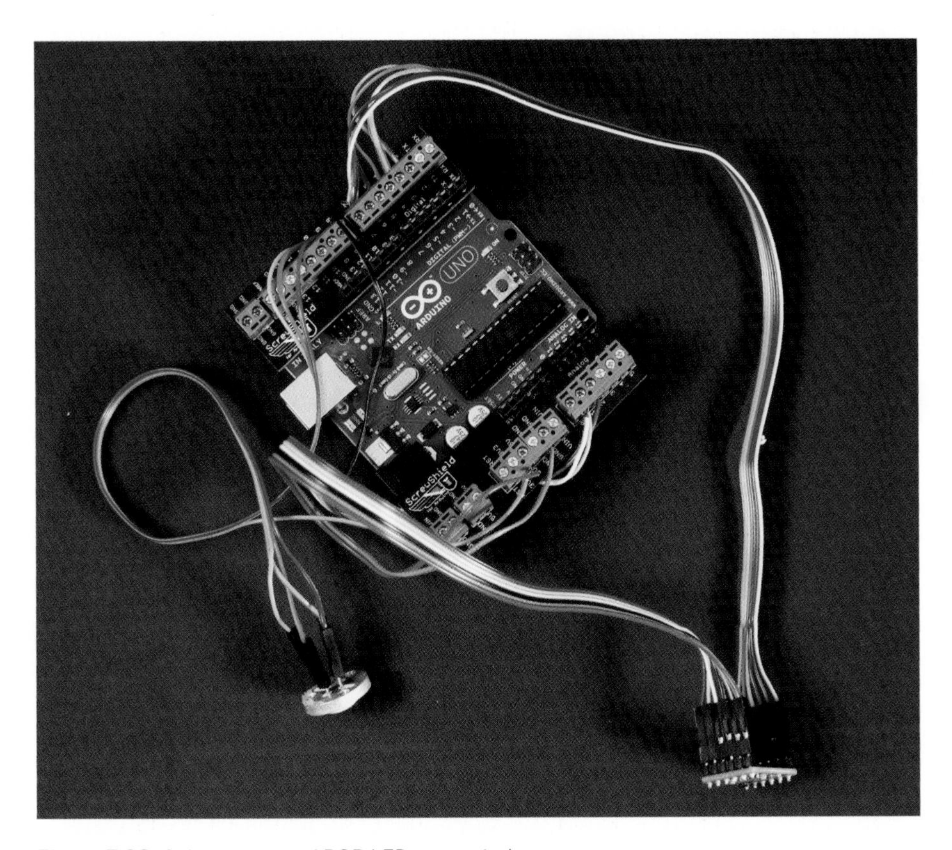

Figure 7-26. *Color sensor and RGB LED connected*

The sketch for the Chameleon Dome is shown in Example 7-11. After you make all the connections (Figure 7-27), run that sketch on your Arduino.

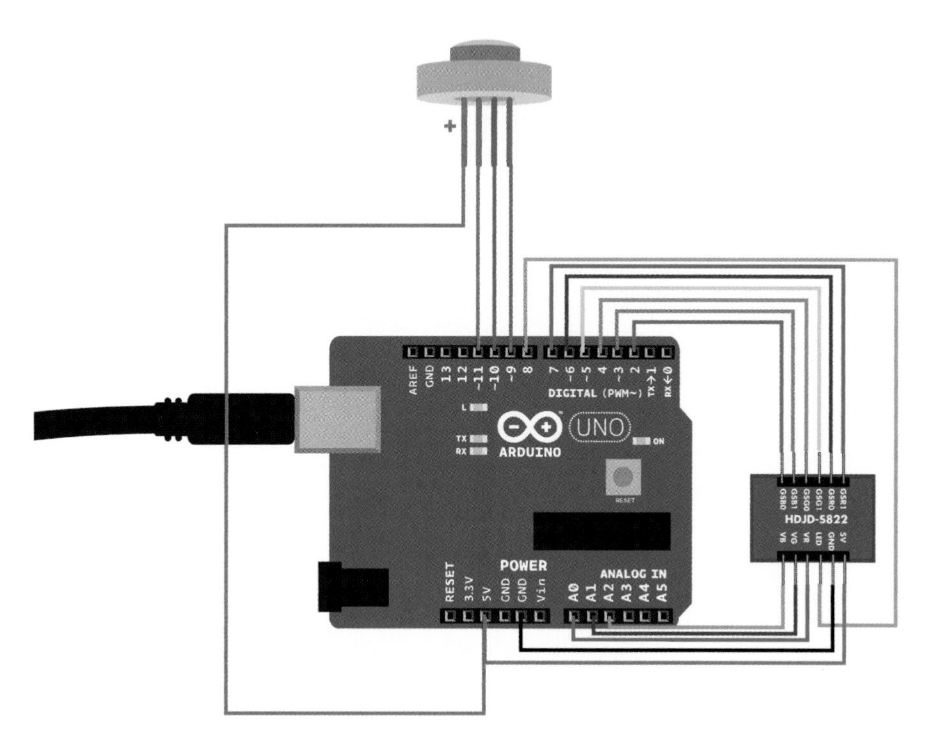

Figure 7-27. *Chameleon Dome connections*

Example 7-11. ***chameleon_cube.ino***

```
// chameleon_dome.ino - cube changes color to match the surface
// (c) BotBook.com - Karvinen, Karvinen, Valtokari

const int gsr1Pin = 7;  // ❶
const int gsr0Pin = 6;
const int gsg1Pin = 5;
const int gsg0Pin = 4;
const int gsb1Pin = 3;
const int gsb0Pin = 2;

const int ledPin = 8;   // ❷

const int redInput = A0;        // ❸
const int greenInput = A1;
const int blueInput = A2;

const int redOutput = 11;       // ❹
const int greenOutput = 10;
const int blueOutput = 9;

int red = -1;   // ❺
int green = -1;
int blue = -1;
```

```
const float newWeight = 0.7;    // ❻

void setup() {
  Serial.begin(115200);
  pinMode(gsr1Pin, OUTPUT);
  pinMode(gsr0Pin, OUTPUT);
  pinMode(gsg1Pin, OUTPUT);
  pinMode(gsg0Pin, OUTPUT);
  pinMode(gsb1Pin, OUTPUT);
  pinMode(gsb0Pin, OUTPUT);
  pinMode(ledPin, OUTPUT);
  pinMode(redOutput, OUTPUT);
  pinMode(greenOutput, OUTPUT);
  pinMode(blueOutput, OUTPUT);

  digitalWrite(ledPin, HIGH);    // ❼

  digitalWrite(gsr1Pin, LOW);
  digitalWrite(gsr0Pin, LOW);
  digitalWrite(gsg1Pin, LOW);
  digitalWrite(gsg0Pin, LOW);
  digitalWrite(gsb1Pin, LOW);
  digitalWrite(gsb0Pin, LOW);
}

void loop() {
  int redValue = analogRead(redInput);  // ❽
  int greenValue = analogRead(greenInput);
  int blueValue = analogRead(blueInput);

  redValue = redValue * 10 / 1.0;
  greenValue = greenValue * 10 / 0.75;
  blueValue = blueValue * 10 / 0.55;

  redValue = map(redValue, 0, 1023, 0, 255);    // ❾
  greenValue = map(greenValue, 0, 1023, 0, 255);
  blueValue = map(blueValue, 0, 1023, 0, 255);

  if(redValue > 255) redValue = 255;    // ❿
  if(greenValue > 255) greenValue = 255;
  if(blueValue > 255) blueValue = 255;

  red = runningAverage(redValue, red);  // ⓫
  green = runningAverage(greenValue, green);
  blue = runningAverage(blueValue, blue);

  Serial.print(red); Serial.print(" ");
  Serial.print(green); Serial.print(" ");
  Serial.print(blue); Serial.println(" ");
  if(red < 200 || green < 180 || blue < 180) {
    green = green - red * 0.3;  // ⓬
    blue = blue - red * 0.3;
  }
```

```
  red = easing(red);    // ⓭
  green = easing(green);
  blue = easing(blue);

  setColor(red,green,blue);     // ⓮

  delay(100);
}

int runningAverage(int input, int old) {
  return newWeight*input + (1-newWeight)*old;    // ⓯
}

int easing(int input) { // ⓰
  float percent = input / 255.0f;
  return 255.0f * percent * percent * percent * percent;
}

int setColor(int r, int g, int b) {     // ⓱
  analogWrite(redOutput, 255-r);        // ⓲
  analogWrite(greenOutput, 255-g);
  analogWrite(blueOutput, 255-b);
}
```

❶ These are the gain pins for each color. This code doesn't use gain, but sets all gain pins LOW. If you decide to use them, see Example 7-8 for more details.

❷ This is the pin for the surface-illuminating LED in the sensor.

❸ These are the input pins to read RGB values.

❹ The output pins for the RGB LED.

❺ Global variables to hold manipulated color values. They are initialized to impossible values to make debugging easier (if you see these values appear later when the code is running, you know something's wrong).

❻ The weighting value to apply to new input (see "Moving Average" on page 187).

❼ Illuminate the surface before taking measurements.

❽ Read the red color. `analogRead()` returns a raw integer value between 0 and 1023.

❾ Map `analogRead()` values (0..1023) to `analogWrite()` values (0..255).

❿ ⓫ Because color values are equalized (e.g., red is multiplied by 10), they could end up higher than the maximum used in the map. This would result in a red value higher than 255. But later you'll be controlling the LED with `analogWrite()`, so you must cap the value at 255.

⓬ Some aesthetic color changes, with values found experimentally.

⓭ Make colors change a little more slowly, instead of just immediately jumping from red to green.

⓮ Set the RGB LED to a calculated color. Now the user can see the result.

⓯ A moving average allows you to smooth out random noise in the input. See "Moving Average" on page 187.

⓰ Easing is an animation term. When Flash or JavaScript animations move things so that they accelerate and decelerate like real objects, that's often done with easing functions. The function here makes values lower than 255 (100%) smaller. Smaller values are affected more. The purpose is to make the output (RGB LED) change slowly and smoothly.

⓱ Change the RGB LED color to the given RGB value. Parameters are integers in the range of 0..255.

⓲ For common anode (common positive), the data pins are negative pins, and the values must be inverted. See "RGB LED Shows Any Color" on page 188.

Dome Building Tips

We found a perfect casing for our Chameleon Dome from IKEA. With some minor hacking, the Solvinden lamp is made for this. Obviously you can use any translucent box or dome that pleases your eye. There are tons of different lamps to choose from, or you could even use a freezer food storage container.

Start by opening Solvinden (Figure 7-28) and pry the bottom cover lid off (Figure 7-29). Remove the original electronics to make space for our gadget as shown in Figure 7-30.

Figure 7-28. *First, remove the dome*

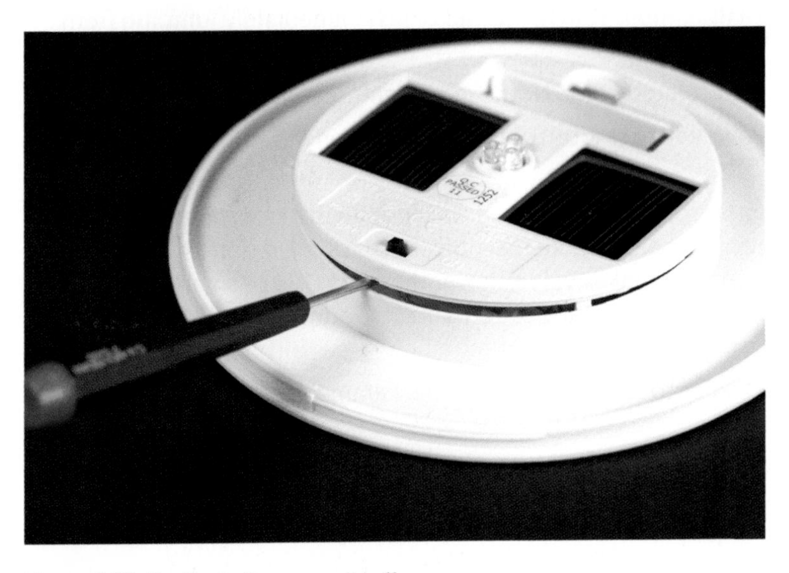

Figure 7-29. *Pry the bottom cover lid off*

Figure 7-30. *Remove the electronics*

There is a plastic stick in the center of the bottom; cut it off (see Figure 7-31). You need to make two holes, one 19 mm for the sensor to see what's below; and one 3 mm to attach the Arduino and the RGB LED (Figure 7-32).

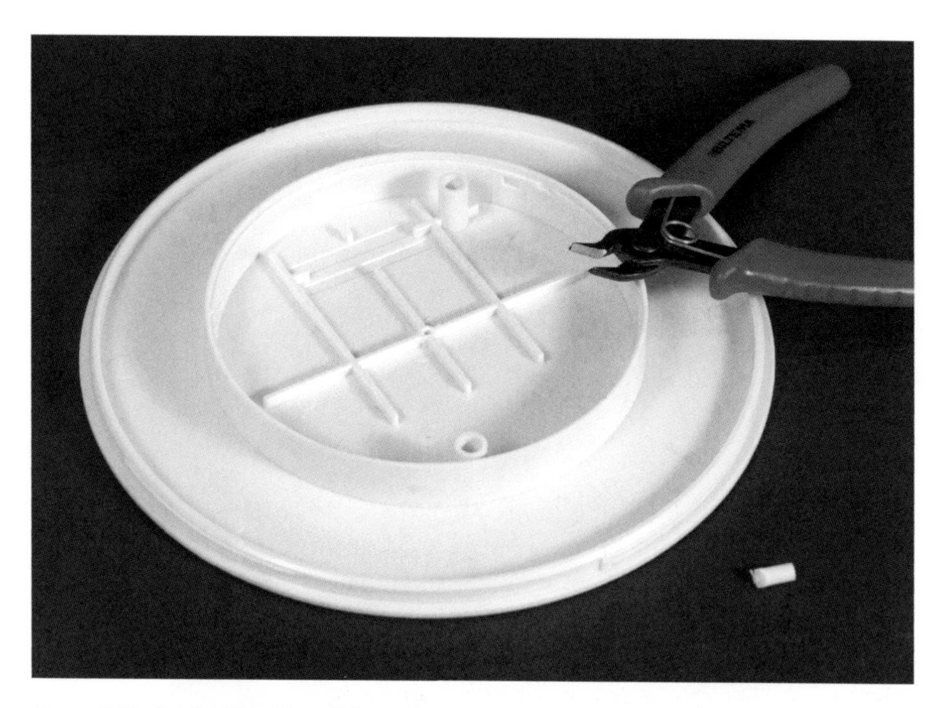

Figure 7-31. *Cut the little stick off from the center*

Figure 7-32. *Drill a 19 mm hole for the sensor and 3 mm hole for a screw to hold the Arduino and RGB LED in place*

Use hot glue to attach the sensor pointing down, and make sure it's in the center of the hole as shown in Figure 7-33.

Figure 7-33. *Cameleon Dome electronics attached*

Use a 3 mm screw to secure Arduino in the bottom (see Figure 7-33). On the top of that same screw holding Arduino, put the RGB LED. Our LED already had holes, but we needed to drill one larger to fit in the 3 mm screw.

Now just put the battery in the clip, turn the power switch on, close the dome, and enjoy your Chameleon Dome (see Figure 7-34).

Figure 7-34. *Dome is ready with new electronics inside*

Now your devices can see the light in many ways: detect the presence and direction of light, measure its intensity, and even its color.

Acceleration

When you tilt your smartphone, the screen probably turns from portrait to landscape. How did the phone know? Smartphones have a built-in accelerometer, and gravity is indistinguishable from acceleration downward, so the downward pull of gravity helps tell the phone which way it's oriented.

Some games are controlled by tilting a phone in the air, using the accelerometer. The popular Wii game console uses an accelerometer for its controllers. In this chapter, you will hack a Wii controller to use its sensors. We used a phone accelerometer for control in the Football Robot from *Make: Arduino Bots and Gadgets*.

Hard disks found in laptops and desktops can now take a lot of punishment. Many are rated to take a shock of 150 g (when off), an acceleration that would immediately kill any human. To brace for impact, some drives will turn themselves off: when the hard disk accelerometer detects it's in free fall, it automatically moves the actuator arm away from the sensitive plates.

Have you ever tried to ride a self-balancing device, like a Segway or a Solowheel? After perhaps a shaky start, it almost feels like a miracle that the device stays upright.

When a self-balancing device detects that it's about to fall over forward, it quickly moves its wheels forward, turning itself upright again. Self-balancers measure angular velocity with a gyroscope. (An accelerometer would gather too many cumulative errors to work in a self-balancing device.)

Acceleration vs. Angular Velocity

Acceleration is the rate at which an object's velocity changes (when it's slowing down or speeding up). Angular velocity measures the rotational speed of an object, as well as the axis that it's rotating around. Depending on your project, you might need acceleration, angular velocity, or even both.

Acceleration is measured in g, as a multiple of the acceleration caused by Earth's gravity. Another commonly used unit of acceleration is meters per second squared (m/s^2). Free-fall acceleration (1 g) is 9.81 m/s^2.

Why are the seconds squared in acceleration? Acceleration is change in speed. If you're using meters per second (m/s) as a unit of speed, then the unit of acceleration (change of speed) is meters per second per second, or m/s^2.

Gyroscopes measure angular velocity, or how fast the sensor is rotating around an axis. For example, a gyroscope might report it's rotating at 10 degrees per second. They are used in self-balancers and airplane gyrocompasses.

Table 8-1. Accelerometer vs. gyroscope

Sensor	Measures	Meaning	Unit	Gravity
Accelerometer	Acceleration	Change of velocity, speeding up or braking	$m/s / s = m/s^2$	Yes, 1 g down
Gyroscope	Angular velocity	Change of angle, spinning	rad/s (SI), often deg/s or RPM	Ignores gravity

Experiment: Accelerate with MX2125

MX2125 is a simple two-axis acceleration sensor (see Figure 8-1). It reports acceleration as a pulse length, making the interface and code simple.

The real, physical world is three dimensional. Objects can go up and down (y), left and right (x), and back and forth (z). A two-axis sensor only measures two of these axes.

The MX2125 only measures up to 3 g per axis. But some sensors can measure extreme acceleration. For example, the maximum measured acceleration of the ADXL377 (200 g), is much more than would kill any human. Thus, it's more than is experienced in a shuttle launch or high-g maneuvers in fighter jets. It could measure an object accelerating faster than a bullet fired from a pistol. When we made an early prototype for an Aalto-1 satellite sun sensor, even the satellite spec did not require acceleration this tough.

It's unlikely that you would need to measure such an extreme acceleration, and it would probably not be possible with a breadboard setup (because the acceleration needed to test would shake your project apart!). The cost is quite minimal, though. However, the wider the area of measured acceleration (from -250 g to +250 g), the less precise the device is.

Figure 8-1. *MX2125 sensor*

Decoding MX2125 Pulse Length

Usually, an accelerometer's conversion formulas are found on data sheets. In this case, it required some more searching.

Finding Data Sheets

Search for the data sheet. The obvious search query is the code name of the component and the word "datasheet." But you may also find the data sheet on the website where you bought the component from (it's usually on the product detail page for the part).

For example, when we searched for "MX2125 datasheet," we found the Parallax breakout board datasheet (*http://bit.ly/Prbb3c*), but not the Memsic datasheet for the MX2125 chip itself. The Parallax datasheet did not contain the required information.

However, the pulse length to g force conversion formula can be found on the Parallax Memsic 2125 Accelerometer Demo Kit (*http://bit.ly/PrblHL*) document. And fortunately, when we later checked the Parallax page, we noticed the data sheet for the actual chip there. We were also able to find the data sheet using a slightly different search term, "MXD2125," which is the part number of the chip on the board. The actual data sheet contained a lot of information not found in the other documents.

MX2125 works by heating a bubble of gas inside the device, and then measuring how the air bubble moves.

When power is on, the MX2125 reports the acceleration on each axis, 100 pulses a second. Consecutive HIGH and LOW signals form a 100 Hz square wave. The more acceleration there is, the more time the wave spends in the HIGH portion, and the less time in the LOW portion. You can read these pulses to determine acceleration.

One full wave contains one HIGH and one LOW. The time taken by one wave (HIGH+LOW) is called period (T). Let's call the time of HIGH part *tHIGH* (time of HIGH).

The *duty cycle* tells you how much of the wave is HIGH. The duty cycle is a percentage, for example 50% (0.50) or 80% (0.80).

```
dutyCycle = tHIGH / T
```

According to the data sheet and other documents, the period T is set to 10 ms by default:

```
dutyCycle = tHIGH / 10 ms
```

Here's the acceleration formula from the data sheet:

```
A = (tHIGH/T-0.50)/20%
```

Or, replacing `tHIGH/T` with `dutyCycle` and `20%` with `.2`:

```
A = (dutyCycle-0.50)/.2
```

Now it can be written as follows (because `x/.2` is `5*x`):

```
A = 5*(dutyCycle-0.50)
```

or:

```
A = 5*(tHIGH/T-0.50)
```

When there is no acceleration (0 g), the duty cycle is 50%:

```
0   = 5*(dutyCycle-0.50)
0/5 = dutyCycle-0.50
0   = dutyCycle-0.50
.50 = dutyCycle
```

At the time we originally checked, the Parallax and Memsic documentation conflicted on the multiplier: the Memsic documentation used 1/20% (5), and Parallax used 1/12.5% (8). In our experiments, we found 1/12.5% (8) to give proper readings with the breakout board from Parallax. And in fact, when we later examined the MXD2125 data sheet from Memsic that was hosted on Parallax's site, both agreed on 1/12.5%. This is why you need to be careful with data sheets you find online: always verify the values with experimentation. So we will use 8 as the multiplier:

```
A = 8*(tHIGH/T-0.50)
```

Because Arduino `pulseIn()` returns the pulse length in microseconds (1 µs = 0.001 ms = 1e-6 s), the formula could be modified to use microseconds:

```
A = 8 * ( tHIGH/(10*1000) - 0.5)
```

The unit of A is then g, which equals 9.81 m/s^2.

For example, if tHIGH is 5,000 µs (5 ms), the duty cycle is

```
dutyCycle = 5 ms / 10 ms = 0.50 = 50%
```

Which equals 0 g:

```
A = 8*(0.50-0.5) = 8*0 = 0        // g
```

To show another example, consider tHIGH of 6250 µs (6.25 ms)

```
A = 8*(6250/10000-0.5) = 8*(0.625-0.5) = 8*0.125 = 1        // g
```

Thus, a 6.25 ms pulse means 1 g acceleration.

Accelerometer Code and Connection for Arduino

Figure 8-2 shows the circuit diagram for Arduino. Wire it up as shown, and load the code from Example 8-1.

Figure 8-2. *MX2125 dual axis accelerometer circuit for Arduino*

Example 8-1. mx2125.ino

```
// mx2125.ino - measure acceleration on two axes using MX2125 and print to serial
// (c) BotBook.com - Karvinen, Karvinen, Valtokari

const int xPin = 8;
const int yPin = 9;


void setup() {
  Serial.begin(115200);
  pinMode(xPin, INPUT);
  pinMode(yPin, INPUT);
```

```
}

void loop() {
  int x = pulseIn(xPin, HIGH);  // ❶
  int y = pulseIn(yPin, HIGH);
  int x_mg = ((x / 10) - 500) * 8;       // ❷
  int y_mg = ((y / 10) - 500) * 8;
  Serial.print("Axels x: ");
  Serial.print(x_mg);
  Serial.print(" y: ");
  Serial.println(y_mg);
  delay(10);
}
```

❶ The length of the pulse tells you the acceleration. `pulseIn()` returns the pulse length in microseconds (μs). 1 μs = 0.001 ms = 0.000001 s = 1e-6 s.

❷ Convert output to *millig*, one-thousandth of the gravity constant *g*.

Accelerometer Code and Connection for Raspberry Pi

Figure 8-3 shows the wiring diagram for the Raspberry Pi. Hook everything up as shown, and then run the code shown in Example 8-2.

Example 8-2. mx2125.py

```
# mx2125.py - print acceleration axel values.
# (c) BotBook.com - Karvinen, Karvinen, Valtokari
import time
import botbook_gpio as gpio

xPin = 24
yPin = 23

def readAxel(pin):
        gpio.mode(pin, "in")
        gpio.interruptMode(pin, "both")
        return gpio.pulseInHigh(pin)     # ❶

def main():
        x_g = 0
        y_g = 0
        while True:
                x = readAxel(xPin) * 1000
                y = readAxel(yPin) * 1000
                if(x < 10):      # ❷
                        x_g = ((x / 10) - 0.5) * 8       # ❸
                if(y < 10):
                        y_g = ((y / 10) - 0.5) * 8
                print ("Axels x: %fg, y: %fg" % (x_g, y_g))      #
                time.sleep(0.5)
```

```
if __name__ == "__main__":
    main()
```

❶ Measure the length of pulse, the time when the pin is HIGH.

❷ Ignore readings that are wildly out of range.

❸ Calculate the acceleration along x axis in g. One g is the acceleration caused by the gravity of earth, 9.81 meters per second squared.

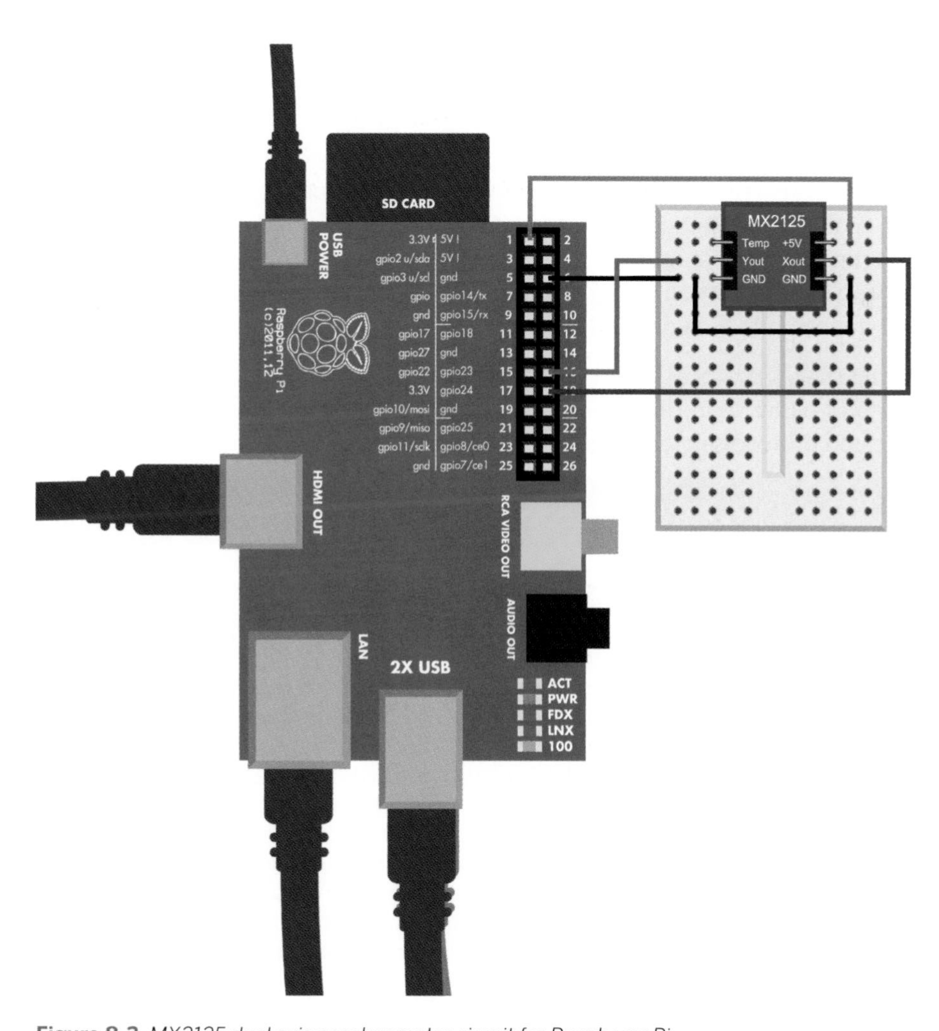

Figure 8-3. *MX2125 dual axis accelerometer circuit for Raspberry Pi*

Experiment: Accelerometer and Gyro Together

When an accelerometer is not moving, it detects gravity and can tell where down is. A gyroscope can tell the orientation reliably, even if you spin it around and around. A gyroscope ignores gravity, though.

Could we combine an accelerometer and gyroscope to get both benefits? Yes.

An IMU (inertial measurement unit) combines multiple sensors and (optionally) some logic to get more precise and reliable motion information. In this experiment, you'll work hands-on with the basic features of the MPU 6050.

In general, IMUs are more expensive and more precise than plain accelerometers and gyros. They also use more advanced protocols to communicate, such as I2C, instead of a simple pulse width signaling protocol.

The MPU 6050 (Figure 8-4) has an accelerometer, gyro, and microcontroller on the same chip. Even though space isn't a premium when you're in the breadboard prototyping stage, it's nice to know that all this functionality fits in a tiny surface-mounted component, just in case you ever run short of circuit real estate. For example, an early prototype of a sun sensor we designed barely fit into a box of about 10 × 10 × 10 cm. The final part had to fit into a very flat 5 mm × 5 mm area on the satellite surface.

The MPU 6050 uses the I2C protocol. Thanks to the *python-smbus* library, Raspberry Pi code is much simpler and easier than the equivalent Arduino code. In general, Raspberry Pi handles complicated protocols in less code than Arduino.

Industry Standard Protocols

Most devices use one of the industry standard protocols instead of inventing their own.

I2C is one of the easiest industry standard protocols. Because it's strictly defined and the protocol includes how data should be encoded and decoded, it's usually the easiest to use. You can find multiple examples of I2C in this chapter.

SPI is also a common industry standard protocol. Because SPI leaves a lot of choices to the implementer, it can be a daunting task to write an interface to a new SPI component. On the other hand, if there is a code example or a reference implementation, it's just a matter of copying and pasting, since someone else has done the hard work for you. You can find an example of using an SPI component without a reference implementation in *http://botbook.com/satellite*.

Serial is often found in disguise: serial over USB, serial over Bluetooth, serial over some jumper wires. The good old serial port you have used with Arduino Serial Monitor is surprisingly common. It solves only part of the problem: serial defines *how* to send characters over wire, but the implementer still has to decide how numbers are encoded and decoded. You can find an example of hacking serial over USB in our book *Make a Mind-Controlled Arduino Robot*.

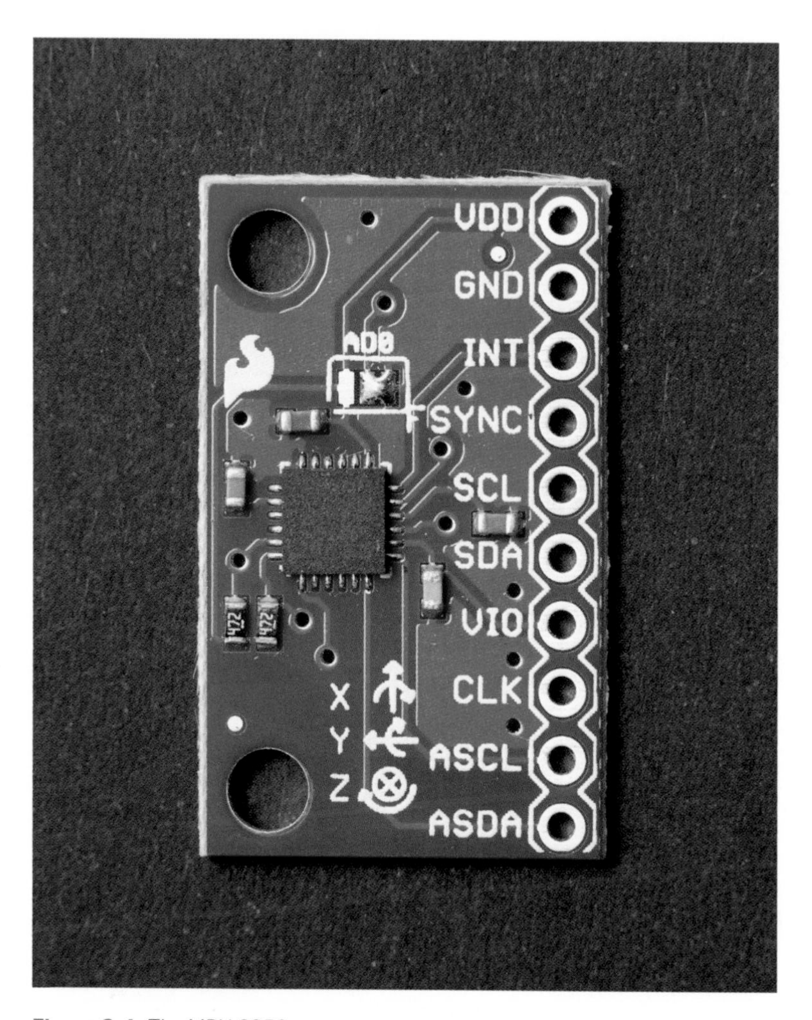

Figure 8-4. *The MPU 6050*

MPU 6050 Code and Connection for Arduino

Figure 8-5 shows the wiring diagram for Arduino. Hook everything up, and then run the code in Example 8-3.

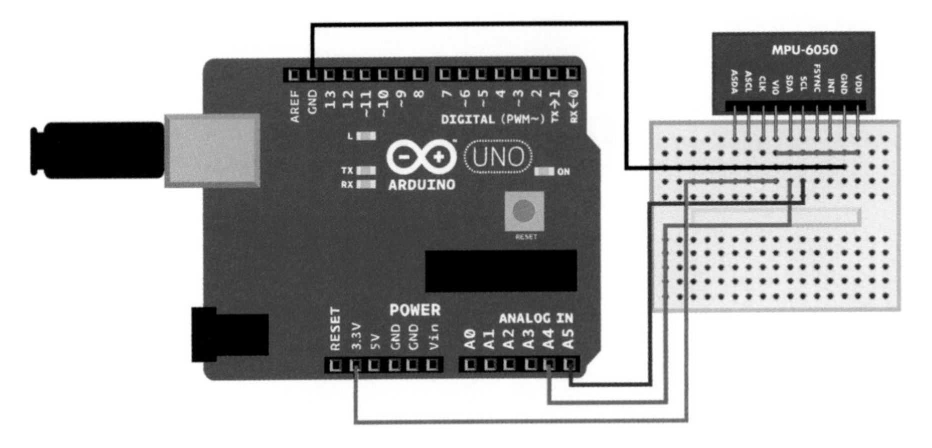

Figure 8-5. *MPU 6050 (accelerometer+gyro) circuit for Arduino*

Difficult code! The code for MPU 6050 contains more difficult programming concepts than most other code examples in this book. If you find endianness, bit shifting, and structs difficult, you can simply use the code and play with the values. You don't need to deeply understand the code to use it.

If you want to understand the code, see the explanations after the code, such as "Hexadecimal, Binary, and Other Numbering Systems" on page 219 and "Bitwise Operations" on page 221.

Example 8-3. mpu_6050.ino

```
// mpu_6050.ino - print acceleration (m/s**2) and angular velocity (gyro, deg/s)
// (c) BotBook.com - Karvinen, Karvinen, Valtokari

#include <Wire.h>       // ❶

const char i2c_address =  0x68;          // ❷

const unsigned char sleep_mgmt = 0x6B;  // ❸
const unsigned char accel_x_out = 0x3B;

struct data_pdu // ❹
{
  int16_t x_accel;      // ❺
  int16_t y_accel;
  int16_t z_accel;
  int16_t temperature;  // ❻
  int16_t x_gyro;       // ❼
  int16_t y_gyro;
  int16_t z_gyro;
};
```

```
void setup() {
  Serial.begin(115200);
  Wire.begin(); // ❽
  write_i2c(sleep_mgmt,0x00);   // ❾
}

int16_t swap_int16_t(int16_t value)     // ❿
{
  int16_t left = value << 8;    // ⓫
  int16_t right = value >> 8;   // ⓬
  right = right & 0xFF; // ⓭
  return left | right;  // ⓮
}

void loop() {
  data_pdu pdu; // ⓯
  read_i2c(accel_x_out, (uint8_t *)&pdu, sizeof(data_pdu));     // ⓰

  pdu.x_accel = swap_int16_t(pdu.x_accel);      // ⓱
  pdu.y_accel = swap_int16_t(pdu.y_accel);
  pdu.z_accel = swap_int16_t(pdu.z_accel);
  pdu.temperature = swap_int16_t(pdu.temperature);      // ⓲
  pdu.x_gyro = swap_int16_t(pdu.x_gyro);
  pdu.y_gyro = swap_int16_t(pdu.y_gyro);
  pdu.z_gyro = swap_int16_t(pdu.z_gyro);

  float acc_x = pdu.x_accel / 16384.0f; // ⓳
  float acc_y = pdu.y_accel / 16384.0f;
  float acc_z = pdu.z_accel / 16384.0f;
  Serial.print("Accelerometer: x,y,z (");
  Serial.print(acc_x,3); Serial.print("g, ");   // ⓴
  Serial.print(acc_y,3); Serial.print("g, ");
  Serial.print(acc_z,3); Serial.println("g)");

  int zero_point = -512 - (340 * 35);   // ㉑
  double temperature = (pdu.temperature - zero_point) / 340.0;  // ㉒
  Serial.print("Temperature (C): ");
  Serial.println(temperature,2);

  Serial.print("Gyro: x,y,z (");
  Serial.print(pdu.x_gyro / 131.0f); Serial.print(" deg/s, ");  // ㉓
  Serial.print(pdu.y_gyro / 131.0f); Serial.print(" deg/s, ");
  Serial.print(pdu.z_gyro / 131.0f); Serial.println(" deg/s)");
  delay(1000);
}

void read_i2c(unsigned char reg, uint8_t *buffer, int size)     // ㉔
{
  Wire.beginTransmission(i2c_address);  // ㉕
  Wire.write(reg);      // ㉖
  Wire.endTransmission(false);  // ㉗
  Wire.requestFrom(i2c_address,size,true);      // ㉘
```

```
  int i = 0;    // 29
  while(Wire.available() && i < size) { // 30
    buffer[i] = Wire.read();    // 31
    i++;
  }
  if(i != size) {       // 32
    Serial.println("Error reading from i2c");
  }

}

void write_i2c(unsigned char reg, const uint8_t data)   // 33
{
  Wire.beginTransmission(i2c_address); // 34
  Wire.write(reg);
  Wire.write(data);
  Wire.endTransmission(true);
}
```

1. *Wire.h* is the Arduino library for the I2C protocol. *Wire.h* comes with the Arduino IDE, so you can just include the library in your code. You don't need to separately install the library or copy any library files manually.
2. The I2C address of the MPU 6050 sensor. One three-wire bus can have many *slaves*. Each slave is recognized by its address. Typically, an I2C bus can have 128 (2^7) slaves. The I2C wires can be only a couple of meters long, so that also puts a practical limit to wire length. The number is represented in hexadecimal, see "Hexadecimal, Binary, and Other Numbering Systems" on page 219 for an explanation of this notation.
3. The registers for commands are from MPU 6050 documentation. If you need a complete list of commands, search the Web for "MPU 6050 data sheet" and "MPU 6050 register map." The numbers are in hex but could be expressed in decimal if you prefer.
4. The *struct* for decoding the answer from the sensor. A struct combines multiple values together. The C struct is only for data, and a struct can't contain any functions. This makes structs different from objects and classes you may be familiar with from other languages. `struct data_pdu` declares a new data type, which you'll later use for declaring variables of type `data_pdu`. This struct has variables that are exactly the same size as the data fields in the protocol used by the sensor. Later, you will read bytes from the sensor directly into the struct. Then you'll go through the variables embedded in the struct to get at the values. Yes, it's a neat trick!

❺ A variable for storing acceleration across the horizontal x-axis. The type `int16_t` is a specifically sized integer defined by avr-libc (the C library used by the Arduino compiler). It's a signed (negative or positive) 16-bit (two-byte) integer. Because the struct is used for decoding raw data from the sensor, the exact sized data types are required.

❻ The sensor also reports temperature. Even if you don't need it, you must have a variable for it so that the `data_pdu` struct ends up being the right size.

❼ Angular velocity around x-axis (roll), read by the gyroscope portion of the MPU 6050.

❽ Initialize I2C communication using the *Wire.h* library.

❾ Wake up the sensor by writing the command `0` to the sleep management register `0x6B`. The MPU 6050 starts out in sleep mode, so this is a required step.

❿ Swap the two bytes in parameter value. MPU 6050 is *big endian*. Arduino is *little endian* like most processors. The endianness must be converted between the platforms. See "Endianness—Typically on the Small Side" on page 224 for more details.

⓫ This new two-byte (16 bit) variable `left` ends up being the rightmost byte of the parameter `value`. After a one-byte (8 bit) left shift (<<), the leftmost byte of `value` is dropped, as it doesn't fit the two bytes of `left`. The right byte of `left` is filled with zeroes in the bit shift.

⓬ This new two-byte (16 bit) variable `right` is now the leftmost byte of `value`. The leftmost byte of `right` is zeroes.

⓭ Zero out the leftmost byte of `right`, just to make sure it's empty. See also "Bit Masking with Bitwise AND &" on page 223

⓮ Combine the left and right bytes. The variable `left` is actually two bytes (16 bits), with the rightmost byte (8 bits) full of zeroes. See also "Bitwise OR |" on page 224

⓯ Create a new variable `pdu` of type `data_pdu`. This is the struct type you created earlier.

⓰ Fill the `pdu` struct with data from the sensor. The first parameter is the register to read (`accel_x_out, 0x3B`). The second parameter is a reference to the `pdu` struct. It is passed as a reference, so that the function can modify the struct itself instead of returning a value. The last parameter is how many bytes to read. Conveniently, you can use the size of the struct to specify the number of bytes to read.

⓱ Convert the number from the sensor's big endian format to little endian used in Arduino.

⓲ You can refer to the variables in the struct with `structname.var`, such as `pdu.temperature`.

⓳ The raw acceleration value is converted to the real-life unit g. The standard gravity g is `9.81 m/s`^{2}. The conversion factor is from the data sheet. To get a floating point (decimal) result, the divider must be floating point.

(20) Print the acceleration to the serial monitor. The unit is g, the acceleration from gravity. 1 g = 9.81 m/s^2.

(21) Calculate the zero point for converting raw measurement to temperature in Celsius, using information from the data sheet. The temperature is from -40 C to +85 C. A raw value of -512 indicates 35 C. From that point, every 1 C change is represented by a raw value change of 340. Thus, to find the 0 C point raw value, you take -512 and subtract the product 340 * 35 (deducting 35 C worth of raw values at 340 per C). The calculation is `-512 - (340 * 35)`, which is -12412. But instead of writing the calculated value -12412, you should write the calculation from the datasheet so that the code is more clear.

(22) Convert raw measurement to temperature in Celsius, using the formula from the data sheet.

(23) Print angular velocity as measured by gyroscope. The raw to degree/s conversion factor 1/131.0 is from the data sheet. To get a floating point (decimal) result, the divider must be floating point, too.

(24) A function to read `size` number of bytes from register `point`. The result is written over the struct, which is represented by the `*buffer` pointer. Because of the pointer, the actual value of the struct is modified, instead of returning a value.

(25) Send the I2C command to the device (the MPU 6050 sensor in the address 0x69).

(26) Specify the register address to read. In this program, `read_i2c()` is only used to read from `accel_x_out` (0x3B).

(27) Keep the connection open, so that you can read data on the next lines.

(28) Request `size` bytes of data from the sensor. Earlier, you stated you want to start from register `accel_x_out` (0x3B). The `true` parameter (the third argument, which is named `stop` in the Arduino documentation) means that the connection is closed after the read ends, which releases the I2C bus for future use.

(29) Declare a new variable for the upcoming `while` loop. The loop variable `i` holds the count of how many bytes have been read. This count `i` equals the number of iterations the loop has run.

(30) Enter the loop only if there are bytes available for reading and you have not yet read all the bytes requested. In the way `read_i2c()` is called in this program, the variable `size` will always be the length of `data_pdu` struct.

(31) Read a byte (8 bits) and store it into the buffer. Considering how `read_i2c()` is called in this program, we can walk through the first iterations. The pointer `*buffer` points to the first byte of `pdu`, which is of struct type `data_pdu`. On the first iteration, `i` is 0, so `buffer[i]` points to the first byte of `pdu`. Because `pdu` was passed to the function with a pointer, the contents of the actual `pdu` (the variable in the main program) are overwritten. No return is needed, so the type of `read_i2c()` is void. On the second iteration, `buffer[1]` points to the second byte of `pdu`. This continues for the whole buffer (`pdu`). When `i == size`, the `while` loop is not re-entered, and execution continues with the code that comes after the `while` loop.

(32) If not enough bytes were available, the loop variable `i` is less than `size`. As `i` was declared outside the loop, it is available to you after the loop.

(33) Write one byte `data` to register `reg` on the sensor.

(34) The address of the sensor comes from the global variable `i2c_address`.

MPU 6050 Code and Connection for Raspberry Pi

Figure 8-6 shows the wiring diagram for Raspberry Pi. Hook it up as shown, and then run the code from Example 8-4.

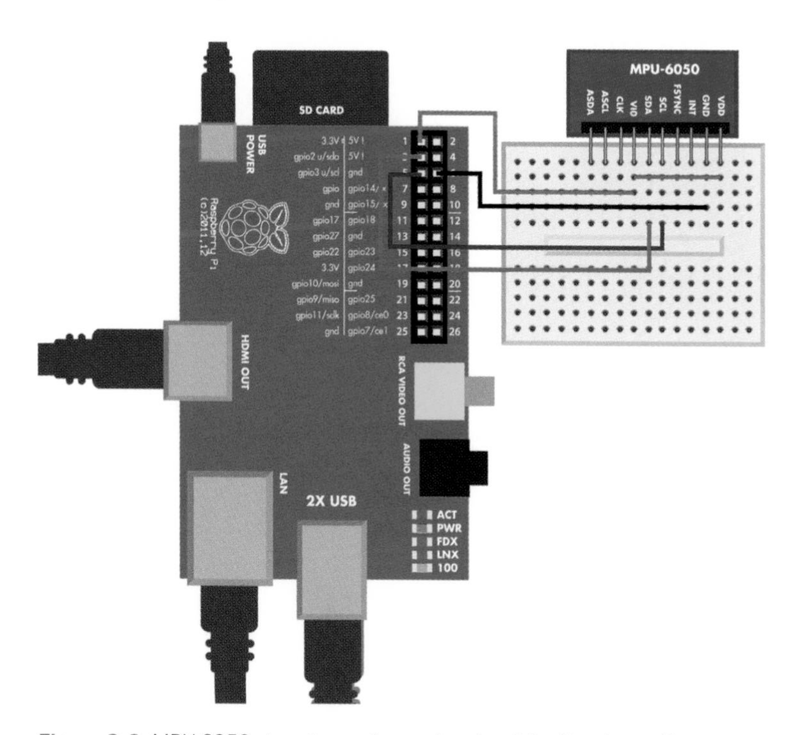

Figure 8-6. *MPU 6050 six-axis accelerometer circuit for Raspberry Pi*

Example 8-4. mpu_6050.py

```
# mpu_6050.py - print acceleration (m/s**2) and angular velocity (gyro, deg/s)
# (c) BotBook.com - Karvinen, Karvinen, Valtokari
import time
import smbus # sudo apt-get -y install python-smbus  # ❶
import struct

i2c_address =  0x68      # ❷
sleep_mgmt = 0x6B        # ❸
accel_x_out = 0x3B       # ❹

bus = None       # ❺
acc_x = 0
acc_y = 0
acc_z = 0
temp = 0
gyro_x = 0
gyro_y = 0
gyro_z = 0


def initmpu():
  global bus    # ❻
  bus = smbus.SMBus(1)  # ❼
  bus.write_byte_data(i2c_address, sleep_mgmt, 0x00)    # ❽

def get_data():
  global acc_x,acc_y,acc_z,temp,gyro_x,gyro_y,gyro_z
  bus.write_byte(i2c_address, accel_x_out)      # ❾
  rawData = ""
  for i in range(14):   # ❿
    rawData += chr(bus.read_byte_data(i2c_address,accel_x_out+i))       # ⓫
  data = struct.unpack('>hhhhhhh', rawData)     # ⓬

  acc_x = data[0] / 16384.0      # ⓭
  acc_y = data[1] / 16384.0
  acc_z = data[2] / 16384.0
  zero_point = -512 - (340 * 35)        # ⓮
  temp = (data[3] - zero_point) / 340.0 # ⓯

  gyro_x = data[4] / 131.0       # ⓰
  gyro_y = data[5] / 131.0
  gyro_z = data[6] / 131.0


def main():
  initmpu()
  while True:   # ⓱
    get_data()  # ⓲
    print("DATA:")
    print("Acc (%.3f,%.3f,%.3f) g, " % (acc_x, acc_y, acc_z))   # ⓳
    print("temp %.1f C, " % temp)
```

```
        print("gyro (%.3f,%.3f,%.3f) deg/s" % (gyro_x, gyro_y, gyro_z))
        time.sleep(0.5) # s # (20)

if __name__ == "__main__":
    main()
```

(1) SMBus implements a subset of the I2C industry standard protocol. The SMBus library makes the Raspberry Pi program much shorter than the Arduino equivalent. The `python-smbus` package must be installed on Raspberry Pi (see "SMBus and I2C Without Root" on page 218 for instructions).

(2) The I2C address of MPU 6050 sensor, found on the data sheet. The number is represented in hexadecimal (see "Hexadecimal, Binary, and Other Numbering Systems" on page 219).

(3) The register address for commands. You can find the register map by searching the Web for "MPU 6050 register map".

(4) The X acceleration register address is the starting address for the values you're interested in: acceleration, temperature, and angular velocity.

(5) Make `bus` a global variable visible to all functions.

(6) To modify the value of a global variable in a function, you must indicate that it's global in the beginning of the function.

(7) Initialize the SMBus (I2C). Store the new object of the SMBus class into the global `bus` variable.

(8) MPU 6050 starts in sleep mode. Wake it up before doing anything with it. The commands to the sensor are given over I2C (SMBus) with the device address, the register, and the value to write to the register.

(9) Request data, starting from the X acceleration address.

(10) Repeat 14 times, with values of `i`, from 0 to 13.

(11) Read the current byte, convert it to ASCII, and add it to `rawData` string.

(12) Convert the `rawData` string to a Python tuple. The format string characters indicate little endian `<`, short signed 2 byte (16 bit) integer `h`.

(13) Convert raw acceleration to real-life g units. The standard gravity g is equal to 9.81 m/s^2.

(14) Calculate the temperature zero point. The Pi does it very quickly, and writing the whole formula here makes code easier to read and typos less likely.

(15) Convert raw temperature to Celsius. To get a floating point result, the divider must also be floating point. Conversion formulas are from the data sheet (Google "MPU 6050 data sheet").

(16) Convert raw angular velocity to real-life units (degrees per second).

(17) The program runs until you press Control-C.

(18) `get_data()` updates global variables, so it doesn't need to return values from the function.

(19) Print acceleration using the format string. The replacement `%.3f` indicates a floating point value with three decimal places.

(20) To let the user read the printed values and avoid taking 100% of CPU time, we add a small delay here.

Find out what walking, running, or skipping does to readings. How about twitching or squirming?

SMBus and I2C Without Root

The Raspberry Pi code uses the Python smbus library for I2C. Luckily, installing software in Linux is a breeze. You can install any software to Raspbian just like you would install it in Debian, Ubuntu, or Mint. Double-click the LXTerminal icon on the left side of your Raspbian desktop. Then:

```
$ sudo apt-get update
$ sudo apt-get install python-smbus
```

To enable I2C support, you'll need to enable the i2c modules. First, make sure they are not disabled. Edit the */etc/modprobe.d/raspi-blacklist.conf* with the command `sudoedit /etc/modprobe.d/raspi-blacklist.conf` and *delete* this line:

```
blacklist i2c-bcm2708
```

Save the file: press Control-X, type y, and then press Enter or Return.

Next, edit the */etc/modules* with the command `sudoedit /etc/modules` and add these two lines:

```
i2c-bcm2708
i2c-dev
```

Save the file: press Control-X, type y, and then press Enter or Return.

To use I2C without needing to be root, create the udev rule file *99-i2c.rules* (shown in Example 8-5) and put it in place. (To avoid typing and inevitable typos, you can download a copy of *99-i2c.rules* file from *http://botbook.com*.)

```
$ sudo cp 99-i2c.rules /etc/udev/rules.d/99-i2c.rules
```

Example 8-5. 99-i2c.rules

```
# /etc/udev/rules.d/99-i2c.rules - I2C without root on Raspberry Pi
# Copyright 2013 http://BotBook.com

SUBSYSTEM=="i2c-dev", MODE="0666"
```

Reboot your Raspberry Pi, open LXTerminal, and confirm that you can see the I2C devices and that the ownership is correct:

```
$ ls -l /dev/i2c*
```

The listing should show two files, and they should list permissions of `crw-rw-rwT`. If not, go over the preceding steps again.

Hexadecimal, Binary, and Other Numbering Systems

The same number can be represented multiple ways. For example, the decimal number 65 is 0x41 in hexadecimal and 0b1000001 in binary. You are probably most familiar with the normal decimal system, where 5+5 is 10.

The different representations are marked with a prefix before the number. Normal decimal numbers don't have a prefix. Hexadecimal numbers start with `0x`, binary numbers start with `0b`, and octal numbers start with `0`.

The different numbering systems are compared in Table 8-2.

Table 8-2. Number representations

Prefix	Representation system	Base	Use	Example	Calculation
	Decimal	10	The normal system	10	$0*10^0 + 1*10^1$
0x	Hexadecimal	16	C and C++ code, datasheets	0xA	$10*16^0$
0b	Binary	2	Low-level protocols, bit-banging	0b1010	$0*2^0 + 1*2^1 + 0*2^2 + 1*2^3$
0	Octal	8	Chmod permissions in Linux	012	$2*8^0 + 1*8^1$

Consider the number 42. It is the exact same number in any representation system, so

```
42 = 0x2a = 0b101010 = 052
```

Only the base number changes. With the familiar, normal decimal system, the base is 10. Starting from the right, you first count ones and then tens.

```
2*1 + 4*10 = 42
```

If there is a big number, such as 1917, the ten is obvious. After ones, it's tens (10), hundreds (10*10) and thousands (10*10*10). You can easily write these numbers as powers:

$$10*10 = 10^2$$
$$10*10*10 = 10^3$$

What about ones? Any number to zeroth power is 1, so it's:

$$10^0 = 1$$

Thus, the number 42 becomes:

$$42 = 4*10^1 + 2*10^0$$

The hex representation of 42 is 0x2A. In hex, numbers bigger than 9 are shown with letters: A=10, B=11 … F=15. Starting from the right, note that A is 10 and count:

```
0x2A = 10*1 + 2*16 = 10 + 32 = 42
```

To play with powers, this can be written

```
10*16^0 + 2*16^1 = 0x2A
```

Try some other numbers. To check your calculations, use the Python console (see "The Python Console" on page 221 or Table 8-3). Can you apply your skills to convert to binary numbers, too?

You can practice working with numbers in the Python console. The number representation (1 == 0x1 == 0b1) is the same in Python, C, and C++. You can run any Python commands in the console:

```
>>> print("Botbook.com")
Botbook.com
>>> 2+2
4
```

Any numbers you enter are converted to decimal system for display:

```
>>> 0x42
66
>>> 66
66
>>> 0b1000010
66
```

There are functions to convert any number to binary, hexadecimal, octal, and ASCII characters:

```
>>> bin(3)
'0b11'
>>> hex(10)
'0xa'
>>> oct(8)
'010'
>>> chr(0x42)
'B'
```

Table 8-3. Example numbers in different representations

Decimal	Hex	Binary	Octal	ASCII
0	0x0	0b0	0	*\0* NUL, terminates string
1	0x1	0b1	01	
2	0x2	0b10	02	
3	0x3	0b11	03	
4	0x4	0b100	04	EOT, end of text, CTRL-D in Linux
5	0x5	0b101	05	
6	0x6	0b110	06	
7	0x7	0b111	07	

Decimal	Hex	Binary	Octal	ASCII
8	0x8	0b1000	010	
9	0x9	0b1001	011	
10	0xA	0b1010	012	*\n* newline
11	0xB	0b1011	013	
16	0x10	0b10000	020	
17	0x11	0b10001	021	
32	0x20	0b100000	040	' ' space, the first printable character
48	0x30	0b110000	060	*0* ASCII zero is not number zero
65	0x41	0b1000001	0101	*A*
97	0x61	0b1100001	0141	*a*
126	0x7e	0b1111110	0176	~ tilde is the last printable ASCII character

Type the command `man ascii` at the terminal to show a manual page showing an ASCII chart in Linux or OS X. For example, you can see that ASCII character A is decimal number 65, hexadecimal 0x41, and octal number 0101. Manual pages are shown in the `less` utility, so pressing space moves forward a page, pressing b moves back, and pressing q quits.

The Python Console

The Python console is started by typing the *python* command at the terminal. In Linux, the terminal (bash, shell) is found in the bottom left Main menu: Accessories→Terminal, or searching Dash for Terminal. In Macintosh, terminal is /Applications/Utilities/Terminal. In Windows, you can start terminal (Applications→Accessories→Command Prompt) or open IDLE from the Start menu (if you've installed Python).

The Python prompt >>> lets you know that your commands will be interpreted as Python code. To finish your Python session, type the `exit()` Python command.

If you want a really neat Python console with tab guessing and interactive help, try *ipython*.

Bitwise Operations

Some sensors send data as a bunch of bits that form a byte, such as *01010000*. To work with data in a binary representation, you must perform *bitwise operations* on them, in which you manipulate a bunch of bits at the same time.

A bit is 1 or 0. One bit can represent a truth value, 0-false or 1-true.

A byte is eight bits, such as 0b1010100. A byte can represent one ASCII character, such as T, 3, or x. It can also represent a number, such as 256 or -127.

Bitwise operations are very low-level functions (you're pretty much working with computer memory in its native format), and it can be hard to read code that uses it, so you should resort to it only when needed. On the other hand, we sometimes meet sensors where binary arithmetic is required just to get the part working. The MPU 6050 integrated motion unit is one such sensor.

The most common binary arithmetic operations are *bit shifting* and binary *Boolean operations*.

To get some confidence with bit operations, practice with them in the Python console ("The Python Console" on page 221). Bit operations work the same way in C and C++ (Arduino), but these languages don't have interactive consoles, so it's easier to play around in Python.

You might already know normal Boolean algebra. "AND" means that both conditions must be true for the whole expression to be considered true. "OR" means either condition can be true for the whole expression to be true.

```
>>> if (True): print "Yes"
...
Yes
>>> if (False): print "Yes"
...     # nothing printed
>>> if (True and True): print "Yes"
...
Yes
```

In Python, True and False must have a capital first letter.

The truth values are there even without "if" construct:

```
>>> True
True
>>> False
False
>>> True and True
True
```

In most languages, 1 is True and 0 is False:

```
>>> 1 and 1
1
>>> 1 and 0
0
```

With bitwise operations, you can do the same thing to every one and zero in a byte.

You can't use the English words "and" or "or" for bitwise operations in Python. Use the characters & (bitwise AND) and | (bitwise OR) instead.

```
>>> 0b0101 & 0b0110
4
```

Maybe the answer is easier to read in binary:

```
>>> bin(0b101 & 0b110)
'0b100'
```

The bitwise AND (&) simply takes each bit and applies the Boolean AND operation to each pair:

```
    0b0101
    0b0110
==========
&   0b0100
```

Starting from the left, 0-false and 0-false is 0-false. 1-true and 1-true is 1-true. 0-false and 1-true is 0-false. 1-true and 0-false is 0-false.

Bit Masking with Bitwise AND &

Bit masking allows you to get just the bits you want. For example, you could get the four leftmost digits of 0b 1010 1010 with the following operation:

```
>>> bin(0b10101010 & 0b11110000)
'0b10100000'
```

Here's how it works:

```
0b 1010 1010    # a number
0b 1111 0000    # the mask
==========================
0b 1010 0000    # & (bitwise AND)
```

AND is true only when both inputs are true (Table 8-4).

Table 8-4. AND truth table

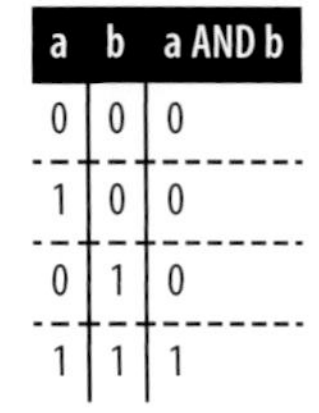

a	b	a AND b
0	0	0
1	0	0
0	1	0
1	1	1

Bitwise OR |

If you build the left part and the right part of a byte separately, how do you put them back together? In Python, you can use an OR (|) operation:

```
>>> left=0b10100000
>>> right=0b1111
>>> bin(left|right)
'0b10101111'
```

Looking at the bits in detail:

```
0b 1010 0000    # left, must have zeroes on the right
0b      1111    # right, zeroes on the left don't matter
======================
0b 1010 1111    # bitwise OR |
```

Bitwise OR "|" is not the same as plus "+". For example, consider 0b1 + 0b1.

Bit Shifting <<

Bit shifting moves bits to the right or left. You can try it in Python:

```
>>> bin(0b0011<<1)
'0b110'
>>> bin(0b0011<<2)
'0b1100'
```

Moving bits over the right edge drops the extra bits:

```
>>> bin(0b0011>>2)
'0b0'
```

Endianness—Typically on the Small Side

Most hardware is *little endian*, just like the numbers you're used to working with every day. Normal numbers are little endian: in the number 1991, thousands come first, then hundreds, tens, and finally ones. The most significant number comes first.

Most computers are little endian too, so that the most significant *byte* comes first. Your workstation and laptop likely run on x86, amd64, or x86_64 architechture, so they are little endian. Arduino is based on the Atmel AVR, and it's little endian. Both Raspberry Pi and Android Linux (the leading smartphone platform) run on ARM in little endian mode.

The number 135 can be stored little endian (typical) or big endian:

```
0b 1000 0111    # little endian, typical, Arduino, Pi, workstation
0b 0111 1000    # big endian, MPU 6050
```

Some devices have atypical endianness. The MPU 6050 sensor is big endian, meaning that the most significant bytes come *last*. When communicating between Arduino and the MPU 6050, the endianness must be swapped between little endian and big endian.

The endianness only matters when doing low-level operations like working with bits and bytes. In high-level programming, you can just assign a value to a floating point variable and get a floating point back. In higher level programming, the environment handles endianness and other details for you.

Experiment: Hacking Wii Nunchuk (with I2C)

Would you like to have an accelerometer, a joystick, and a button—all in a cheap package? Look no further, because the Nunchuk controller for the Wii gaming console is all that. If you want to go even cheaper, there are cheap compatible copies available, too.

The Wii Nunchuk also teaches an important hacking lesson: engineers are human. And just like the rest of us humans, engineers want to work with tried and true protocols, where libraries and tools are available. The Wii has its own proprietary connector, and originally, very little documentation. But under the surface, it's the standard I2C protocol. As you saw earlier ("Industry Standard Protocols" on page 208), I2C is our favorite industry standard short-range communication protocol.

Nintendo produces the Wii Nunchuk in large batches, which keeps quality up and prices down. The wide availability of Nunchuk also means there is a lot of example code and documentation available. You don't even need to cut the cord, as there is even a WiiChuck adapter you can push into Nunchuk's connector to connect it to Arduino or a breadboard (Figure 8-7).

Figure 8-7. *Nunchuk connected to Arduino with WiiChuck adapter*

The Wii Nunchuk can use the SMBus protocol to communicate. It is a simplified protocol, by virtue of it being a subset of the industry standard I2C protocol, which makes it even more clear how the communication should happen.

Nunchuk Code and Connection for Arduino

Figure 8-8 shows the connection diagram for Arduino. Wire things up as shown, and run the sketch shown in Example 8-6.

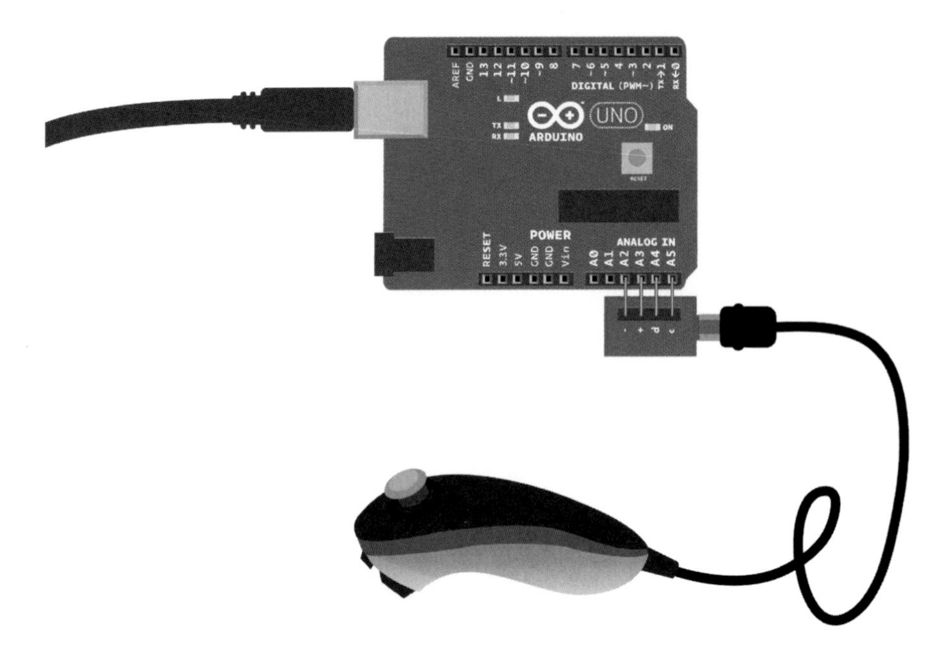

Figure 8-8. *WiiChuck circuit for Arduino*

Example 8-6. wiichuck_adapter.ino

```
// wiichuck_adapter.ino - print joystick, accelerometer and button data to serial
// (c) BotBook.com - Karvinen, Karvinen, Valtokari


#include <Wire.h>

const char i2c_address = 0x52;

unsigned long lastGet=0; // ms
int jx = 0, jy = 0, accX = 0, accY = 0, accZ = 0, buttonZ = 0, buttonC = 0;     // ❶

void setup() {
  Serial.begin(115200);
  Wire.begin();
  pinMode(A2, OUTPUT);
```

```
  pinMode(A3, OUTPUT);
  digitalWrite(A2, LOW);        // ❷
  digitalWrite(A3, HIGH);       // ❸
  delay(100);
  initNunchuck();       // ❹

}

void loop() {
  if(millis() - lastGet > 100) {        // ❺
    get_data(); // ❻
    lastGet = millis(); // ❼
  }
  Serial.print("Button Z: ");
  Serial.print(buttonZ);        // ❽
  Serial.print(" Button C: ");
  Serial.print(buttonC);
  Serial.print(" Joystick: (x,y) (");
  Serial.print(jx);       // ❾
  Serial.print(",");
  Serial.print(jy);
  Serial.print(") Acceleration (x,y,z) (");
  Serial.print(accX);   // ❿
  Serial.print(",");
  Serial.print(accY);
  Serial.print(",");
  Serial.print(accZ);
  Serial.println(")");

  delay(10); // ms
}

void get_data() {
  int buffer[6]; // ⓫
  Wire.requestFrom(i2c_address, 6);     // ⓬
  int i = 0;     // ⓭
  while(Wire.available()) {     // ⓮
    buffer[i] = Wire.read();    // ⓯
    buffer[i] ^= 0x17;  // ⓰
    buffer[i] += 0x17;  // ⓱
    i++;
  }
  if(i != 6) {  // ⓲
    Serial.println("Error reading from i2c");
  }
  write_i2c_zero();     // ⓳

  buttonZ = buffer[5] & 0x01;   // ⓴
  buttonC = (buffer[5] >> 1) & 0x01;    // ㉑
  jx = buffer[0];       // ㉒
  jy = buffer[1];
  accX = buffer[2];
  accY = buffer[3];
  accZ = buffer[4];
```

```
}

void write_i2c_zero() {
  Wire.beginTransmission(i2c_address);
  Wire.write(0x00);
  Wire.endTransmission();
}

void initNunchuck()
{
  Wire.beginTransmission(i2c_address);
  Wire.write(0x40);
  Wire.write(0x00);
  Wire.endTransmission();
}
```

❶ Declare global variables. Because Arduino functions can't easily return multiple values, data is passed from function to function using global variables.

❷ Use analog pin A2 as ground (GND, 0 V, LOW). This way, WiiChuck can be pushed into the Arduino pin header without needing any breadboard or jumper wires.

❸ Use A3 as +5 V power (HIGH, positive, VCC).

❹ Initialize the Wii Nunchuk by calling `initNunchuck()`, which sends 0x40 and 0x00 over I2C.

❺ Get data every 100 ms. This is a common program pattern to perform something every *x* milliseconds. The `millis()` function returns the Arduino's uptime (how long it's been turned on) in milliseconds. The `lastGet` variable holds the time since the last time `get_data()` was called, in milliseconds of uptime. (In the first iteration, `lastGet` is 0). The difference between `millis()` and `lastGet` is the time since `get_data()` was last called. If more time than 100 ms has passed, Arduino executes the block below.

❻ As `get_data()` updates globals, it doesn't need to return a value.

❼ Update the time of the last call to `get_data()`.

❽ Button states are 0-down, 1-up.

❾ Both joystick axes (`jx`, `jy`) are raw values from 30 to 220.

❿ All accelerometer axes (`accX`, `accY`, `accZ`) are raw values from 80 to 190.

⓫ Declare a new array of six integers.

⓬ Request six bytes of data from the Nunchuk.

⓭ The loop variable `i` will contain the number of bytes processed.

⓮ Enter the loop only if there are bytes to read.

⓯ Read one byte and store it into the current cell in `buffer[]`. On the first iteration, this is `buffer[0]`. On the last iteration, this is `buffer[5]`.

⓰ Perform an XOR on the current value with 0x17, and replace the current value with the result (this is called an *in place* operation). Exclusive or (XOR) is a Boolean operation similar to OR, but it's true only when either a or b is true, but not when both are true.

⓱ Add 0x17 to the current value.

⓲ If the number of bytes read is not six, print a warning.

⓳ Ask for another reading by sending 0x00 over I2C.

⓴ Get the last bit of the last byte. The last byte is `buffer[5]`. The last bit is extracted with bit masking; the code performs a bitwise AND on byte against 0b 0000 0000 0000 0001. See "Bit Masking with Bitwise AND &" on page 223 and Table 8-5. Button state is 0-down, 1-up.

㉑ Get the second-to-last bit of the last byte. The second-to-last bit is moved last with bit shifting `>> 1`. Then the last bit is extracted like in the previous line. See "Bit Shifting <<" on page 224.

㉒ The `jx` joystick x axis is just a full byte. The rest of the joystick and accelerometer axes are read in similar fashion.

Table 8-5. Nunchuk 6-byte data block

Byte	Use
0	Joystick X
1	Joystick Y
2	Accelerometer X
3	Accelerometer Y
4	Accelerometer Z
5	ZCxxyyzz: Buttons "Z" and "C", accelerometer precision

Nunchuk Code and Connection for Raspberry Pi

Figure 8-9 shows the connection diagram for Raspberry Pi. Hook it up, and run the program shown in Example 8-7.

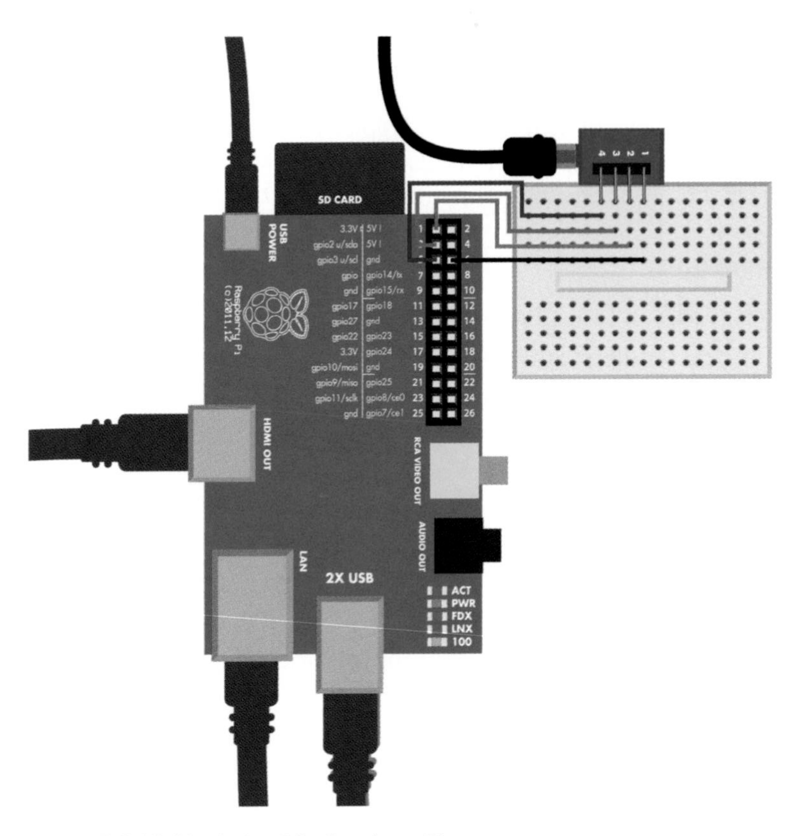

Figure 8-9. *WiiChuck circuit for Raspberry Pi*

Example 8-7. wiichuck_adapter.py

```
# wiichuck_adapter.py - print Wii Nunchuck acceleration and joystick
# (c) BotBook.com - Karvinen, Karvinen, Valtokari

import time
import smbus # sudo apt-get -y install python-smbus     # ❶

bus = None
address = 0x52     # ❷

z = 0     # ❸
c = 0
joystick_x = 0
joystick_y = 0
ax_x = 0
ax_y = 0
ax_z = 0

def initNunchuck():
    global bus
    bus = smbus.SMBus(1) # ❹
```

```
    bus.write_byte_data(address, 0x40, 0x00)    # ❺

def send_request():
    bus.write_byte(address, 0x00)    # ❻

def get_data():
    global bus, z, c, joystick_x, joystick_y, ax_x, ax_y, ax_z
    data = [0]*6
    for i in range(len(data)): # ❼
        data[i] = bus.read_byte(address)
        data[i] ^= 0x17
        data[i] += 0x17


    z = data[5] & 0x01    # ❽
    c = (data[5] >> 1) & 0x01    # ❾

    joystick_x = data[0]
    joystick_y = data[1]
    ax_x = data[2]
    ax_y = data[3]
    ax_z = data[4]
    send_request()

def main():
    initNunchuck()
    while True:
        get_data()
        print("Button Z: %d Button C: %d joy (x,y) (%d,%d) \
                acceleration (x,y,z) (%d,%d,%d)" \
                % (z,c,joystick_x,joystick_y,ax_x, ax_y, ax_z))
        time.sleep(0.1)


if __name__ == "__main__":
    main()
```

❶ The python-smbus library must be installed on Raspberry Pi (see "SMBus and I2C Without Root" on page 218).

❷ The address of your Wii Nunchuk. The value is in hex. See "Hexadecimal, Binary, and Other Numbering Systems" on page 219 for more information on hex values.

❸ Global variable for one of the buttons.

❹ Create a new object of class `SMBus`, and store it to the new variable `bus`. The constructor `SMBus()` takes one parameter, the device number. Number 1 means the file `/dev/i2c-1`. This device number is common with modern Raspberry Pi boards. If you have an old revision 1 Raspberry Pi board, you might need to use number 0 for `/dev/i2c-0` instead.

❺ The Nunchuk must be initialized with I2C commands. `bus.write_byte_data(addr=0x52, cmd=0x40, val=0x00)` sends the Nunchuk the command 0x40 with value 0x00.

❻ You can ask the Nunchuk for the next set of values with the null character 0x00, which is equivalent to *\0* or just 0.

❼ Read six bytes of data. This contains the data block described in Table 8-5. The rest of the function will decode the data.

❽ Get "Z" button status: 1 for down, 0 for up. The bit mask is just one bit, `0b1`. When used with bitwise AND (`&`), it is the rightmost bit, and everything left of it is considered zero. So bitwise AND with `0b1` simply returns the rightmost bit. See also "Bit Masking with Bitwise AND &" on page 223.

❾ Get the "C" button status, 1-down or 0-up. To get the second-to-last bit, move the second-to-last bit to the last place (`x >> 1`) and use a bitmask to get just the last bit.

Test Project: Robot Hand Controlled by Wii Nunchuk

Control a robot hand with the Nunchuk. As you can already read acceleration and joystick position, you can simply turn servos according to these numbers. Add mechanics, and you've got a Nunchuk-controlled robot hand.

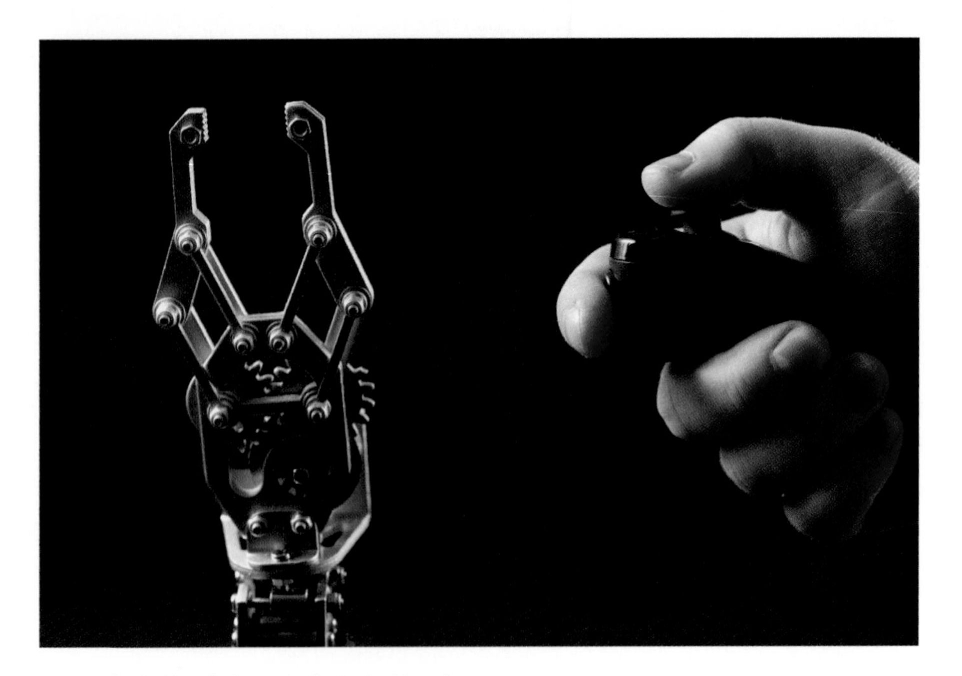

Figure 8-10. *Nunchuk-controlled robot hand*

What You'll Learn

In the *Robot Hand* project, you'll learn how to:

- Use the accelerometer and a mechanical joystick with outputs.
- Combine servos for complex movement.

You'll also refresh your skills on servo control and filtering noise with running averages (see "Servo Motors" on page 115 and "Moving Average" on page 187).

Start with just the servos (Figure 8-11). Once you get some movement, you can continue with hand mechanics.

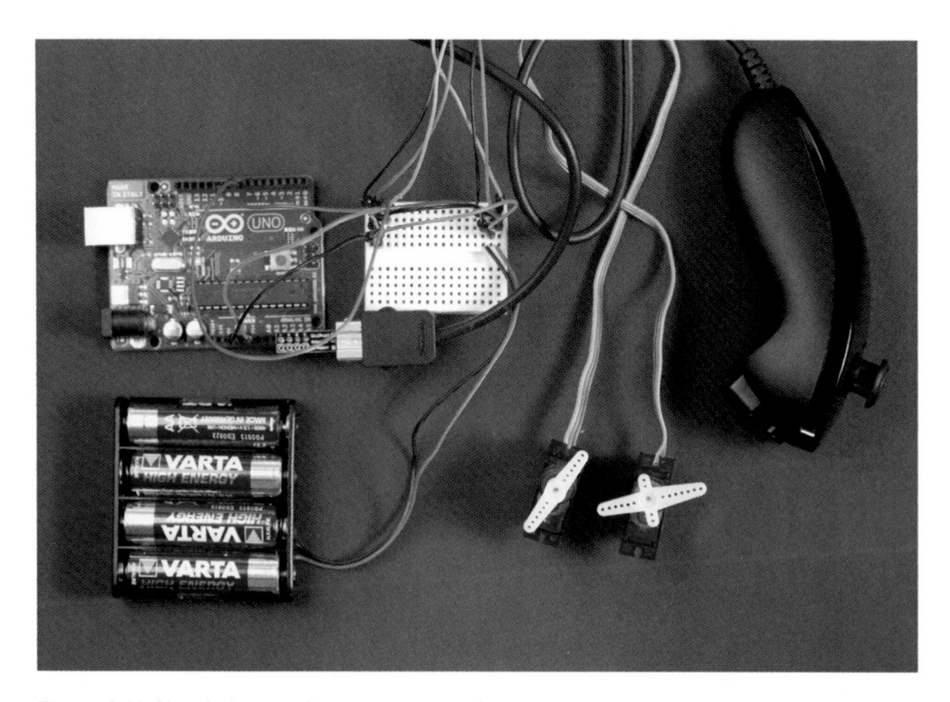

Figure 8-11. *Nunchuk controls two servos with Arduino*

Figure 8-12 shows the wiring diagram. Hook everything up as shown, and then run the code shown in Example 8-8.

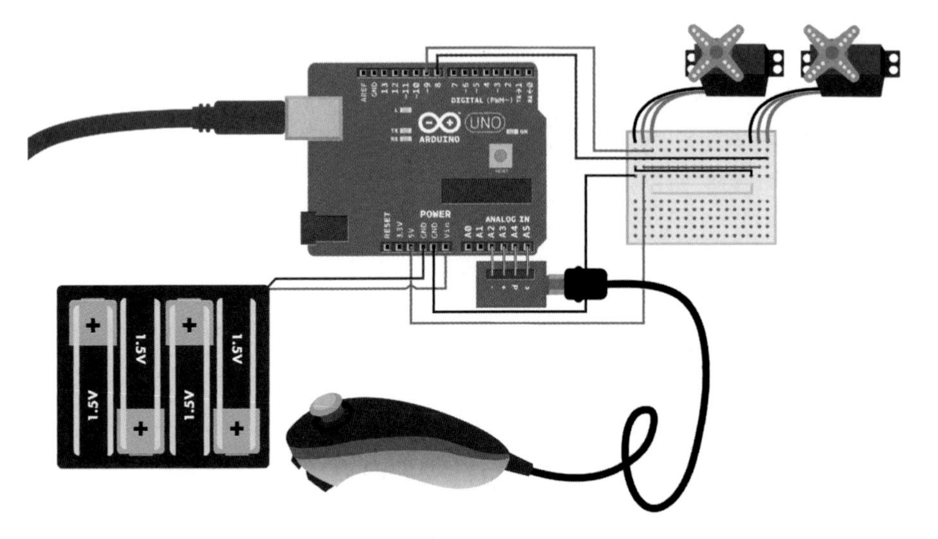

Figure 8-12. *Claw circuit for Arduino*

For more reading on Wii Nunchuk with Arduino, see "Nunchuk Code and Connection for Arduino" on page 226.

Example 8-8. *wiichuck_adapter_claw.ino*

```
// wiichuck_adapter_claw.ino - control robot hand with Nunchuck
// (c) BotBook.com - Karvinen, Karvinen, Valtokari

#include <Wire.h>

const int clawPin = 8;
const int armPin = 9;
int armPos=0, clawPos=0;
float wiiP = 0.0;          // ❶
float wiiPAvg = 0.0;       // ❷
int lastarmPos = 350;

const char i2c_address = 0x52;
int jx = 0, jy = 0, accX = 0, accY = 0, accZ = 0, buttonZ = 0, buttonC = 0;

void setup() {
  Serial.begin(115200);

  // Nunchuck
  Wire.begin();
  pinMode(A2, OUTPUT);
  pinMode(A3, OUTPUT);
  digitalWrite(A2, LOW);
  digitalWrite(A3, HIGH);
```

```
  delay(100);
  initNunchuck();

  // Servos
  pinMode(clawPin, OUTPUT);
  pinMode(armPin, OUTPUT);
}

void loop() {
  get_data();   // ❸

  wiiP = (accZ-70.0)/(178.0-70.0);      // ❹
  if (accY>120 && accZ>100) wiiP=1;
  if (accY>120 && accZ<100) wiiP=0;
  if (wiiP>1) wiiP=1;
  if (wiiP<0) wiiP=0;
  wiiPAvg = runningAvg(wiiP, wiiPAvg);  // ❺
  armPos = map(wiiPAvg*10*1000, 0, 10*1000, 2200, 350);

  clawPos = map(jy, 30, 220, 1600, 2250);       // ❻

  pulseServo(armPin, armPos);   // ❼
  pulseServo(clawPin, clawPos);

  printDebug();
}

float runningAvg(float current, float old) {
  float newWeight=0.3;
  return newWeight*current + (1-newWeight)*old; // ❽
}

// servo

void pulseServo(int servoPin, int pulseLenUs)   // ❾
{
  digitalWrite(servoPin, HIGH);
  delayMicroseconds(pulseLenUs);
  digitalWrite(servoPin, LOW);
  delay(15);
}

// i2c

void get_data() {
  int buffer[6];
  Wire.requestFrom(i2c_address,6);
  int i = 0;
  while(Wire.available()) {
    buffer[i] = Wire.read();
    buffer[i] ^= 0x17;
    buffer[i] += 0x17;
    i++;
  }
```

```
  if(i != 6) {
    Serial.println("Error reading from i2c");
  }
  write_i2c_zero();

  buttonZ = buffer[5] & 0x01;
  buttonC = (buffer[5] >> 1) & 0x01;
  jx = buffer[0];
  jy = buffer[1];
  accX = buffer[2];
  accY = buffer[3];
  accZ = buffer[4];

}

void write_i2c_zero() {
  Wire.beginTransmission(i2c_address);
  Wire.write((byte)0x00);
  Wire.endTransmission();
}

void initNunchuck()
{
  Wire.beginTransmission(i2c_address);
  Wire.write((byte)0x40);
  Wire.write((byte)0x00);
  Wire.endTransmission();
}

// debug

void printDebug()
{
  Serial.print("accZ:");
  Serial.print(accZ);
  Serial.print("        wiiP:");
  Serial.print(wiiP);
  Serial.print("        wiiPAvg:");
  Serial.print(wiiPAvg);
  Serial.print("        jy:");
  Serial.print(jy);
  Serial.print("        clawPos:");
  Serial.println(clawPos);
}
```

❶ `wiiP` holds the Wii tilt as a percentage value. Back is 0.0 and forward is 1.0.

❷ Running average of `wiiP`.

❸ Reading the Wii Nunchuk through i2c requires a 20 ms delay between reads. The two calls to `pulseServo()` later in `loop()` provide this delay.

❹ Calculate a percentage from raw values read from Wii. This formula is similar to the built-in `map()` function.

❺ To filter out random spikes, use the average of a couple of the last samples for z-axis acceleration.

❻ Take the raw joystick value (30 to 220) and map it to servo pulse (1500 to 2400). For example, the joystick value 30 maps to the servo pulse length of 1500 μs.

❼ Send one pulse to the servo controlling the arm. To keep the servo turning, you must send a continuous stream of these pulses to the servo.

❽ To get a running average over multiple values, you use a weighted average. This way, only one previous data point needs to be stored, but older values can still affect the average.

❾ Send one pulse to the servo. For reliable servo control, this function should be called about 50 times a second. See "Servo Motors" on page 115.

Adding Hand Mechanics

You already know how to control servo motors with Wii Nunchuk, and it's easy to adapt this to commercial robot arms and hands. There are plenty of choices available, and if you have strong mechanical skills you can even build your own. Our robot arm was purchased from *http://dx.com*.

Whichever arm or hand you choose, it's a good idea to mount it firmly on a solid base to keep it upright and prevent it from tipping over. We attached our arm to a thick wood base with screws (see Figure 8-13).

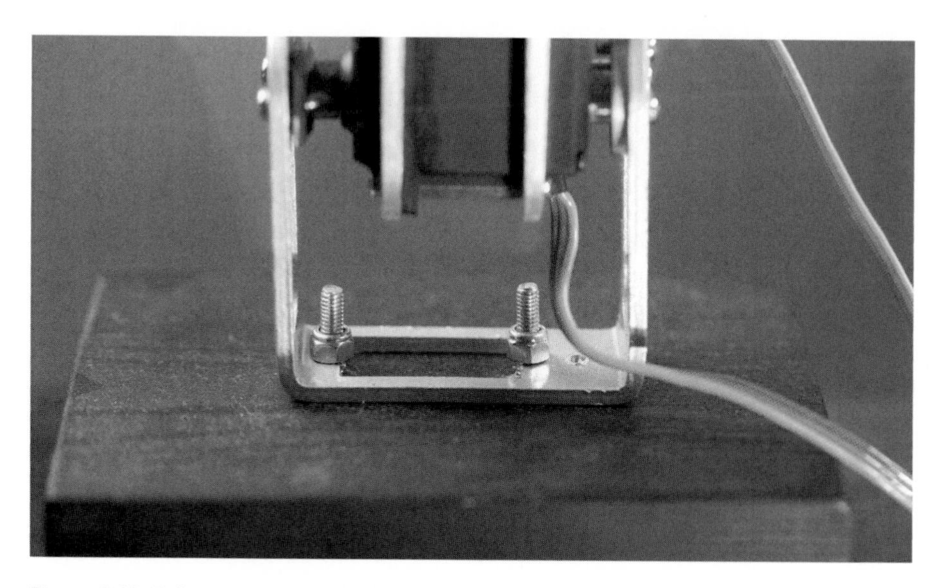

Figure 8-13. *Robot arm screwed to base*

All the code is finished already (see Example 8-8). Connect the servos so that you control the gripper with the Nunchuk thumbstick, and arm movement with the accelerometer. See Figure 8-14 for the final result.

You can now measure acceleration and angular velocity. You know how your device is oriented and which way it's spinning. You can measure just one of these, or use an integrated mobile unit to measure both if this is required in your project.

To use any of the sensors, you probably noticed you can use code examples in a cookbook fashion. But if you took the harder route and learned the bitwise operations to understand how the code works, you can pat yourself on the back. These skills are useful if you work in a new, difficult sensor without any code examples.

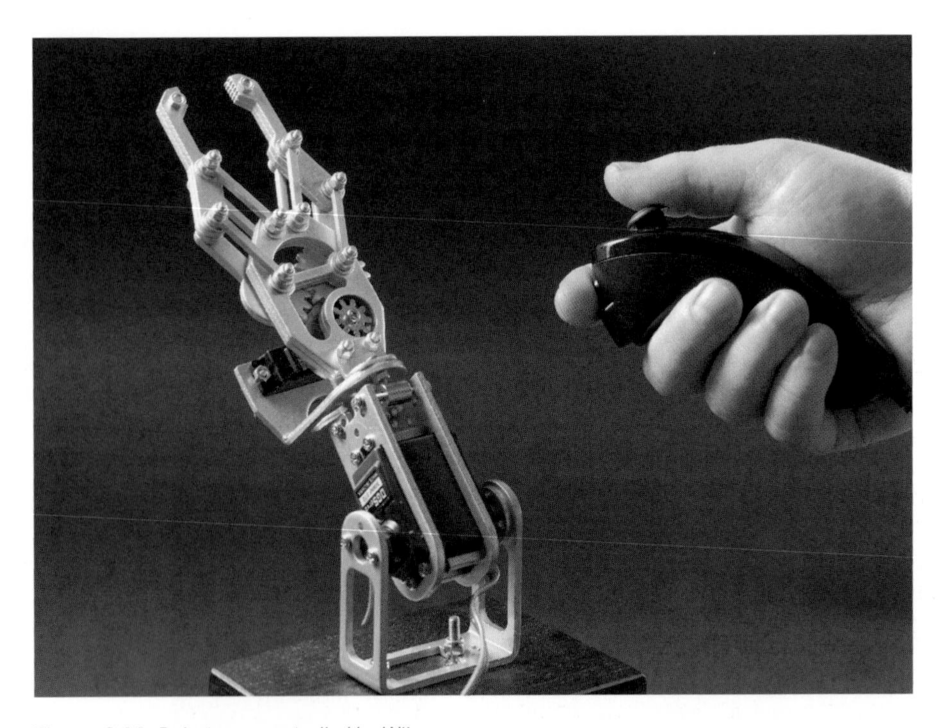

Figure 8-14. *Robot arm controlled by Wii*

If you can stop playing with your robot hand project for a moment, it's time to identify things. In the next chapter, you'll learn to read fingerprints and RFID tags.

8 Identity

Who are you? What object is sitting on top of the sensor? Identification sensors identify people and objects. In this chapter, you'll learn to identify objects with RFID and people with fingerprints and keypads.

RFID (radio frequency identification) has long been hailed as the next big thing in identification. In reality, it's slowly been becoming part of everyday life. RFID has largely replaced bar codes in warehouses, but bar codes are still used with individual consumer packages of products. RFID can be read from a distance, and it can store more data than a bar code. For example, a bar code identifies a product as "1 liter of Foobar, Inc. milk." An RFID tag can also tell you that this specific carton was carton number 12,209,312, uniquely identifying it.

Biometric identification is getting commonplace. Many laptops have a fingerprint reader to protect you from bystanders "shoulder-surfing" as you type your password. Most governments want to store fingerprints and digitally readable facial images of every citizen with the excuse of using them in passports. The digital passports also have a standard for storing an iris image, even though that's not used yet. DNA markers are already used for identifying criminals.

The benefit of biometric identification is that it's always with you. The biggest downside is that once it's copied by the adversary, it's not possible to change it. For example, you can't change your fingerprints. In practice, many cheap biometric sensors can be easily misled. There have also been claims that even professional, manual fingerprint identification is not as reliable as has been assumed.

Keypad is likely the most common method for identification. Just think how many times a week you type your PIN into different devices.

Keypad

A keypad is a quick way to type in some numbers. If you used a cell phone when they still had keypads, you might remember that keypads work well with one-handed use, and some people could even type numbers (or text!) without looking at the phone. You have similar keypads in television remote controls and microwave oven controls. Keypad are also used for password entry. You have probably used one with an ATM, numeric door lock, or a burglar alarm.

This experiment uses inexpensive numeric keypad, *http://dx.com* part number 149608, "DIY 4 x 4 16-Key Numeric Keypad - Black," shown in Figure 9-1. The wiring used is similar to many other cheap number pads.

Figure 9-1. *16-key numeric keypad*

For Raspberry Pi, you could just use a normal USB keyboard or a USB number pad.

Under each key, vertical and horizontal wires cross. When you press a key, it connects the wires at each crossing.

To find out which key is pressed, you need to take one of the vertical columns LOW, but leave all the other columns HIGH. If one of the horizontal lines goes LOW, the key that was pressed is the one where the HIGH lines cross. If there is no match, test the rest of the columns until you find a key that was pressed. The code keeps testing columns, returning to the first column to check for more keypresses.

Keypad Code and Connection for Arduino

Figure 9-2 shows the connections for the keypad and Arduino. Wire it up as shown, and run the sketch shown in Example 9-1.

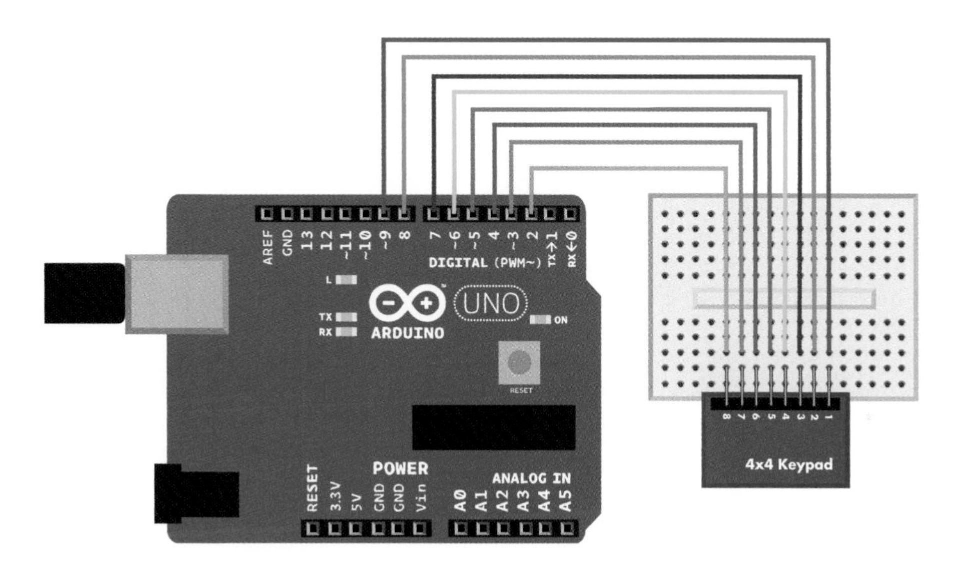

Figure 9-2. *Keypad Arduino connections*

Arduino has internal pull-up resistors. When a digital pin is in INPUT mode, digitalWrite(pin, HIGH) connects it to +5 V through 20 kOhm resistor.

Example 9-1. keypad.ino

```
// keypad.ino - read 16-key numeric keypad (dx.com sku 149608)
// (c) BotBook.com - Karvinen, Karvinen, Valtokari

const int count = 4;    // ❶
char keymap[count][count] = {   // ❷
  {'1', '2', '3', 'A'},
  {'4', '5', '6', 'B'},
  {'7', '8', '9', 'C'},
  {'*', '0', '#', 'D'}
};
```

```
const char noKey = 'n';
byte columns[count] = {9, 8, 7, 6};        // ❸
byte rows[count] = {5, 4, 3, 2};
unsigned int lastReadTime;
unsigned int bounceTime = 30; // ms

void setup()
{
  Serial.begin(115200);
  lastReadTime = millis();

  for(int i = 0; i < count; i++)
  {
    pinMode(rows[i], INPUT);
    digitalWrite(rows[i], HIGH); // pull up     // ❹

    pinMode(columns[i], INPUT);
  }
}

void loop()
{
  char key = getKeyPress();
  if(key != noKey) {
    Serial.print(key);
  }
  delay(100);
}
// This does not support multiple presses. first one is
// returned
char getKeyPress()
{
  char foundKey = noKey;
  if((millis() - lastReadTime) > bounceTime) {  // ❺
    //Pulse columns and read row pins
    for(int c = 0; c < count; c++) {
      //Start pulse
      pinMode(columns[c], OUTPUT);       // ❻
      digitalWrite(columns[c], LOW);
      //Read rows
      for(int r = 0; r < count; r++) {
        if(digitalRead(rows[r]) == LOW) {        // ❼
          //Find right character
          foundKey = keymap[r][c];       // ❽
        }
      }
      digitalWrite(columns[c], HIGH);
      pinMode(columns[c], INPUT);        // ❾
      if(foundKey != noKey)
      {
        break;
      }
    }
    lastReadTime = millis();
```

```
  }
  return foundKey;
}
```

❶ The count of keys vertically and horizontally. It's a 4 by 4 keypad.

❷ Map the position of the key (e.g., column 0, row 1) to the key number ("4"). The keymap is a two-dimensional array.

❸ Map columns (e.g., column 2) to Arduino digital pins (D7).

❹ Pull up each row pin (D6, D7, D8, D9), so that when they are not connected, they go HIGH. Using the internal pull-up saves you a bunch of external pull-up resistors!

❺ Check that the difference between uptime, as measured by millis(), and the time of the last read is more than 30 ms. This is the typical program pattern for checking that enough time has passed. In this program, the main program loop() never calls this function more often than once in 100 ms, due to the `delay()` in `loop()`. The check is useful when you start using getKeyPress() in your own programs outside this example.

❻ Turn the current (vertical) column LOW. Other columns are left HIGH.

❼ If a row is LOW…

❽ … then the pressed key is in the intersection of the column you set LOW and the row that went LOW.

❾ In our tests, a column in OUTPUT mode seemed to interfere with the readings. Setting it to INPUT mode sets it to high impedance (off), thus allowing correct readings.

Keypad Code and Connection for Raspberry Pi

Figure 9-3 shows the wiring diagram for Raspberry Pi. Hook everything up as shown, and run the program in Example 9-2.

Try USB keyboards and USB number pads with Raspberry Pi. Just like any keyboard, a USB number pad works automatically when plugged in. You can use `raw_input()` or pyGame to read keypresses, just like on your normal laptop or desktop.

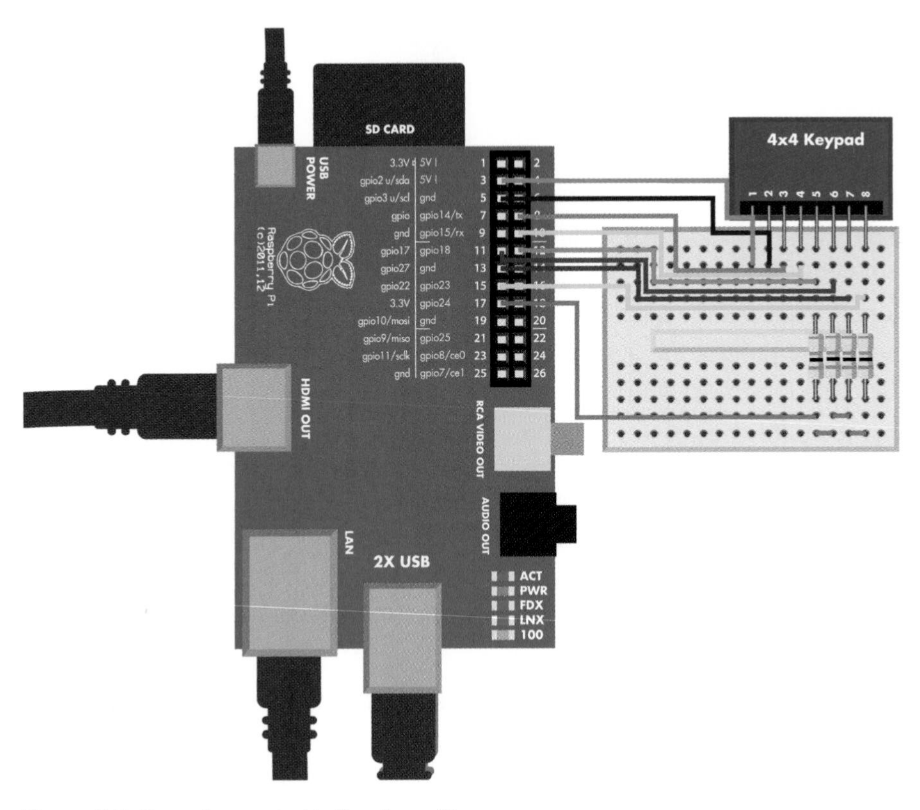

Figure 9-3. *Keypad connected to Raspberry Pi*

Example 9-2. *keypad.py*

```
# keypad.py - read 16-key numeric keypad (dx.com sku 149608)
# (c) BotBook.com - Karvinen, Karvinen, Valtokari


import time
import botbook_gpio as gpio     # ❶

keymap = []
keymap.append(['1', '2', '3', 'A'])     # ❷
keymap.append(['4', '5', '6', 'B'])
keymap.append(['7', '8', '9', 'C'])
keymap.append(['*', '0', '#', 'D'])

columns = [2, 3, 14, 15]        # ❸
rows = [18, 17, 27, 22]
lastReadTime = None
bounceTime = 0.03 # s

def initializeKeyPad():
        for x in range(len(rows)):
                gpio.mode(rows[x], 'in')
```

```
            gpio.mode(columns[x], 'in')

def getKeyPress():
        global lastReadTime
        foundKey = None
        if((time.time() - lastReadTime) > bounceTime):  # ❹
                #pulse columns and read pins
                for c in range(len(columns)):
                        gpio.mode(columns[c], 'out')
                        gpio.write(columns[c], gpio.LOW)        # ❺

                        for r in range(len(rows)):
                                if gpio.read(rows[r]) == gpio.LOW:      # ❻
                                        foundKey = keymap[r][c] # ❼

                        gpio.write(columns[c], gpio.HIGH)
                        gpio.mode(columns[c], 'in')     # ❽
                        if not foundKey == None:
                                break   # ❾
                lastReadTime = time.time()
        return foundKey

def main():
        global lastReadTime
        initializeKeyPad()
        lastReadTime = time.time()
        while True:
                key = getKeyPress()
                if not key == None:
                        print(key)
                time.sleep(0.1) # s

if __name__ == "__main__":
        main()
```

❶ Make sure there's a copy of the *botbook_gpio.py* library in the same directory as this program. You can download this library along with all the example code from *http://botbook.com*. See "GPIO Without Root" on page 19 for information on configuring your Raspberry Pi for GPIO access.

❷ Map the key's physical position (e.g., row 1, column 0) to a key label ("4").

❸ Map a column number (e.g., 2) to a gpio pin (gpio 14).

❹ Check that the keypad is not read more often than once every 30 ms. The check is useful when you start using getKeyPress() in your own programs. In this example, the main() function calls getKeyPress() so rarely that the check is redundant. The `time.time()` function returns the time as seconds since the Unix epoch, such as 1376839738.068395. This is compared with the time of the last keypress read, also stored as seconds since the epoch. The epoch is 00:00:00 UTC on January 1, 1970.

❺ Take one column (vertical) LOW. Others are left HIGH.

❻ If a row became LOW, other rows are pulled HIGH by the physical pull-up resistor, as you can see in Figure 9-3.

❼ The key that was pressed is at the intersection of LOW column and LOW row.

❽ Put the column back to "in" mode, so that it's high impedance (off) and doesn't affect the measurements.

❾ Break out of the for loop (because a pressed key was found).

Environment Experiment: Revealing Fingerprints

How can you bypass a keypad, short of breaking it?

You could try all the combinations, but that's a lot of trial and error. For example, there are a lot of three-character permutations in a 16-key pad.

```
16 * 16 * 16 = 16**3 =  4096
```

Can you see which keys are pressed most? In a keypad that's been used a lot, the paint or sticker on frequently used keys could be worn.

You probably have a new keypad. Press the same code a couple of times. Try, if you can, to see grease on well-used keys by using bright light (Figure 9-4).

Figure 9-4. *Try to see grease on well-used keys by using bright light*

What if you pulverize graphite from a pencil or take some beauty powder and spread that on keys? What if you clean the keys before typing the combination? Does it matter if your fingers are greasy?

Latent (hidden) fingerprints can be captured with cyanoacrylate superglue vapor. Even though this is a very efficient technique for glass and plastic, cyanoacrylate fumes are unhealthy and must be handled in a fume hood. Also, superglue annoyingly sticks to everything.

Did you find the three most used keys? Now there are fewer to try, as you just have to find the correct order of these three keys.

```
3 * 3 * 3 = 3**3 = 27
```

Fingerprint Scanner GT-511C3

A fingerprint scanner is a more sophisticated identification device than the keypad. It is unique and won't get lost or forgotten. Also you don't have to worry about someone shoulder surfing as you type in a password.

The GT-511C3 fingerprint scanner (Figure 9-5) uses serial pins to communicate with Arduino or Raspberry Pi. Its protocol is described in the data sheet, available by searching the Web for "GT-511C3 data sheet" and from the distributor's (SparkFun) catalog page (*https://www.sparkfun.com/products/11792*).

The scanner can store the fingerprints into its own memory, making it easier to use from Arduino.

The GT-511C3 fingerprint scanner communicates at 9600 bit/s over serial port. The scanner is sometimes slow to respond, so you should configure the serial port for an extra-long timeout with `setTimeout()`, as shown in the next example program.

The protocol consists of *packages*. This code stores fingerprint images in the scanner, so only command packages are used (see Table 9-1). The header and device are always the same, and checksum is always calculated the same way. The part that changes is your command and parameter.

Table 9-1. Command package of the GT-511C3 fingerprint sensor

Purpose	Example contents	Comment
Header	0x55 0xAA	Same for all command packages
Device ID	1	Always the same, fixed on device
Parameter	0	Turn off
Command	0x12	Control LED (CMD_LED)
Checksum		Sum of bytes

Figure 9-5. *Fingerprint scanner GT-511C3*

The microcontroller (master) sends a command package to the scanner (slave). Commands include the following:

- Light up an LED.
- Delete all fingerprints from scanner memory.
- Enroll 1 (scan and store a new fingerprint to slot one).
- Identify (scan a fingerprint and tell if it's one of the stored fingerprints).

The scanner responds with a command package. The response command is either ACK (acknowledge, OK, success) or NACK (negative acknowledge, failure, error). The parameter in the answer contains additional information, such as the following:

- ID of recognized fingerprint
- The fact that fingerprint is unknown
- Error code

The most common error codes are 100F (connection already opened) and 100A (memory is empty).

You can get a lot done with the code in this chapter. But if you need more details about the fingerprint sensor, search the Web for "GT-511C3 datasheet."

Fingerprint scanners are vulnerable to fake fingers. Gelatin, like that found in gummy bears, is an easy material to get started with. Other materials include Scotch tape, Play-Doh, and printed transparency with Blu-Tack. The choice of material for fake prints depends on the technique used in the target fingerprint scanner. Chaos Computer Club has published information on extracting fingerprints (http://bit.ly/Prbqew) from objects and using them for creating fake fingerprints.

Fingerprint Sensor Code and Connection for Arduino Mega

Use the Arduino Mega for easier debugging. Mega has multiple serial ports, so you can connect the fingerprint scanner to one serial and use another for USB code upload and debugging.

If you want a cheaper option, try the Arduino Leonardo. Although it looks a lot like an Arduino Uno, the Leonardo has the advantage of having two serial ports. Unlike the Uno, the RX/TX pins (pins 0 and 1) and the USB/Serial port are not tied together. With the Leonardo, `Serial` refers to the USB/Serial port, and `Serial1` refers to pins 0 (RX) and 1 (TX). If you use the Leonardo instead of the Mega, you will need to change the example code and the circuit wiring to use Serial1 (pins 0 and 1) for the fingerprint scanner instead of Serial3 (pins 15 and 14).

Figure 9-6 shows the connection diagram for Arduino. Wire it up as shown, and then run the sketch shown in Example 9-3.

Difficult code! This code is more difficult than some other code examples in this book: it uses pointers to convert between a struct and a byte buffer. You don't have to completely understand it to use it. You could just build the circuit, upload the code, and enjoy your fingerprint reader. But if you want to understand it, read on.

Arduino uses 5 V, but fingerprint sensor uses 3.3 V. Excessive voltage could destroy the sensor, reduce its usable age, or cause incorrect readings. In this project, you'll use a voltage divider to reduce the voltage given by the Arduino TX pin. All you need are two resistors.

You don't need a voltage divider on the sensor's TX pin, because it has a maximum of 3.3 V—much less than Arduino's RX pin. Arduino will still recognize 3.3 V as HIGH, because it's more than half of 5 V.

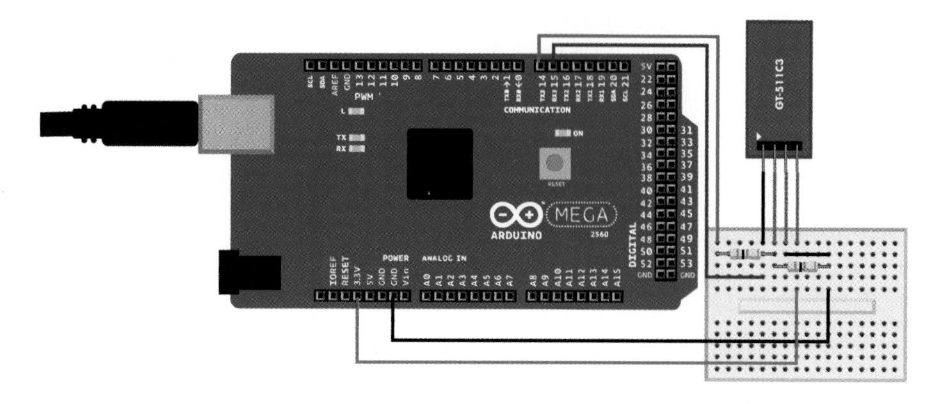

Figure 9-6. *Connections for the fingerprint scanner with Arduino*

Example 9-3. ***fingerprint_scanner.ino***

```
// fingerprint_scanner.ino - learn and recognize fingerprints with GT-511C3
// (c) BotBook.com - Karvinen, Karvinen, Valtokari
// Requires Arduino Mega for extra serial port

const byte STX1 = 0x55; // ❶
const byte STX2 = 0xAA;

const word CMD_OPEN = 0x01;       // ❷
const word CMD_CLOSE = 0x02;
const word CMD_LED = 0x12;
const word CMD_GET_ENROLL_COUNT = 0x20;
const word CMD_ENROLL_START = 0x22;
const word CMD_ENROLL_1 = 0x23;
const word CMD_ENROLL_2 = 0x24;
const word CMD_ENROLL_3 = 0x25;
const word CMD_IS_FINGER_PRESSED = 0x26;
const word CMD_DELETE_ALL = 0x41;
const word CMD_IDENTIFY = 0x51;
const word CMD_CAPTURE_FINGER = 0x60;

const word ACK = 0x30;  // ❸
const word NACK = 0x31;  //Error

struct package {        // ❹
  byte header1;
  byte header2;
  word deviceID;
  unsigned long param;
  word cmd;
  word checksum;
};

const int SIZE_OF_PACKAGE = 12;
```

```
/*
To calculate checksum we add all bytes in pdu together.
*/
word calcChecksum(struct package *pkg) {        // ❺
  word checksum = 0;
  byte *buffer = (byte*)pkg;    // ❻
  for(int i=0; i < (sizeof(struct package) - sizeof(word)); i++)
  {
    checksum += buffer[i];
  }
  return checksum;
}


int sendCmd(word cmd, int param) {       // ❼
  struct package pkg;
  pkg.header1 = STX1;
  pkg.header2 = STX2;
  pkg.deviceID = 1;       // ❽
  pkg.param = param;
  pkg.cmd = cmd;
  pkg.checksum = calcChecksum(&pkg);
  //Serial.println("Sending command");
  byte *buffer = (byte*)&pkg;    // ❾

  int bytesSent = Serial3.write(buffer, sizeof(struct package));

  if(bytesSent != sizeof(struct package)) {
    Serial.println("Error communicating");
    return -1;
  }

  int bytesReceived = 0;
  char recvBuffer[SIZE_OF_PACKAGE];      // ❿
  struct package *recvPkg = (struct package*) recvBuffer;        // ⓫

  bytesReceived = Serial3.readBytes(recvBuffer, sizeof(struct package));        // ⓬
  if(bytesReceived != SIZE_OF_PACKAGE) {
    Serial.println("Error communicating");
    return -1;
  }

  if( recvPkg->header1 != STX1 || recvPkg->header2 != STX2) {   // ⓭
    Serial.println("Header error!");
    return -1;
  }

  if(recvPkg->checksum != calcChecksum(recvPkg)) {
    Serial.println("Checksum mismatch error!");
    return -1;
  }
  if(recvPkg->cmd == NACK) {
    Serial.println("NACK - Cmd error!");
```

```
    Serial.print("Error: ");
    Serial.println(recvPkg->param,HEX);
    return -1;
  }

  return recvPkg->param;
}

//All custom codes here as they may use variables defined in protocol implementation.
void setup() {
  Serial.begin(115200); // ⓮
  Serial3.begin(9600);  // ⓯
  Serial3.setTimeout(10*1000); // ms    // ⓰
}

void flashLed(int time) {
  sendCmd(CMD_LED, 1);
  delay(time);
  sendCmd(CMD_LED, 0);

}

void loop() {
  Serial.println("Sending open command");
  sendCmd(CMD_OPEN, 0);
  //Delete all fingerprints on start for testing purpose only.
  if(sendCmd(CMD_DELETE_ALL, 0) >= 0) {
    //Flash LED 3 times for victory dance and to indicate that we are ready for enrolling.
    flashLed(500);
    delay(500);
    flashLed(500);
    delay(500);
    flashLed(500);
  }

  Serial.println("Starting capture");

  int id = 0;
  id = sendCmd(CMD_GET_ENROLL_COUNT, 0);          // ⓱
  sendCmd(CMD_LED, 1);
  sendCmd(CMD_ENROLL_START, id);
  Serial.println("Press finger to start enroll");
  int ret = 0;
  WaitForFinger(false);
  Serial.println("Capturing finger");
  ret = sendCmd(CMD_CAPTURE_FINGER, 1); // ⓲
  if(ret < 0) {
    EnrollFail();
    return;
  }
  Serial.println("Remove finger");

  sendCmd(CMD_ENROLL_1, 0);
  WaitForFinger(true);
```

```
Serial.println("Press finger again");

WaitForFinger(false);
ret = sendCmd(CMD_CAPTURE_FINGER, 1);
if(ret < 0) {
  EnrollFail();
  return;
}
Serial.println("Remove finger");

sendCmd(CMD_ENROLL_2, 0);
WaitForFinger(true);
Serial.println("Press finger again");

WaitForFinger(false);
ret = sendCmd(CMD_CAPTURE_FINGER, 1);
if(ret < 0) {
  EnrollFail();
  return;
}
Serial.println("Remove finger");

ret = sendCmd(CMD_ENROLL_3, 0);
if(ret != 0) {
  EnrollFail();
  return;
}
WaitForFinger(true);
flashLed(500);
delay(500);
flashLed(500);
delay(500);
Serial.println("Enroll completed");
Serial.println("Press finger for identify");
sendCmd(CMD_LED, 1);

// Identify
WaitForFinger(false);
ret = sendCmd(CMD_CAPTURE_FINGER, 1); // ⓳
if(ret < 0) {
  IdentFail();
  return;
}
ret = sendCmd(CMD_IDENTIFY, 0);        // ⓴
if(ret >= 0 && ret < 200) {
  Serial.print("ID found: ");
  Serial.println(ret);
  flashLed(500);
  delay(500);
  flashLed(500);
  delay(500);
  flashLed(500);
  delay(500);
  flashLed(500);
```

```
    delay(500);
    flashLed(500);
    delay(500);
    flashLed(500);
    delay(500);
  } else {
    Serial.println("ID not found");
  }
  sendCmd(CMD_CLOSE,0);
  delay(100000);

}

void WaitForFinger(bool bePressed) {
  delay(500);
  if(!bePressed) {

    while(sendCmd(CMD_IS_FINGER_PRESSED, 0) > 0) {
      delay(200);
    }
  } else {
    while(sendCmd(CMD_IS_FINGER_PRESSED, 0) == 0) {
      delay(200);
    }
  }
}
// Flash LED 3 times for failure
// and close device.
void IdentFail() {
  Serial.println("Ident failed!");
  flashLed(500);
  delay(500);
  flashLed(500);
  delay(500);
  flashLed(500);
  delay(500);
  sendCmd(CMD_CLOSE, 0);
}
// Flash LED 4 times for failure
// and close device.
void EnrollFail() {
  Serial.println("Enroll failed!");
  flashLed(500);
  delay(500);
  flashLed(500);
  delay(500);
  flashLed(500);
  delay(500);
  flashLed(500);
  sendCmd(CMD_CLOSE, 0);
}
```

❶ Command package header. As this code doesn't read the fingerprint images, data packets are not used.

❷ Commands as described in the data sheet. 0x is the hexadecimal prefix (see "Hexadecimal, Binary, and Other Numbering Systems" on page 219).

❸ Possible return values: ACK (acknowledge) meaning success, NACK (negative acknowledge) meaning failure.

❹ Command package structure. The exact length of variables will come into play later, when you use the struct to decode received data. See Table 9-1.

❺ The checksum is the sum of all bytes.

❻ To process every byte, the `*pkg` must be converted to a `byte` buffer.

❼ `sendCmd()` sends one command and returns the parameter from the response. It returns -1 for any errors.

❽ The device ID is fixed on the device side. It must still be accounted for, so that the raw bytes converted from this struct will be correct.

❾ Convert the command package struct to raw bytes for transmitting.

❿ The receive buffer is the exact same length as the command package struct.

⓫ The `*recvPkg` pointer sees the receive buffer as a struct. You can later use this pointer to easily access the variables, as in `recvPkg->cmd`.

⓬ Fill the receive buffer with bytes, not worrying about what they mean yet.

⓭ Using the struct package pointer, the individual values of struct are easily accessible. This would not have been possible with a plain byte buffer.

⓮ This first serial port is serial over USB. You can access it from within the Arduino IDE with Tools→Serial Monitor. Debugging is much easier when you can send output to the serial monitor.

⓯ The fingerprint scanner is connected to Serial3. This connection is available on Arduino Mega but not on Arduino Uno.

⓰ The scanner is sometimes slow, so we extend the timeout to 10 seconds.

⓱ Find the first free slot.

⓲ Parameter 1-slow capture, 0-quick image.

⓳ Scan the finger to be identified.

⓴ Identify the last scanned finger. The return value of `sendCmd()` will be the id of the finger (1, 2 or 3) or 0 for an unidentified finger.

Fingerprint Sensor Code and Connection for Raspberry Pi

To use the serial port in Raspberry Pi, you must first release it from use as a login terminal. See "Enabling the Serial Port in Raspberry Pi" on page 320. Wire up the scanner as shown in Figure 9-7, and then run the code in Example 9-4.

To access the serial port from Python, you need to install the PySerial library by running the command:

```
sudo apt-get update && sudo apt-get install python-serial
```

The connection for Raspberry Pi is simple—even simpler than the connection for Arduino. Raspberry Pi uses 3.3 V for HIGH, the same as the fingerprint sensor. As the voltage is already correct, you don't need resistors to build a voltage divider with Raspberry Pi.

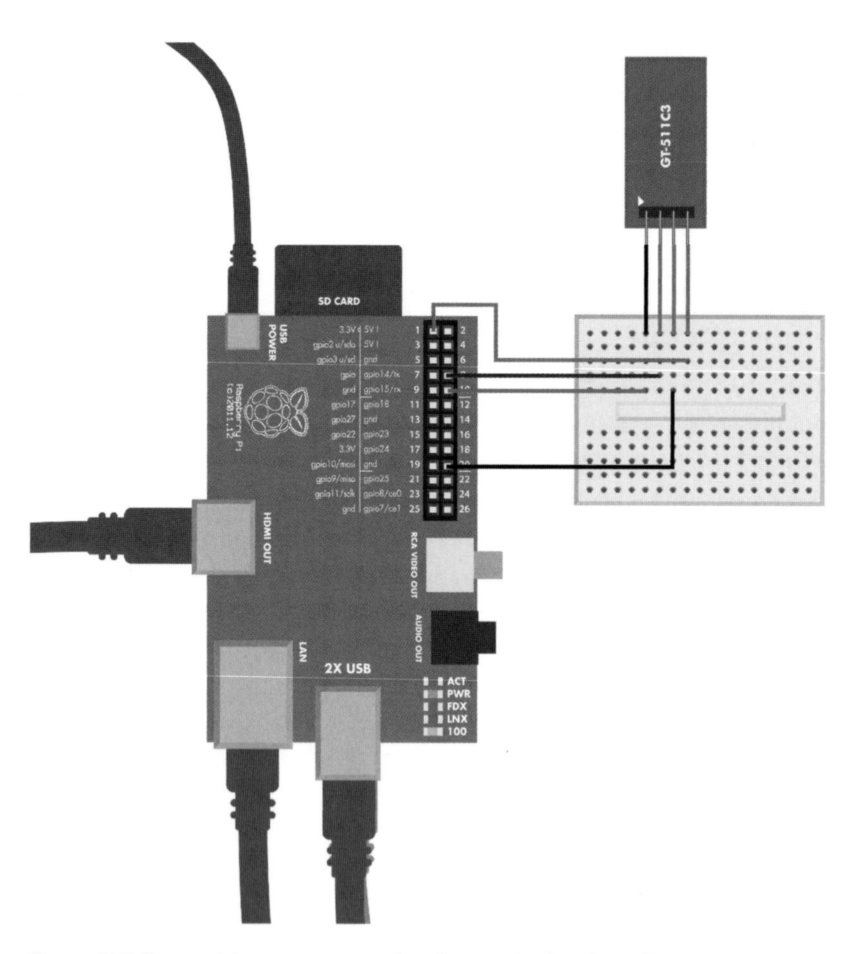

Figure 9-7. *Fingerprint scanner connection diagram for Raspberry Pi*

Example 9-4. ***fingerprint_scanner.py***

```python
# fingerprint_scanner.py - learn and recognize fingerprints with GT-511C3
# (c) BotBook.com - Karvinen, Karvinen, Valtokari

import time
import serial
import struct
```

```
STX1 = 0x55     # 1
STX2 = 0xAA

CMD_OPEN =                              0x01    # 2
CMD_CLOSE =                     0x02
CMD_LED =                               0x12
CMD_GET_ENROLL_COUNT =  0x20
CMD_ENROLL_START =              0x22
CMD_ENROLL_1 =                  0x23
CMD_ENROLL_2 =                  0x24
CMD_ENROLL_3 =                  0x25
CMD_IS_FINGER_PRESSED = 0x26
CMD_DELETE_ALL =                0x41
CMD_IDENTIFY =                  0x51
CMD_CAPTURE_FINGER =    0x60

ACK = 0x30      # 3
NACK = 0x31

port = None

def calcChecksum(package):      # 4
        checksum = 0
        for byte in package:
                checksum += ord(byte)
        return int(checksum)

def sendCmd(cmd, param = 0):    # 5
        package = chr(STX1)+chr(STX2)+struct.pack('<hih', 1, param, cmd)       # 6
        checksum = calcChecksum(package)
        package += struct.pack('<h',checksum)   # 7

        sent = port.write(package)

        if(sent != len(package)):
                print "Error communicating"
                return -1

        recv = port.read(sent)  # 8
        recvPkg = struct.unpack('cchihh',recv)  # 9

        if recvPkg[4] == NACK:
                print("error: %s" % recvPkg[3])
                return -2
        time.sleep(1)
        return recvPkg[3]

def startScanner():
        print("Open scanner communications")
        sendCmd(CMD_OPEN)

def stopScanner():
        print("Close scanner communications")
```

```
        sendCmd(CMD_CLOSE)


def led(status = True):
        if status:
                sendCmd(CMD_LED,1)
        else:
                sendCmd(CMD_LED,0)

def enrollFail():
        print("Enroll failed")
        led(False)
        stopScanner()

def identFail():
        print("Ident failed")
        led(False)
        stopScanner()

def startEnroll(ident):
        sendCmd(CMD_ENROLL_START,ident)

def waitForFinger(state):
        if(state):
                while(sendCmd(CMD_IS_FINGER_PRESSED) == 0):
                        time.sleep(0.1)
        else:
                while(sendCmd(CMD_IS_FINGER_PRESSED) > 0):
                        time.sleep(0.1)

def captureFinger():
        return sendCmd(CMD_CAPTURE_FINGER)

def enroll(state):
        if state == 1:
                return sendCmd(CMD_ENROLL_1)
        if state == 2:
                return sendCmd(CMD_ENROLL_2)
        if state == 3:
                return sendCmd(CMD_ENROLL_3)

def identifyUser():
        return sendCmd(CMD_IDENTIFY)

def getEnrollCount():
        return sendCmd(CMD_GET_ENROLL_COUNT)

def removeAll():
        return sendCmd(CMD_DELETE_ALL)

def main():
        print("Remove all identities from scanner")
        startScanner()
        removeAll()
```

```
        led()
        print("Start enroll")
        newID = getEnrollCount()
        print(newID)

        startEnroll(newID)
        print("Press finger to start enroll")
        waitForFinger(False)
        if captureFinger() < 0:
                enrollFail()
                return
        enroll(1)
        print("Remove finger")
        waitForFinger(True)


        print("Press finger again")
        waitForFinger(False)
        if captureFinger() < 0:
                enrollFail()
                return
        enroll(2)
        print("Remove finger")
        waitForFinger(True)

        print("Press finger again")
        waitForFinger(False)
        if captureFinger() < 0:
                enrollFail()
                return

        if enroll(3) != 0:
                enrollFail()
                return

        print("Remove finger")
        waitForFinger(True)

        print("Press finger again to identify")
        waitForFinger(False)
        if captureFinger() < 0: # ➓
                identFail()
                return
        ident = identifyUser()
        if(ident >= 0 and ident < 200): # ⓫
                print("Identity found: %d" % ident)
        else:
                print("User not found")
        led(False)
        stopScanner()

if __name__ == "__main__":
        try:
                if port == None:
```

```
                    port = serial.Serial(
                            "/dev/ttyAMA0",
                            baudrate=9600,
                            timeout=None)   # ⓬
            main()
    except Exception, e:
            print e
            port.close()
    finally:
            port.close()
```

❶ Command package header (see Table 9-1).

❷ The commands from the data sheet. 0x indicates a hexadecimal number (see "Hexadecimal, Binary, and Other Numbering Systems" on page 219 for more information).

❸ An ACK (acknowledge) reply means success.

❹ Each command package has a checksum. The checksum is the sum of bytes.

❺ The `sendCmd()` function sends a command package and reads the response. It returns the parameter in the response, or a negative number for any errors.

❻ Pack the Python variables into raw bytes for transmission. The header bytes STX1 and STX2 are concatenated. The `param` and `cmd` values are packed with `struct.pack`: little endian (<), short signed two-byte integer (`h`), and normal four-byte integer (`i`).

❼ The calculated checksum is appended to the raw bytes. The `pack()` function converts checksum to little endian (<) and short signed two-byte integer (`h`).

❽ Read the response. The command packages are always the same length, so you can use the length `sent` as the number of bytes to read.

❾ Unpack the raw data to a tuple. The parameters to `struct.pack()` are: normal one-byte char (`c`), short int (`h`), and normal int (`i`).

❿ Given the command `CMD_CAPTURE_FINGER` (0x60), the scanner returns a negative return value for an error. No identification is attempted yet. The `captureFinger()` function simply sends this command and returns its value.

⓫ The `identifyUser()` function returns the id for recognized finger (1, 2, 3) or a negative number for unknown finger.

⓬ The scanner communicates at 9600 bits/second. In Linux, serial ports are raw device files. The scanner can be slow to answer, so the timeout needs to be disabled (which is what we did here) or set to a long value. The timeout is specified in seconds (you can use floating point for fractions of seconds).

RFID with ELB149C5M Electronic Brick

Radio frequency identification (RFID) offers cheap, unique object identification from a distance. It's already used a lot in warehouses, and as prices go down, it's making its way to consumer packages. Some pets have microchips in the neck so that they can be identified even if they lose their collar tags. In Finland, some libraries use RFID instead of EAN bar codes.

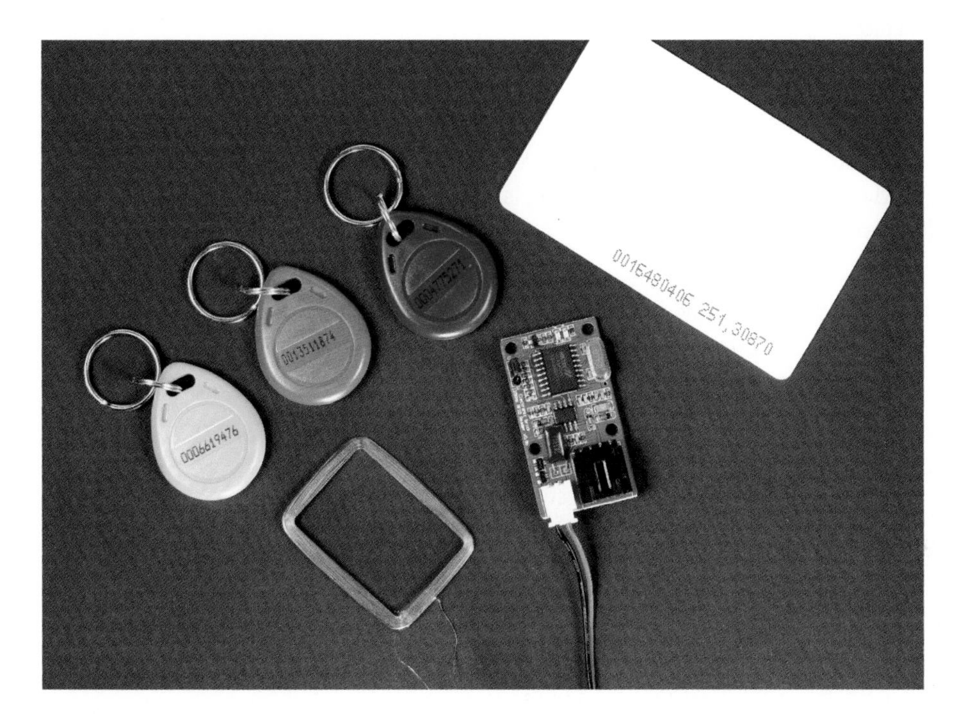

Figure 9-8. *The ELB149C5M brick*

There are multiple RFID standards that differ on price, reading distance, security, amount of stored data, and popularity. In this project, you'll play with the ELB149C5M Electronic Brick. It reads uem4100 standard tags at 125 kHz.

Once you've built the circuit and powered the sensor, hold the card to the sensor. A green light blinks when the sensor reads the card.

The ELB149C5M sensor can use one of two protocols, serial port or *Wiegand protocol*. As serial port is much more common and familiar, the experiment here uses serial port. You can find the description of the protocol by searching for "ELB149C5M data sheet."

As of this writing, the ELB149C5M Electronic Brick was listed as being discontinued, and it was getting difficult to find it for sale. However, Seeedstudio has released a newer module, the Grove 125KHz RFID Reader (http://bit.ly/PrbsmD). It includes a connector that is designed for its Grove System (http://bit.ly/Prbw5Z), but you can connect the module directly to an Arduino Mega with jumper wires: black goes to GND, red to +5 V, and yellow to RX3 (pin 15) on the Mega. White should remain unconnected.

To select serial mode, the UW jumper on the sensor is set to U for UART serial port. The serial port uses 9600 N81 TTL: 9600 bit/s speed with typical settings of no verify bit (`N`), 8 data bits (`8`), and one stop bit (`1`). The signal levels are TTL/UART, so you can just connect it to the Arduino pins with jumper wire.

You can usually connect TTL directly to Arduino or Raspberry PI. RS232 connects to serial port in older computers and should not be connected directly to Raspberry Pi or Arduino.

TTL (transistor-transistor logic) uses LOW for 0 bit and HIGH for 1 bit. For TTL, LOW is 0 V (GND) and HIGH is typically 3.3 V or 5 V.

The old-fashioned RS232 serial port uses voltages between -25 V and +25 V. To connect RS232 to Arduino or Raspberry Pi, you would need a conversion chip like the MAX 232.

The RFID reader initiates communication by sending a packet. The packet contains the static identifying number of the RFID tag (see Table 9-2).

The RFID reader sends ASCII characters in an odd way, such that every pair of characters represents a hexadecimal number. For example, the two byte string "3E" represents one byte hex code 0x3E (62).

For more information on hexadecimal numbers, see "Hexadecimal, Binary, and Other Numbering Systems" on page 219.

After the microcontroller verifies the packet checksum, you know you have a valid card number that you can use. In this experiment, the program prints the number on the serial monitor.

Table 9-2. 14 character packet from Electronic Brick RFID reader

Purpose	Length in ASCII chars	Length in bytes	Comment
Start	1 char	not decoded	0x02 - STX, ASCII start of text
Card number	10 char	5 B	1 B manufacturer, 4 byte id
Checksum	2 char	1 B	bitwise XOR between bytes
End	1 char	not decoded	0x03 - ETX, ASCII end of text

RFID Code and Connection for Arduino Mega

For easier debugging, this project uses Arduino Mega. Mega has multiple serial ports, so you can use the serial monitor (Tools→Serial Monitor in the Arduino IDE) at the same time the RFID reader is attached. With the Uno, you would have only one serial port, which would make prototyping extremely slow and frustrating. Our code example uses two serial ports and will not work on Uno without modifications.

Wire up the Arduino as shown in Figure 9-9, then run the sketch listed in Example 9-5.

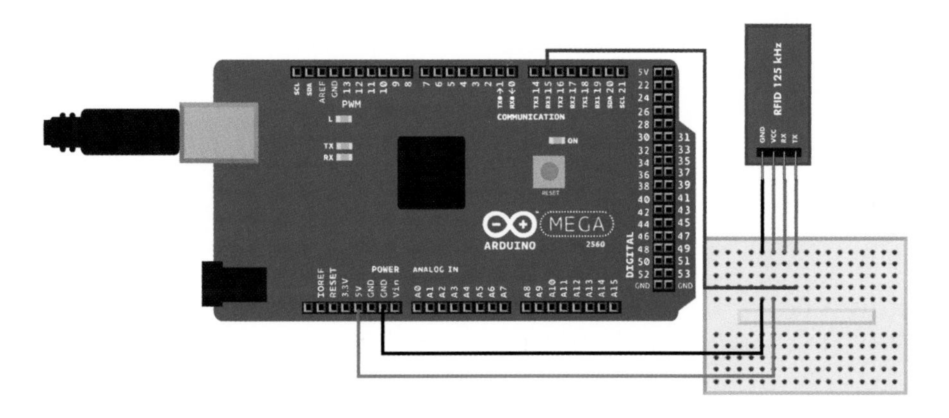

Figure 9-9. *Connection diagram for Arduino and RFID reader*

Example 9-5. rfid_reader.ino

```
// rfid_reader.ino - read 125 kHz RFID tags with ELB149C5M electronic brick
// (c) BotBook.com - Karvinen, Karvinen, Valtokari
// Requires Arduino Mega for extra serial port


int bytesRead = 0;        // ❶
char buffer[13];          // ❷

void setup() {
  Serial.begin(115200); // computer
  Serial3.begin(9600);  // RFID reader  // ❸
}
```

```
void loop() {
  char recv;
  if(Serial3.available() > 0) { // ❹
    recv = Serial3.read();
    if(recv == 0x02) {  // ❺
      bytesRead = 0;
      Serial.println("Start reading tag");
    } else if(bytesRead == 12 && recv == 0x03) {        // ❻
      Serial.println();
      String data = buffer;
      byte checksum = 0;
      byte chk = toLong(data.substring(10, 12));
      long id = toLong(data.substring(4, 10));
      for(int i = 0; i < 10; i=i+2) {
        checksum ^= toLong(data.substring(i, i+2));     // ❼
      }
      Serial.print(id); // ❽
      if(checksum == chk) {     // ❾
        Serial.println(" Card ok");
      } else {
        Serial.println(" Checksum error!");
      }
    } else {
      buffer[bytesRead] = recv;
      bytesRead++;
      Serial.print(recv);
    }
  }
  delay(10);
}

long toLong(String data) {
  char buf[20];
  data = "0x"+data;
  data.toCharArray(buf, 19);
  return strtol(buf, NULL, 0);

}
```

❶ bytesRead will contain the number of bytes handled in the current packet. You'll use it to select the correct element within buffer[].

❷ Initialize a 14-byte buffer for storing the packet from the reader. The count starts from zero, buffer[0], buffer[1], ... buffer[13].

❸ The extra serial ports of the Arduino Mega make debugging convenient.

❹ Try to read bytes from the serial stream only if there is actually some data available.

❺ 0x02 indicates the start of text. Ignore any existing content in the buffer and start from the beginning. See Table 9-2.

❻ 0x03 indicates the end of the text. If we've read 12 bytes...

❼ ...calculate the checksum. The checksum is calculated as a *bitwise XOR* of every byte of the card id. The inplace bitwise XOR `a ^= b` means: `a = a^b`, where `^` is bitwise exclusive OR.

❽ Print the id to the serial port.

❾ Indicate whether the calculated checksum matches the one in the packet.

RFID Code and Connection for Raspberry Pi

To use the serial port in Raspberry Pi, you must first release it from use as a login terminal. See "Enabling the Serial Port in Raspberry Pi" on page 320. Figure 9-10 shows the connection diagram for the RFID reader and Raspberry Pi. Wire it up as shown, and then run the code from Example 9-6.

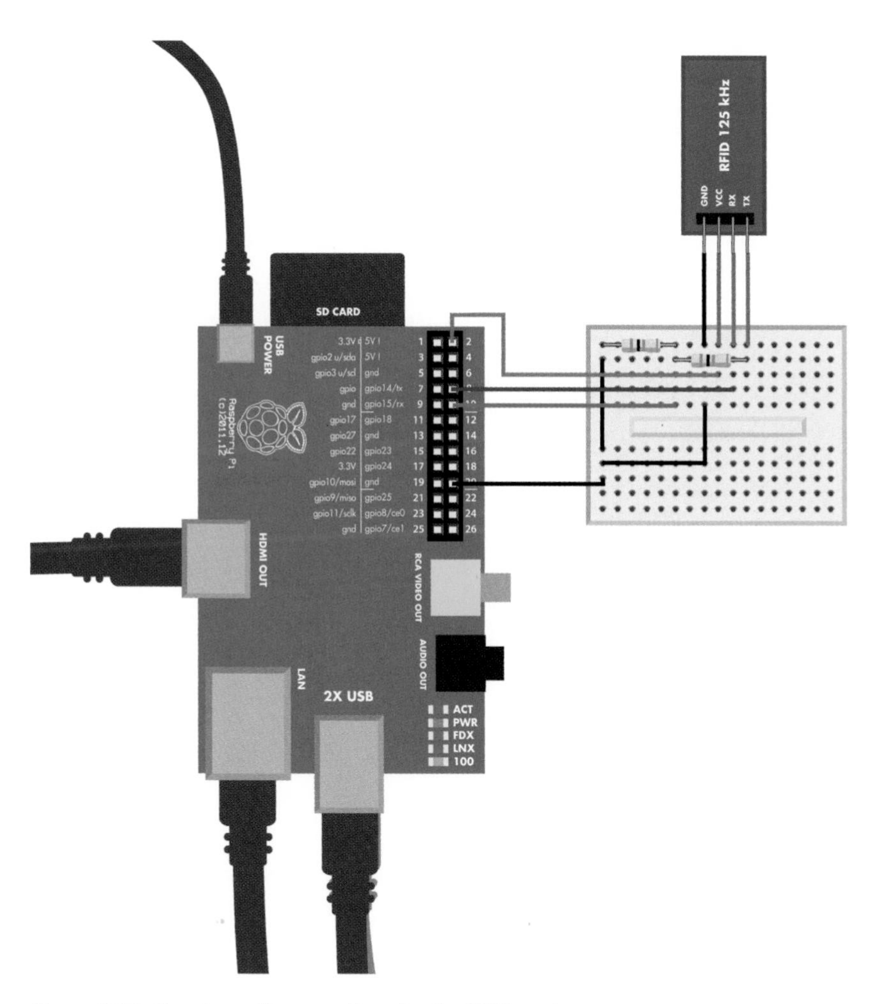

Figure 9-10. *Raspberry Pi connections for the RFID reader*

Example 9-6. rfid_reader.py

```
# rfid_reader.py - read 125 kHz RFID tags with ELB149C5M electronic brick
# (c) BotBook.com - Karvinen, Karvinen, Valtokari

import time
import serial   # ❶
import struct

port = None

def main():
        global port
        bytesRead = -1  # ❷
        buff = [0x00]*12        # ❸
        print("Ready to receive tag")
        while True:     # ❹
                recv = port.read()      # ❺
                if(ord(recv) == 0x02):  # ❻
                        bytesRead = 0
                        print("Start reading tag")
                elif(bytesRead == 12 and ord(recv) == 0x03):    # ❼
                        print("Checking tag")
                        data = ""       # ❽
                        checksum  = 0x00
                        for x in 0, 2, 4, 6, 8, 10:
                                hexString = ''.join( buff[ x : x+2 ] )  # ❾
                                translatedByte = int(hexString, 16)
                                data += chr(translatedByte)     # ❿
                                checksum = checksum ^ translatedByte    # ⓫
                        cardData = struct.unpack(">cic", data)  # ⓬

                        if checksum != 0:
                                print "Checksum calculation failed"
                        print cardData[1]       # ⓭
                else:   # ⓮
                        buff[bytesRead] = recv  # ⓯
                        bytesRead += 1


if __name__ == "__main__":
        if port == None:
                port = serial.Serial("/dev/ttyAMA0", baudrate=9600, timeout=None)
                port.flushInput()
        main()
```

❶ Import the Python PySerial library.

❷ Initialize `bytesRead` to an impossible value that will never come up if everything is working right. This helps with debugging (if you see this value in your results, you know something is wrong).

❸ Initialize `buff` to zeroes. The table multiplication creates a table of 12 elements with a zero value in each element. The start of text character (0x2) and end of text character (0x3) are not stored in the buffer. See Table 9-2.

❹ The program will run until you press Control-C.

❺ Read one byte from serial. The `read()` function is a *blocking function*, so it will automatically wait until there is data available to read.

❻ 0x02 indicates the start of text; this means it's time to discard any old data that was previously read and start reading a new packet. See Table 9-2. The `ord()` function returns the integer corresponding to the character supplied to it. For 8-bit characters, this will map to its ASCII value.

❼ 0x03 indicates the end of text. If the length is 12, then we have read a whole tag.

❽ The `data` variable will store the raw bytes.

❾ Currently, `buff` contains ASCII representation of the hex codes of the bytes. For example: `buff = ['3', 'E', '0', '0', 'F', 'B', '7', '8', '8', 'D', '3', '0']`. Yes, that is weird. Each two-ASCII-character pair is converted to byte, e.g., the string "3E" is converted to 0x3E (i.e., 62).

❿ Append the byte data to variable `data`. It's a common programming pattern to both put the packet into a variable and calculate the checksum in the same loop so you don't have to perform a loop twice.

⓫ The checksum is each byte bitwise XOR'ed. There is a special trick for comparing the calculated checksum to the checksum in the packet. The last XOR operation is between the calculated checksum (e.g., 0x73) and the checksum in the packet (hopefully 0x73). Any number XORed with itself is zero (`0x73 XOR 0x73 == 0x0`). This way, you don't need to convert the weird hex ASCII twice.

⓬ Unpack the values into a tuple. The values have different length in bytes, so `unpack()` is needed. The parameters `>cic` for unpack are big endian (>), one byte char (c), 4 byte signed integer (`i`) and another char (c).

⓭ Print the second cell of the tuple, which is the 10 byte card number.

⓮ If the byte we read is not the start or end byte, it's probably the body, so...

⓯ ...add the byte we just read to `buff`, which holds all the collected ASCII characters.

Test Project: Ancient Chest from the Future

Marry a chest and fingerprint sensor. You'll get a box that opens only with your finger. You can even authorize your friends' fingers, as well.

Figure 9-11. *Ancient Chest from the Future*

What You'll Learn

In the *Ancient Chest* project, you'll learn how to:

- Use the fingerprint sensor to control a lock.
- Split your code into multiple files.
- Build a simple lock with a servo.
- Package your project with ancient material for style.

Figure 9-16 shows the wiring diagram for the chest.

Operating the Chest

Are you authorized to open the box? The fingerprint scanner glows blue. Press your finger on it, and—beep, whir—the lock opens. Enjoy the contents of the *Ancient Chest from the Future*.

Inside the chest, there are two buttons: add and reset. These buttons are needed only to change who is authorized to open the box. While authorizing fingers, the chest communicates with beeps.

Click the chest reset button to erase all trusted fingerprints. (The chest reset button is not the same as the Arduino on-board reset button.)

Click add to authorize a finger. Your finger is scanned three times. If scanning fails, five beeps tell you to press the add button and try again. The first successful scan makes one beep, the second makes two beeps, and the third makes three beeps. Now that you've authorized your finger, you can authorize more fingers or just go ahead and play with the box. To authorize another finger, press the add button.

An authorized finger can lock the chest. Close the lid and press your finger on the blue fingerprint scanner. Beep, whir—the chest locks. Press your finger over the blue reader again, and—beep, whir—the chest opens.

The Box

If you don't happen to have a 19th century box to modify like we did, you'll need to adjust these instructions a little bit. First make a hole for the fingerprint sensor (see Figure 9-12). We did this by drawing the sensor outline on the box, drilling holes in the corners of the outline, and filing off the rest.

Figure 9-12. *Hole for the sensor*

We used a very simple locking mechanism. A servo arm in the bottom part locks the box by going over a top bracket (see Figure 9-13). For the bracket, we used a part from a Meccano toy, but any small L-bracket from the hardware store will do (see Figure 9-14).

Figure 9-13. *Lock servo*

Figure 9-14. *Lock bracket*

Figure 9-15. *Everything packed inside the chest*

Ancient Chest Code and Connection for Arduino

Figure 9-16 shows the wiring diagram for the Ancient Chest. Wire it up as shown, and then run the sketch shown in Example 9-7.

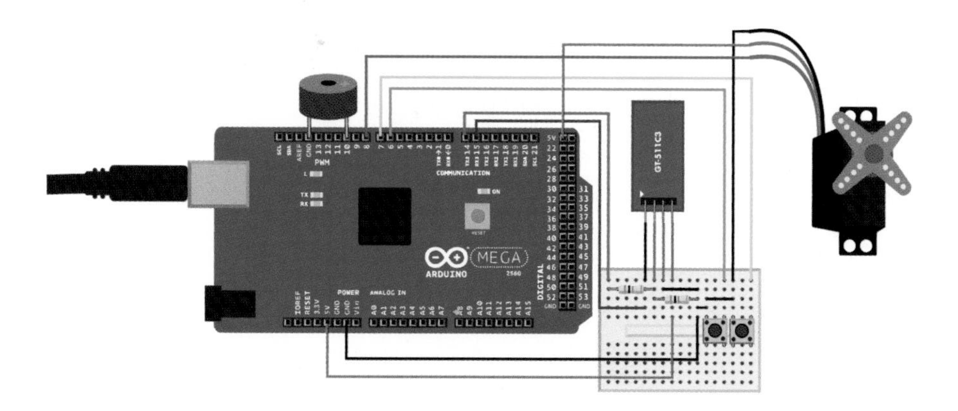

Figure 9-16. *Wiring up the Ancient Chest*

The fingerprint scanner is explained in "Fingerprint Scanner GT-511C3" on page 247, and its Arduino code is in "Fingerprint Sensor Code and Connection for Arduino Mega" on page 249. Servo motors are explained in "Servo Motors" on page 115. The comments after Example 9-7 just explain the new concepts introduced in this example.

Example 9-7. ancient_chest.ino

```
// ancient_chest.ino - fingerprint unlocks chest
// (c) BotBook.com - Karvinen, Karvinen, Valtokari

const byte STX1 = 0x55;
const byte STX2 = 0xAA;

const word CMD_OPEN = 0x01;
const word CMD_CLOSE = 0x02;
const word CMD_LED = 0x12;
const word CMD_GET_ENROLL_COUNT = 0x20;
const word CMD_ENROLL_START = 0x22;
const word CMD_ENROLL_1 = 0x23;
const word CMD_ENROLL_2 = 0x24;
const word CMD_ENROLL_3 = 0x25;
const word CMD_IS_FINGER_PRESSED = 0x26;
const word CMD_DELETE_ALL = 0x41;
const word CMD_IDENTIFY = 0x51;
const word CMD_CAPTURE_FINGER = 0x60;

const word ACK = 0x30;
const word NACK = 0x31;

struct package {
  byte header1;
  byte header2;
  word deviceID;
  unsigned long param;
  word cmd;
  word checksum;
};

const int SIZE_OF_PACKAGE = 12;

const int lockPin = 8;
const int resetButtonPin = 7;
const int addButtonPin = 6;
const int speakerPin = 10;

float lowPeep = 220;
float highPeep = 440;
int closed = 2000;
int opened = 1000;
int state = 0;

word calcChecksum(struct package *pkg) {
```

```
  word checksum = 0;
  byte *buffer = (byte*)pkg;
  for(int i=0; i < (sizeof(struct package) - sizeof(word)); i++)
  {
    checksum += buffer[i];
  }
  return checksum;
}

int sendCmd(word cmd, int param) {
  struct package pkg;
  pkg.header1 = STX1;
  pkg.header2 = STX2;
  pkg.deviceID = 1;
  pkg.param = param;
  pkg.cmd = cmd;
  pkg.checksum = calcChecksum(&pkg);
  byte *buffer = (byte*)&pkg;

  int bytesSent = Serial3.write(buffer, sizeof(struct package));

  if(bytesSent != sizeof(struct package)) {
    Serial.println("Error communicating");
    return -1;
  }

  int bytesReceived = 0;
  char recvBuffer[SIZE_OF_PACKAGE];
  struct package *recvPkg = (struct package*) recvBuffer;


  bytesReceived = Serial3.readBytes(recvBuffer, sizeof(struct package));
  if(bytesReceived != SIZE_OF_PACKAGE) {
    Serial.println("Error communicating");
    return -1;
  }

  if( recvPkg->header1 != STX1 || recvPkg->header2 != STX2) {
    Serial.println("Header error!");
    return -1;
  }

  if(recvPkg->checksum != calcChecksum(recvPkg)) {
    Serial.println("Checksum mismatch error!");
    return -1;
  }
  if(recvPkg->cmd == NACK) {
    Serial.println("NACK - Cmd error!");
    Serial.print("Error: ");
    Serial.println(recvPkg->param,HEX);
    return -1;
  }

  return recvPkg->param;
```

```
}

void wave(int pin, float frequency, int duration)
{
  float period=1/frequency*1000*1000; // microseconds (us)
  long int startTime=millis();
  while(millis()-startTime < duration) {
    digitalWrite(pin, HIGH);
    delayMicroseconds(period/2);
    digitalWrite(pin, LOW);
    delayMicroseconds(period/2);
  }
}

void pulseServo(int servoPin, int pulseLenUs)
{
  digitalWrite(servoPin, HIGH);
  delayMicroseconds(pulseLenUs);
  digitalWrite(servoPin, LOW);
  delay(15);
}

void peep(int count, float frequency)
{
  for(int i = 0; i < count; i++) {
    wave(speakerPin, frequency, 400);
    delay(400);
  }
}

void enrollFinger() {
  int id = 0;
  int ret = 0;
  id = sendCmd(CMD_GET_ENROLL_COUNT, 0);
  sendCmd(CMD_ENROLL_START, id);
  peep(1,lowPeep);
  WaitForFinger(false);
  ret = sendCmd(CMD_CAPTURE_FINGER, 1);
  if(ret < 0) {
    peep(5,highPeep);
    return;
  }
  sendCmd(CMD_ENROLL_1, 0);
  peep(1,highPeep);
  WaitForFinger(true);

  WaitForFinger(false);
  ret = sendCmd(CMD_CAPTURE_FINGER, 1);
  if(ret < 0) {
    peep(5,highPeep);
    return;
  }
  sendCmd(CMD_ENROLL_2, 0);
  peep(2,highPeep);
```

```
  WaitForFinger(true);

  WaitForFinger(false);
  ret = sendCmd(CMD_CAPTURE_FINGER, 1);
  if(ret < 0) {
    peep(5,highPeep);
    return;
  }
  sendCmd(CMD_ENROLL_3, 0);
  peep(3,highPeep);
  WaitForFinger(true);
}

void WaitForFinger(bool bePressed) {
  delay(500);
  if(!bePressed) {

    while(sendCmd(CMD_IS_FINGER_PRESSED, 0) > 0) {
      delay(200);
    }
  } else {
    while(sendCmd(CMD_IS_FINGER_PRESSED, 0) == 0) {
      delay(200);
    }
  }
}

void setup() {
  Serial.begin(115200);
  Serial3.begin(9600);
  Serial3.setTimeout(10*1000);
  delay(100);
  sendCmd(CMD_OPEN, 0);
  sendCmd(CMD_LED, 1);
  pinMode(resetButtonPin, INPUT);
  pinMode(addButtonPin, INPUT);
  pinMode(lockPin, OUTPUT);
  pinMode(speakerPin, OUTPUT);
  digitalWrite(resetButtonPin, HIGH);
  digitalWrite(addButtonPin, HIGH);
  for(int i = 0; i < 20; i++) {
    pulseServo(lockPin, closed);
  }
}

void loop() {
  if (digitalRead(resetButtonPin) == LOW) {     // ❶
      if(sendCmd(CMD_DELETE_ALL, 0) >= 0) {
        peep(5,lowPeep);
      } else {
        peep(2,lowPeep); // Already empty
      }
  }
  if (digitalRead(addButtonPin) == LOW) {       // ❷
```

```
      enrollFinger();
    }
    if(sendCmd(CMD_GET_ENROLL_COUNT, 0) == 0)      // ❸
    {
      delay(100);
      return;
    }

    if(sendCmd(CMD_IS_FINGER_PRESSED, 0) == 0) {
        sendCmd(CMD_CAPTURE_FINGER, 1);
        int ret = sendCmd(CMD_IDENTIFY, 0);
        if(ret >= 0 && ret < 200) {       // ❹
          if(state == 0) {
            peep(1,highPeep);
            for(int i = 0; i < 20; i++) {
              pulseServo(lockPin, opened);       // ❺
            }
            state = 1;
            Serial.println("Open");
          } else {
            Serial.println("Close");
            peep(1,lowPeep);
            for(int i = 0; i < 20; i++) {
              pulseServo(lockPin, closed);
            }
            state = 0;
          }
        } else {
          peep(5,lowPeep);
        }
        WaitForFinger(true);
    }
  }
```

❶ The chest reset button removes all stored fingerprints. The chest reset button is not to be confused with the Arduino reset button.

❷ The add button stores fingerprints that are allowed to open the chest.

❸ If there are no fingerprints authorized, then there is no point in scanning and checking fingers.

❹ We found an authorized finger!

❺ Open the lock. The `opened` is just a constant holding the value for a 1000 µs (1 ms) pulse. When sent repeatedly, this pulse turns the servo to its minimum angle.

Who or What Is It?

Now you know who (or what) your device is talking to: to identify objects, you can put an RFID sticker on them. To identify people, you can use fingerprints.

As you close your secrets to the ancient chest, it's time to move on to electromagnetism.

Electricity and Magnetism

10

Can you feel the power radiate through your body? Of course not, even though electromagnetic radiation from power lines and cell phones is all around you. Electricity powers all your gadgets, especially the ones you build from this book.

Hall sensors detect a magnetic field. They come in different varieties: some just detect the presence of a magnet, while others can tell you the strength of the magnetic field in *teslas*.

Voltage and current sensors work like a multimeter, measuring electricity passing through them. They can easily measure power that would otherwise break a microcontroller's analog input pin.

An electronic compass can tell you where magnetic north is. Better ones combine an accelerometer, so that they can point to north even when tilted.

The sun's north and south poles change places about every 11 years. As of this writing, we're waiting for the flip to happen at any week now. However, you don't have to worry about the Earth's poles flipping on you. The last time that happened was 780,000 years ago.

Experiment: Voltage and Current

In this experiment, you'll use the AttoPilot to measure voltage of a battery pack.

Are the batteries running out, and how much power is your robot's motor draining? *AttoPilot Compact DC Voltage and Current Sense* measures voltage and current. It measures big voltage and current, and then reports it with analog output.

AttoPilot can measure a lot of power. The most powerful model is rated for 50 V and 180 A. Power (P) is a function of voltage (U) and current (I):

```
P = UI = 50 V * 180 A = 9000 VA = 9000 W = 9 kW
```

You are likely to also see this expressed with watts (W) as a function of voltage (V) and current (A):

```
W = VA = 50 V * 180 A = 9000 W = 9 kW
```

The 50 V/ 180 A model of the AttoPilot sensor can handle 9 kilowatts. You can't. Stay safe and use only sensible voltages, like batteries and USB power.

Table 10-1. AttoPilot voltage and current sense, conversion factors

Model	Voltage U, output/measured	Current I, output/measured	Comment
13.6 V / 45 A	242.3 mV / V	73.20 mV / A	Used in this experiment
50 V / 90 A	63.69 mV / V	36.60 mV / A	
50 V / 180 A	63.69 mV / V	18.30 mV / A	9 kW

Because such power can't be safely handled in most projects, we used the 13 V / 45 A model for this test.

```
P = UI = 13 V * 45 A = 585 W     # model used here
```

The AttoPilot has two analog outputs. One reports current; the other reports voltage. The maximum output is 3.3 V—considerably less than the maximum 50 V measured. The conversion factors for the 13.6 V / 45 A model used here are calculated from the component's maximum specs. Using 13.6 V / 45 A model as an example, here's the formula for current:

```
3.3 V output / 45 A measured = 73.3 mV output / A measured
```

But in your code, you'll switch things around:

```
45 A measured/ 3.3 V output = 13.6363 A measured / V output
```

So when you want to calculate the measured current, you can multiply the voltage you read from the current sense output by 13.6363:

```
0.05 V output * 13.6363 ≈ .6818 A = 681.8 mA
```

And here's the formula for voltage:

```
3.3 V output / 13.6 V measured = 242.6 mV output / V measured
```

To express this in a way that's useful for calculating measured voltage in your code, look at it this way:

```
13.6 V measured / 3.3 V output = 4.1212 V measured / V output
```

When you want to calculate the measured voltage, you can multiply the voltage you read on the voltage sense output by 4.1212:

```
1.213 V output * 4.1212 ≈ 5 V
```

AttoPilot Code and Connection for Arduino

Figure 10-1 shows the connections for Arduino. Wire it up as shown, and then run the sketch shown in Example 10-1.

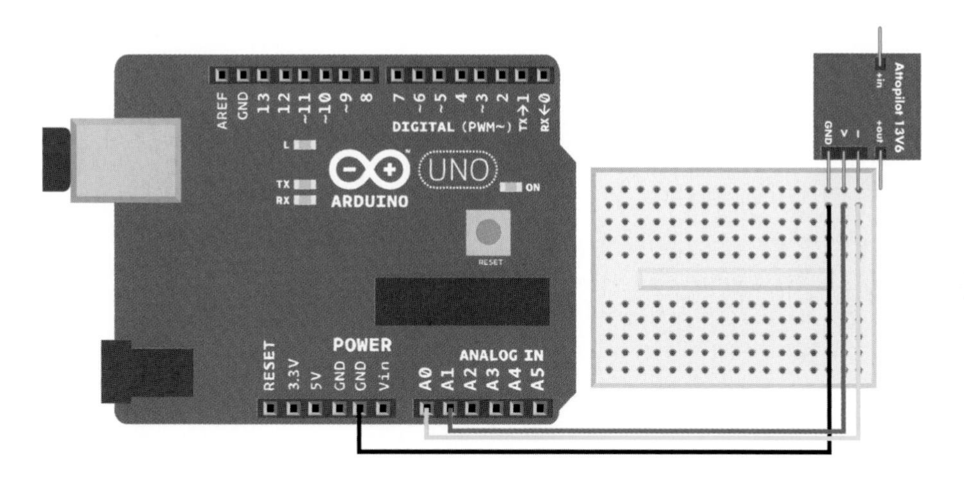

Figure 10-1. *AttoPilot connections for Arduino*

Example 10-1. attopilot_voltage.ino

```
// attopilot_voltage.ino - measure current and voltage with Attopilot 13.6V/45A
// (c) BotBook.com - Karvinen, Karvinen, Valtokari


int currentPin = A0;
int voltagePin = A1;

void setup()
{
  Serial.begin(115200);
  pinMode(currentPin, INPUT);
  pinMode(voltagePin, INPUT);
}

float current()
{
  float raw = analogRead(currentPin);
  Serial.println(raw);
  float percent = raw/1023.0;   // ❶
  float volts = percent*5.0;    // ❷
  float sensedCurrent = volts * 45 / 3.3;        // A/V  // ❸
  return sensedCurrent; // A    // ❹
}
```

```
float voltage()
{
  float raw = analogRead(voltagePin);
  float percent = raw/1023.0;
  float volts = percent*5.0;
  float sensedVolts = volts * 13.6 / 3.3;        // V/V // ❺
  return sensedVolts;   // V
}

void loop()
{
  Serial.print("Current: ");
  Serial.print(current(),4);
  Serial.println(" A");
  Serial.print("Voltage: ");
  Serial.print(voltage());
  Serial.println(" V");
  delay(200); // ms
}
```

❶ The maximum reading of `analogRead()` is 1023, so we calculate the reading as a percentage of that maximum.

❷ Five volts is the maximum voltage of Arduino's `analogRead()`. Five volts corresponds to a raw value of 1023.

❸ The conversion factor is 45 A measured / 3.3 V output ≈ 13.7 A/V. See also Table 10-1.

❹ Return current in amperes. It's always a good idea to include the unit in your comments.

❺ The voltage conversion factor is `V measured max / V output max`.

AttoPilot Code and Connection for Raspberry Pi

Figure 10-2 shows the connections for the Raspberry Pi and AttoPilot. Hook everything up as shown, and then run the program in Example 10-2.

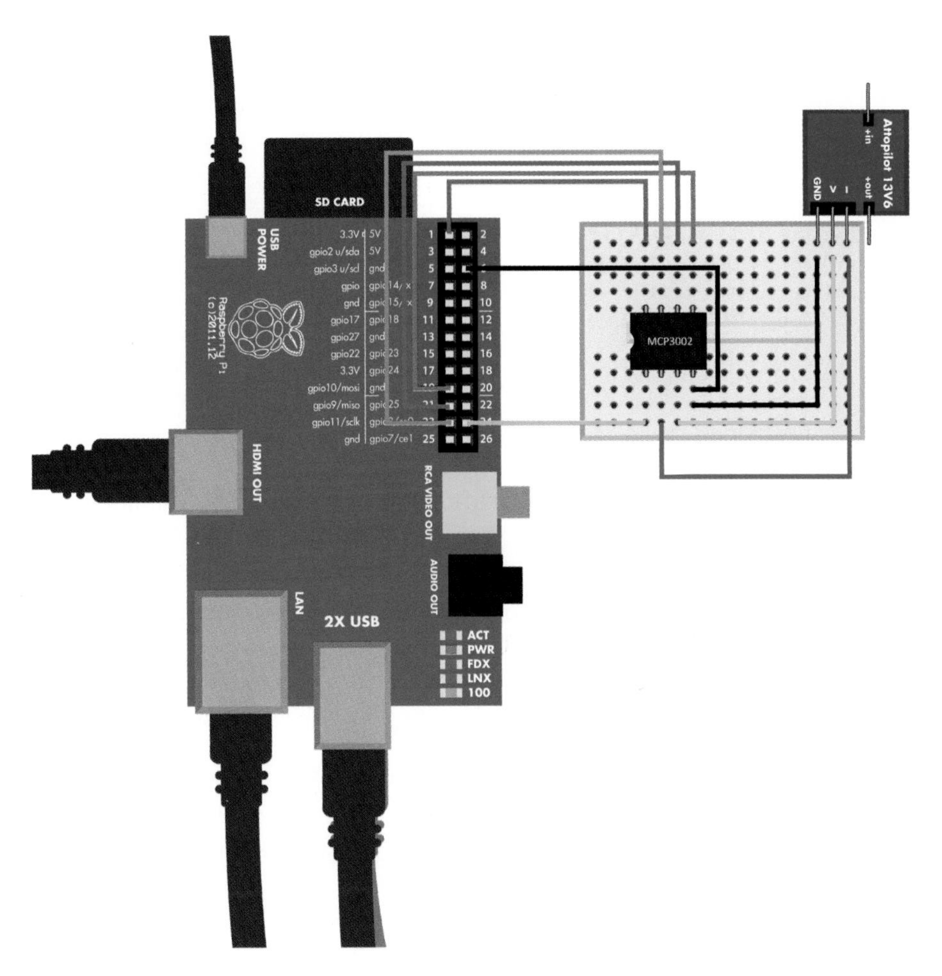

Figure 10-2. *Raspberry Pi connections for the AttoPilot*

Example 10-2. ***attopilot_voltage.py***

```
# attopilot_voltage.py - measure current and voltage with Attopilot 13.6V/45A
# (c) BotBook.com - Karvinen, Karvinen, Valtokari
import time
import botbook_mcp3002 as mcp    # (1)

def readVoltage():
        raw = mcp.readAnalog(0,1)          # (2)
        percent = raw / 1023.0   # (3)
        volts = percent * 3.3    # (4)
        sensedVolts = volts * 13.6 / 3.3          # V/V   # (5)
        return sensedVolts       # V

def readCurrent():
        raw = mcp.readAnalog(0,0)
        percent = raw / 1023.0
```

```
        volts = percent * 3.3
        sensedCurrent = volts * 45.0 / 3.3      # A/V # ❻
        return sensedCurrent    # A

def main():
        while True:
                voltage = readVoltage()
                current = readCurrent()
                print("Current %.2f A" % current)
                print("Voltage %.2f V" % voltage)
                time.sleep(0.5) # s

if __name__ == "__main__":
        main()
```

❶ Import the library for the MCP3002 analog-to-digital converter chip. The library *botbook_mcp3002.py* must be in the same directory as this program (*attopilot_voltage.py*).

❷ Read the second channel. The AttoPilot voltage and current sensing outputs are connected to different channels.

❸ 1023 is the maximum value of `readAnalog()`. It corresponds to 3.3 V.

❹ The maximum GPIO voltage for the Raspberry Pi 3.3 V.

❺ Conversion factors are calculated just like with Arduino. The voltage conversion factor is `V measured max / V output max`.

❻ The conversion factor is 45 A measured / 3.3 V output ≈ 13.7 A/V. See also Table 10-1.

Experiment: Is It Magnetic?

A Hall effect sensor measures a magnetic field. Hall effect sensors are used in bike speedometers, where a magnet on the wheel helps the sensor count revolutions.

A magnetic field causes electrons to divert from their straight path, causing a voltage change in a conductor. This is the Hall effect.

The sensor reports a magnetic field as a voltage. This voltage can be read just like an analog resistance sensor, using `analogRead()` or `botbook_mcp3002.readAnalog()`.

This effect is implemented in sensors from a variety of manufacturers. We used the KY-024 Magnetic Detecting Sensor Module (part number 232563) from *http://dx.com*, shown in Figure 10-3.

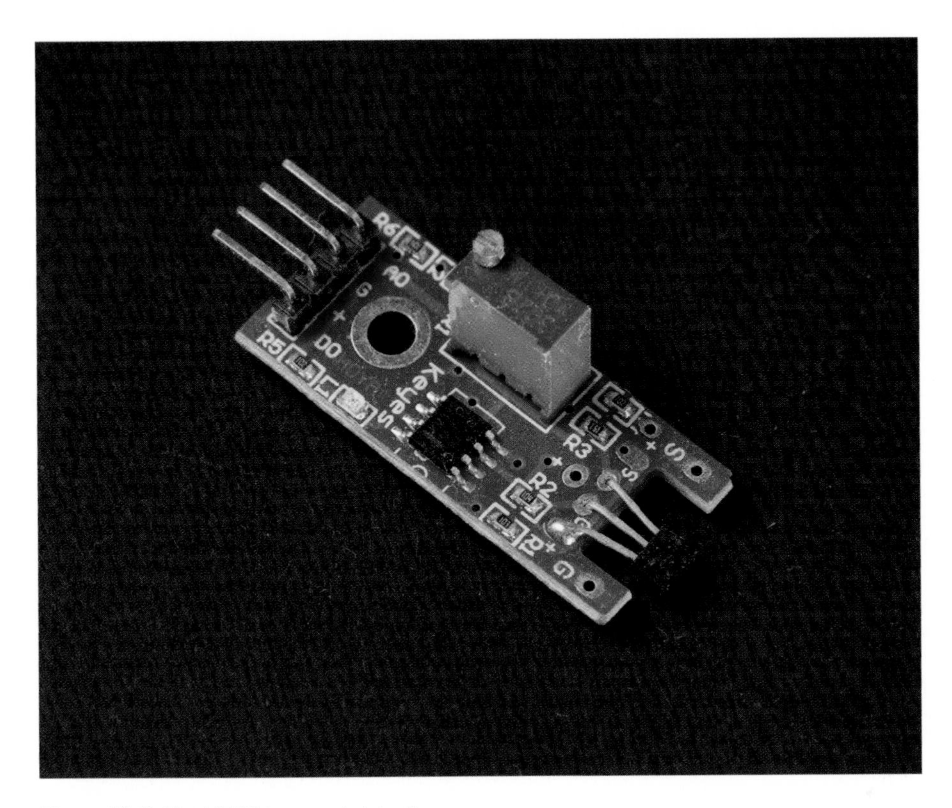

Figure 10-3. *The KY-024 magnet detecting sensor*

Hall Effect Sensor Code and Connection for Arduino

Figure 10-4 shows the connections for Arduino. Wire it up as shown, and then run the sketch shown in Example 10-3.

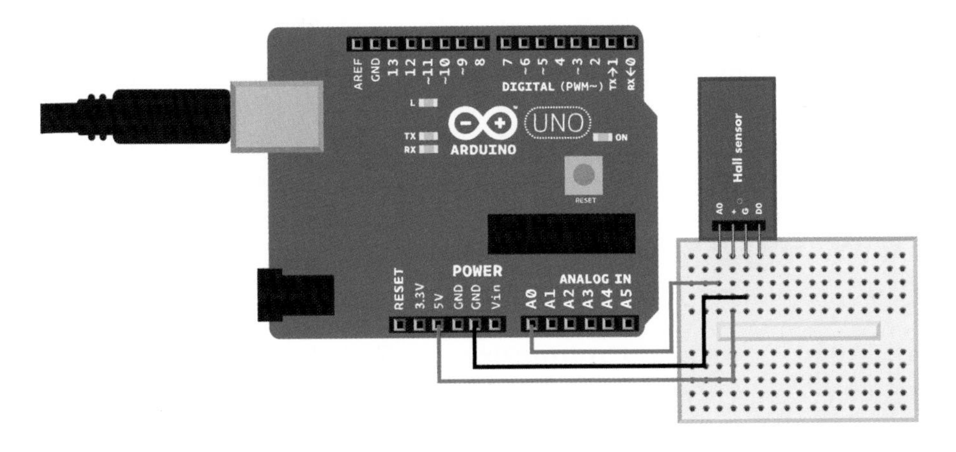

Figure 10-4. *Arduino connections for the Hall effect sensor*

Example 10-3. hall_sensor.ino

```
// hall_sensor.ino - print raw value and magnets pole
// (c) BotBook.com - Karvinen, Karvinen, Valtokari

const int hallPin = A0;
int rawMagneticStrength = -1;   // ❶
int zeroLevel = 527;    // ❷

void setup() {
  Serial.begin(115200);
  pinMode(hallPin, INPUT);
}

void loop() {
  rawMagneticStrength = analogRead(hallPin);    // ❸
  Serial.print("Raw strength: ");
  Serial.println(rawMagneticStrength);
  int zeroedStrength = rawMagneticStrength - zeroLevel;
  // If you know your Hall sensor's conversion from
  // voltage to gauss then you can do it here
  // zeroedStrength * conversion
  Serial.print("Zeroed strength: ");
  Serial.println(zeroedStrength);
  if(zeroedStrength > 0) {
    Serial.println("South pole");
  } else if(zeroedStrength < 0) {
    Serial.println("North pole");
  }
  delay(600); // ms
}
```

❶ Initialize to a value that wouldn't come up as a result from reading the sensor. That way, you know that if you see this value in your running code, there's a problem you need to debug.

❷ This is the raw `analogRead()` value when there is no magnetic field. Our sensor reports 527 when there is no magnetic field. The manufacturer's specification gives 500 (raw), presumably for 5 V logic level. If you see different values in the Arduino Serial Monitor when you have no magnetic field, you may need to adjust the `zeroLevel` variable.

❸ The Hall sensor works just like an analog resistance sensor, even though it's not a resistor.

Hall Effect Sensor Code and Connection for Raspberry Pi

Figure 10-5 shows the connections for Raspberry Pi. Hook them up as shown, and then run the code shown in Example 10-4.

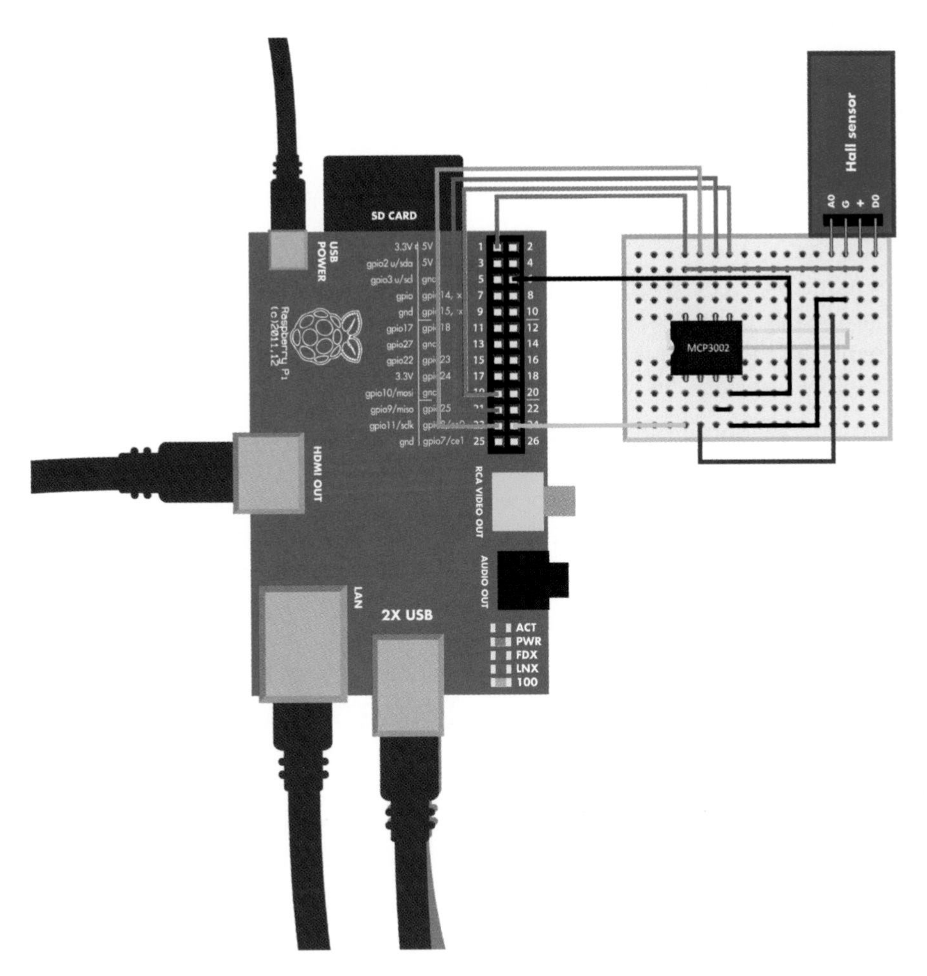

Figure 10-5. *Raspberry Pi/Hall effect sensor connections*

Example 10-4. hall_sensor.py

```
# hall_sensor.py - print raw value and magnets pole
# (c) BotBook.com - Karvinen, Karvinen, Valtokari
import time
import botbook_mcp3002 as mcp    # ❶

zeroLevel = 388 # ❷

def main():
  while True:
        rawMagneticStrength = mcp.readAnalog()
        print("Raw strength:  %i " % rawMagneticStrength)
        zeroedStrength = rawMagneticStrength - zeroLevel
        print("Zeroed strength: %i " % zeroedStrength)
        if(zeroedStrength > 0):
                print("South pole")
```

```
        elif(zeroedStrength < 0):
                print("North pole")
        time.sleep(0.5)

if __name__ == "__main__":
    main()
```

❶ The library (*botbook_mcp3002.py*) must be in the same directory as this program. You must also install the *spidev* library, which is imported by *botbook_mcp3002*. See the comments in the beginning of *botbook_mcp3002/botbook_mcp3002.py* or "Installing SpiDev" on page 56.

❷ `zeroLevel` is the raw output of `readAnalog()` when no magnetic field affects the sensor. Our test gave us 388 for our sensor. The manufacturer's specifications give a raw value of 500 under a 5 V logic level, which would give a raw value of 330 for `zeroLevel` at 3.3 V: `500 / (3.3/5) = 330`. If you're seeing a different raw value in the program output when no magnet is present, you may need to change the value of the `zeroLevel` variable.

Experiment: Magnetic North with LSM303 Compass-Accelerometer

The LSM303 compass-accelerometer (Figure 10-6) gives you the direction of magnetic north. With the accelerometer, it can correct this reading for its orientation.

If you have experience orienteering, you know to hold your compass horizontally when setting the map or taking a bearing. But with this sensor, your device can find north even sideways.

Compass sensors, such as LSM303, are sensitive to external interference. Keep the compass away from power cables and big pieces of metal.

The LSM303 board we used doesn't have any markings for north. To mark it yourself, turn the board so that the text "SA0" is oriented so you can read it left to right. North is to the right edge of the board when the text `SA0` is properly oriented.

In the output values, a heading of 0 is north.

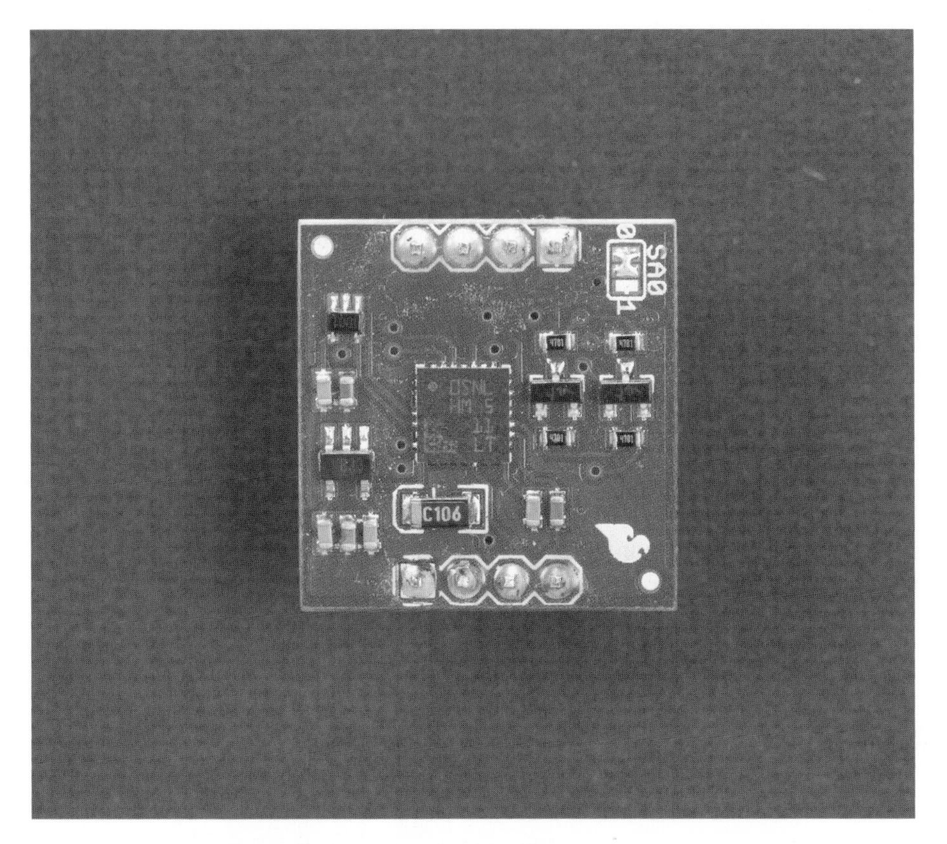

Figure 10-6. *SparkFun's LSM303 compass-accelerometer*

Don't have a compass handy? In the Northern Hemisphere, satellite dishes point south. Most satellite dishes, like the ones for TV, are statically mounted. That means that they point to geostationary satellites that orbit the earth at exactly the same speed the earth is turning. So the satellites are orbiting east above the equator, and satellite dishes point toward the equator.

Calibrate Your Module

The compass must be calibrated to get correct values. You can try it out without calibrating, and you should see values. However, you'll get correct values only after calibration.

Here are the calibration steps:

- Build the circuit for your chosen platform (Arduino or Raspberry Pi).
- Run the program to see that you get some values.
- Change `runningMode = 0` in the code to put the device into calibration mode. In calibration mode, the program will show only minimum and maximum values for each axis.

- Set maximum values initially to zero. For Raspberry Pi, change `magMax[]` and `magMin[]`. For Arduino, change each of the six `magMax_x`, `magMax_y` … `magMin_z`.
- Wave and turn the device around, maybe drawing a figure-eight shape in the air. Continue until the values stop changing. If you can't get any higher or lower value, you have found min and max.
- Hard-code the values into your code. These are the same maximum values you zeroed earlier, `magMax[]` and `magMin[]` or `magMax_x`...

Once the calibration is done, change `runningMode = 0` to use the sensor normally and see headings.

Try running the code and calibrating the device. Once you have played with the compass, you can read about implementation details (if you want) in "LSM330 Protocol" on page 299.

LSM303 Code and Connection for Arduino

Figure 10-7 shows the circuit for Arduino. Hook everything up as shown in the figure, and then run the sketch shown in Example 10-5.

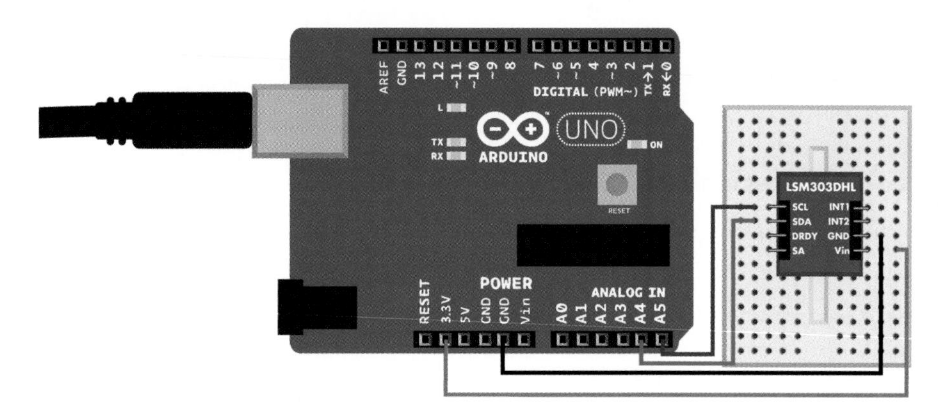

Figure 10-7. *LSM303 compass-accelerometer circuit for Arduino*

Remember to calibrate the module (see "Calibrate Your Module" on page 289).

Example 10-5. *lsm303.ino*

```
// lsm303.ino - normal use and calibration of LSM303DLH compass-accelerometer
// (c) BotBook.com - Karvinen, Karvinen, Valtokari


#include <Wire.h>

const char accelerometer_address = 0x30 >> 1;    // ❶
```

```
const char magnetometer_address = 0x3C >> 1;

const int runningMode = 0;        // ❷

float magMax_x = 0.1; float magMax_y = 0.1; float magMax_z = 0.1;        // ❸
float magMin_x = -0.1; float magMin_y = -0.1; float magMin_z = -0.1;

float acc_x = 0; float acc_y = 0; float acc_z = 0; // ❹
float mag_x = 0; float mag_y = 0; float mag_z = 0;

int heading = 0;

void setup() {
  Serial.begin(115200);
  Wire.begin(); // ❺
  Serial.println("Initialize compass");
  initializelsm();       // ❻
  delay(100);
  Serial.println("Start reading heading");
}

void loop() {
  updateHeading();       // ❼
  if(runningMode == 0) {        // ❽
    magMax_x = max(magMax_x, mag_x);
    magMax_y = max(magMax_y, mag_y);
    magMax_z = max(magMax_z, mag_z);
    magMin_x = min(magMin_x, mag_x);
    magMin_y = min(magMin_y, mag_y);
    magMin_z = min(magMin_z, mag_z);
    Serial.print("Max x y z: ");
    Serial.print(magMax_x); Serial.print(" ");
    Serial.print(magMax_y); Serial.print(" ");
    Serial.print(magMax_z); Serial.print(" ");
    Serial.print("Min x y z: ");
    Serial.print(magMin_x); Serial.print(" ");
    Serial.print(magMin_y); Serial.print(" ");
    Serial.print(magMin_z); Serial.println(" ");
  } else {
    calculateHeading();
    Serial.println(heading);    // ❾
  }
  delay(100); // ms
}


void initializelsm() {
    write_i2c(accelerometer_address, 0x20, 0x27);        // ❿
    write_i2c(magnetometer_address, 0x02, 0x00);        // ⓫
}

void updateHeading() {
  updateAccelerometer();
  updateMagnetometer();
```

```
}

void updateAccelerometer() {    // ⓬
  Wire.beginTransmission(accelerometer_address);
  Wire.write(0x28 | 0x80);      // ⓭
  Wire.endTransmission(false);
  Wire.requestFrom(accelerometer_address, 6, true);     // ⓮
  int i = 0;
  while(Wire.available() < 6) {
    i++;
    if(i > 1000) {
      Serial.println("Error reading from accelerometer i2c");
      return;
    }
  }
  uint8_t  axel_x_l = Wire.read();      // ⓯
  uint8_t  axel_x_h = Wire.read();
  uint8_t  axel_y_l = Wire.read();
  uint8_t  axel_y_h = Wire.read();
  uint8_t  axel_z_l = Wire.read();
  uint8_t  axel_z_h = Wire.read();

  acc_x = (axel_x_l | axel_x_h << 8) >> 4; // ⓰
  acc_y = (axel_y_l | axel_y_h << 8) >> 4;
  acc_z = (axel_z_l | axel_z_h << 8) >> 4;

}

void updateMagnetometer() { // ⓱
  Wire.beginTransmission(magnetometer_address);
  Wire.write(0x03);
  Wire.endTransmission(false);
  Wire.requestFrom(magnetometer_address, 6, true);
  int i = 0;
  while(Wire.available() < 6) {
    i++;
    if(i > 1000) {
      Serial.println("Error reading from magnetometer i2c");
      return;
    }
  }
  uint8_t  axel_x_h = Wire.read();
  uint8_t  axel_x_l = Wire.read();
  uint8_t  axel_y_h = Wire.read();
  uint8_t  axel_y_l = Wire.read();
  uint8_t  axel_z_h = Wire.read();
  uint8_t  axel_z_l = Wire.read();

  mag_x = (int16_t)(axel_x_l | axel_x_h << 8);
  mag_y = (int16_t)(axel_y_l | axel_y_h << 8);
  mag_z = (int16_t)(axel_z_l | axel_z_h << 8);
}
//Heading to north
void calculateHeading() {
```

```
  // Up unit vector
  float dot = acc_x*acc_x + acc_y*acc_y + acc_z*acc_z;  // ⓲
  float magnitude = sqrt(dot);  // ⓳
  float nacc_x = acc_x / magnitude;     // ⓴
  float nacc_y = acc_y / magnitude;
  float nacc_z = acc_z / magnitude;

  // Apply calibration
  mag_x = (mag_x - magMin_x) / (magMax_x - magMin_x) * 2 - 1.0; // ㉑
  mag_y = (mag_y - magMin_y) / (magMax_y - magMin_y) * 2 - 1.0;
  mag_z = (mag_z - magMin_z) / (magMax_z - magMin_z) * 2 - 1.0;

  // East
  float ex = mag_y*nacc_z - mag_z*nacc_y;        // ㉒
  float ey = mag_z*nacc_x - mag_x*nacc_z;
  float ez = mag_x*nacc_y - mag_y*nacc_x;
  dot = ex*ex + ey*ey + ez*ez; // ㉓
  magnitude = sqrt(dot);
  ex /= magnitude;
  ey /= magnitude;
  ez /= magnitude;      // ㉔


  // Project
  float ny = nacc_z*ex - nacc_x*ez;     // ㉕

  float dotE = -1 * ey; // ㉖
  float dotN = -1 * ny;

  // Angle
  heading = atan2(dotE, dotN) * 180 / M_PI;      // ㉗

  heading = round(heading);
  if (heading < 0) heading +=360; // ㉘
}

void write_i2c(char address, unsigned char reg, const uint8_t data)
{
  Wire.beginTransmission(address);
  Wire.write(reg);
  Wire.write(data);
  Wire.endTransmission();
}
```

❶ The addresses of the accelerometer and magnetometer need to be bit shifted (see "Bitwise Operations" on page 221).

❷ The normal operation of this sketch (printing compass headings) is `runningMode=1`. To calibrate the compass, see "Calibrate Your Module" on page 289.

❸ Initial calibration data helps you to get some readings even before you find the exact values for your device (with `runningMode=0`).

❹ Variables for the current heading are initialized to zero.

❺ *Wire.h* is the standard I2C library for Arduino.

❻ Call the initialization function.

❼ This function updates global variables, so there is no return value needed.

❽ Calibration mode prints only max and min.

❾ Normal operation prints the acceleration- (tilt-) corrected heading.

❿ Enable the accelerometer with default values. The values are taken from the data sheet for this module.

⓫ Enable the magnetometer and set it to continuous conversion.

⓬ Read the sensor according to the protocol described in the data sheet, and then update the global variables.

⓭ 0x80 is 0b10000000: the most significant bit (MSB) is 1, and the other seven bits are zero. A bitwise XOR of 0x80 and any other number results in the MSB being changed to 1. 0x28 is the `OUT_X_L_A` register, from the component's data sheet.

⓮ Read in six bytes.

⓯ Each raw value is split over two bytes, the low part (`axel_x_l`) and the high, more significant part (`axel_x_h`).

⓰ Each raw value is 8 + 4 = 12 bits. The high byte contains the most significant 8 bits. The low part contains the last 4 bits. The last 4 bits of the low byte (axel_x_l) are ignored. See also "Bitwise Operations" on page 221.

⓱ The magnetometer (compass) raw values are read similarly to `updateAccelerometer()`.

⓲ Take a dot product of the vector…

⓳ …to calculate its magnitude (length).

⓴ Divide each dimension by length to get a vector whose length is one. So you have an up-pointing vector (`nacc_x`, `nacc_y`, `nacc_z`).

㉑ Apply calibration data.

㉒ East is 90 degrees from north. It's also 90 degrees from up. A *cross product* vector calculation gives you the vector that's perpendicular to these two other vectors.

㉓ East is normalized, just as you normalized up.

㉔ After normalization, the vector (`ex`, `ey`, `ez`) has a length of one and points east. The meaning of the /= symbol is similar to +=. Saying `foo /= 2` is the same as `foo = foo/2`.

㉕ Use a cross product to find the north vector in the gravity horizontal plane.

㉖ Project the north vector N from the NE plane to the XY plane.

㉗ Find the angle between the y-axis and projected N vector. Convert radians to degrees (`deg=rad/(2*pi)*360`).

㉘ Wrap around, so that the compass heading is between 0 and 360 degrees.

LSM303 Code and Connection for Raspberry Pi

Figure 10-8 shows the circuit for Raspberry Pi. Hook everything up as shown, and then run the code listed in Example 10-6.

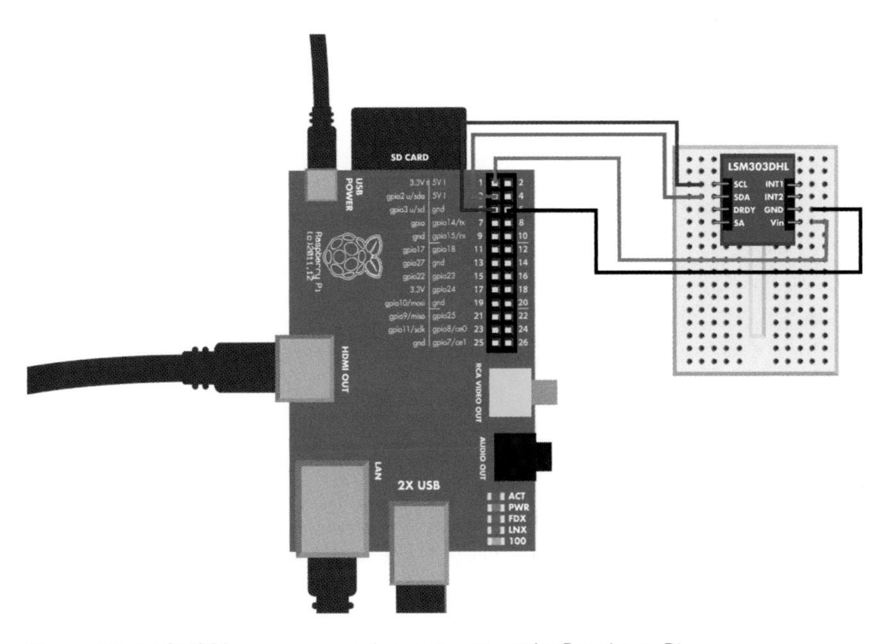

Figure 10-8. *LSM303 compass-accelerometer circuit for Raspberry Pi*

Remember to calibrate the module (see "Calibrate Your Module" on page 289).

Example 10-6. *lsm303.py*

```
# lsm303.py - normal use and calibration of LSM303DLH compass-accelerometer
# (c) BotBook.com - Karvinen, Karvinen, Valtokari
import time
import smbus # sudo apt-get -y install python-smbus
import struct
import math

accelerometer_address = 0x30 >> 1
magnetometer_address = 0x3C >> 1

calibrationMode = True  # ❶
```

```
magMax = [ 0.1, 0.1, 0.1 ]       # (2)
magMin = [ 0.1, 0.1, 0.1 ]

acc = [ 0.0, 0.0, 0.0 ] # (3)
mag = [ 0.0, 0.0, 0.0 ]

heading = 0

def initlsm():
  global bus
  bus = smbus.SMBus(1)
  bus.write_byte_data(accelerometer_address, 0x20, 0x27)        # (4)
  bus.write_byte_data(magnetometer_address, 0x02, 0x00)  # (5)

def updateAccelerometer():
  global acc    # (6)
  bus.write_byte(accelerometer_address, 0x28 | 0x80)     # (7)
  rawData = ""
  for i in range(6):
    rawData += chr(bus.read_byte_data(accelerometer_address, 0x28+i)) # (8)

  data = struct.unpack('<hhh',rawData) # (9)
  acc[0] = data[0] >> 4 # (10)
  acc[1] = data[1] >> 4
  acc[2] = data[2] >> 4

def updateMagnetometer():         # (11)
  global mag
  bus.write_byte(magnetometer_address, 0x03)
  rawData = ""
  for i in range(6):
    rawData += chr(bus.read_byte_data(magnetometer_address, 0x03+i))

  data = struct.unpack('>hhh',rawData)
  mag[0] = data[0]
  mag[1] = data[1]
  mag[2] = data[2]


def calculateHeading():
  global heading, acc, mag
  #normalize
  normalize(acc)        # (12)

  #use calibration data
  mag[0] = (mag[0] - magMin[0]) / (magMax[0] - magMin[0]) * 2.0 - 1.0    # (13)
  mag[1] = (mag[1] - magMin[1]) / (magMax[1] - magMin[1]) * 2.0 - 1.0
  mag[2] = (mag[2] - magMin[2]) / (magMax[2] - magMin[2]) * 2.0 - 1.0

  e = cross(mag, acc)   # (14)
  normalize(e)  # (15)

  n = cross(acc,e)      # (16)
```

```
  dotE = dot(e,[0.0, -1.0, 0.0])          # ⓱
  dotN = dot(n,[0.0, -1.0, 0.0])

  heading = round(math.atan2(dotE, dotN) * 180.0 / math.pi)       # ⓲
  if heading < 0:          # ⓳
    heading += 360         # ⓴

def normalize(v):          # ㉑
  magnitude = math.sqrt(dot(v,v))
  v[0] /= magnitude
  v[1] /= magnitude
  v[2] /= magnitude

def dot(v1, v2):           # ㉒
  return v1[0]*v2[0] + v1[1]*v2[1] + v1[2]*v2[2]

def cross(v1, v2):         # ㉓
  vr = [0.0, 0.0, 0.0]
  vr[0] = v1[1] * v2[2] - v1[2] * v2[1]
  vr[1] = v1[2] * v2[0] - v1[0] * v2[2]
  vr[2] = v1[0] * v2[1] - v1[1] * v2[0]
  return vr

def main():
  initlsm()
  while True:
    updateAccelerometer()        # ㉔
    updateMagnetometer()

    if calibrationMode: # ㉕
      magMax[0] = max(magMax[0], mag[0])
      magMax[1] = max(magMax[1], mag[1])
      magMax[2] = max(magMax[2], mag[2])
      magMin[0] = min(magMin[0], mag[0])
      magMin[1] = min(magMin[1], mag[1])
      magMin[2] = min(magMin[2], mag[2])
      print("magMax = [ %.1f, %.1f, %.1f ]" % (magMax[0], magMax[1], magMax[2]))
      print("magMin = [ %.1f, %.1f, %.1f ]" % (magMin[0], magMin[1], magMin[2]))
    else:
      calculateHeading()         # ㉖
      print(heading)
    time.sleep(0.5)

if __name__ == "__main__":
  main()
```

❶ When you set `calibrationMode` to True, it runs through the calibration process instead of its normal operation. See "Calibrate Your Module" on page 289 for more information.

❷ We provide you with sample calibration values, so you can try your compass without calibration (but you should still calibrate it).

❸ The most recently read and the calculated sensor values are stored in the global variables `acc`, `mag`, and `heading`.

❹ Enable the accelerometer using the control codes we learned from the data sheet.

❺ Enable the magnetometer.

❻ To update a global variable, it must be explicitly declared as global in the beginning of a function.

❼ Send a message where first bit is 1 (0x80 == 0b1000000) and the ending is `OUT_X_L_A` register address (0x28 == 0b 10 1000). An exclusive OR (XOR) operation combines these to 0b10101000. See "Bitwise Operations" on page 221 for more information on bitwise operations.

❽ Read six bytes, starting from the accelerometer address 0x28.

❾ Unpack integer values from the bytes you read from the sensor (`rawData`). Unpack parameters are the following: little endian (<), short (2 byte, 16 bit) integer (h). `Struct.unpack()` is needed so that you can define the exact length of the integers you read in.

❿ `data[0]` is 16 bits, where the value is the first 12 bits. The last 4 bits should be ignored. Bit shifting to the right achieves this. See also "Bitwise Operations" on page 221.

⓫ `updateMagnetometer()` reads values similarly to `updateAccelerometer()`.

⓬ Normalize the acceleration vector; that is, convert it to a unit vector. It points up, but after normalization its length is one.

⓭ Apply calibration data.

⓮ East `e` is 90 degrees from north (`mag`) and 90 degrees from up (`acc`).

⓯ After normalization, `e` is a unit vector (length one), pointing East.

⓰ The new `n` vector is now perpendicular to both gravity up and east. At this point, NE plane is aligned (exactly at level) with gravity horizontally. It's not yet aligned with the device's horizontal XY plane.

⓱ Use a *dot product* to project the north vector to the device's horizontal XY plane. East vector is needed because the whole NE plane controls the projection.

⓲ The angle between Y and north is the *heading*. It's converted from radian (where a whole circle is 2*pi) to degrees (where a whole circle is 360 degrees) by simple division.

⓳ If the heading is below zero…

⓴ …wrap around, so that compass headings are always between 0 and 360 degrees.

㉑ Vector normalization creates a unit vector. The unit vector has length (magnitude) one, and points to the same direction as the original vector.

㉒ The vector dot product formula is from a math textbook. This program uses dot product to calculate the magnitude (`length`) of a vector. Magnitude `mag` of vector `v` is `sqrt(dot(v))`.

23. Vector cross product formula is also from a math textbook. In this program, you use it to find a vector perpendicular to two other vectors. That is, for `c=cross(a,b)`, `c` is perpendicular (90 degrees) from both `a` and `b`.
24. `updateAccelerometer()` and `updateMagnetometer()` work by modifying global values, so they don't return a value.
25. Setting `calibrationMode` to `0` causes the program to run through its calibration process.
26. This is the normal operating mode, printing tilt corrected compass headings in degrees.

LSM330 Protocol

Before you get tempted to read this section, build the project first and run the code! You can use the code without bothering with how it works. An elaborate way to say this is "without spending time on the implementation details." But when you're curious, come on back to this section.

To report measured values, LSM303 uses the industry standard I2C protocol. As I2C is quite strictly defined, it's usually easier than other similar protocols like SPI.

LSM303 is an *I2C slave*, so the microcontroller (Arduino or Raspberry Pi) is the *master*. The master initiates communication by asking for values. I2C communication consists of simply sending and reading values that are specified in the component's data sheet. You can find the data sheet for this component by searching the Web for "LSM303DLH data sheet."

If you want to understand the protocol, I2C communication was explained in more detail in Figure 8-3. Just like many I2C codes, the code for the compass uses hex codes (0xA == 10) and bit shifting (`0b01 << 1 == 0b10`). These concepts are explained in the "Hexadecimal, Binary, and Other Numbering Systems" on page 219 and "Bitwise Operations" on page 221.

Compass Heading Calculation

The sensor already knows where north is. It gives this value as a three-dimensional vector `n = [nx, ny, nz]`. The compass sensor is really three-dimensional, and this north-pointing vector is already correct. So why do you need to do this math?

Vector mathematics ahead. The sensor will work even if you don't learn vectors, but following the explanation requires some vector knowledge.

Humans usually want a compass heading, a number between 0 deg and 360 deg. A compass heading is two dimensional. The compass heading tells how much we have turned right from north. The turn is measured by degrees, growing clockwise, as you turn right.

To convert the three-dimensional north vector to degrees, you use the up vector from the accelerometer.

All vectors are relative to the device, so that Y points in the device-forward direction. Thus, the vector [0, 1, 0] would point directly device forward.

Table 10-2. Variables and definitions for LM330

Arduino	Raspberry Pi	Values	Explanation
heading	heading	int, 0 deg .. 360 deg	The angle between north and forward (y). How much have we turned right?
mag_x, mag_y...	mag[]	signed integer, list of signed integer	Three-dimensional vector pointing north
acc_x, acc_y...	acc[]	signed integer, list of signed integer	Three-dimensional vector pointing up (gravity up)
[0, 0, 1]			A vector pointing device up (Z up)
[0, 1, 0]			A vector pointing forward (Y forward)
[1, 0, 0]			A vector pointing to the right (X right)

Both Arduino and Raspberry Pi codes use the same kind of logic. For clarity, we'll use Python here, but the Arduino code works in the same way and uses similar variable names.

First, read up and north from the sensor. Get these vectors:

- `acc` (up gravitywise, 3d)
- `mag` (north, 3d)

The angle between these vectors can be anything.

Calibration is then applied to `mag`, the north-pointing vector.

Next you need to calculate the east vector. This vector is perpendicular to both of the vectors `mag` and `acc`. So there is a 90-degree angle between `acc` (gravity up) and east, and there is a 90-degree angle between mag and east. Thus, you can calculate the east vector with a vector cross product.

```
e = cross(mag, acc)
e = normalize(e)
```

After normalization, e is a unit vector (length 1) pointing east. This east is perpendicular to gravity up.

Now the north (`n`) vector and east (`e`) vector form an NE plane. This NE plane is likely still in an angle to the device-horizontal XY plane.

To get the `n` and `e` planes aligned to gravity horizontal,

```
n = cross(acc, e)
```

Now you have north `n` and east `e` in the gravity-horizontal plane.

The compass heading must be calculated for the user. The heading tells how many degrees to the right the device has turned from the north. From the device, a vector is projected to the NE plane. Then you can calculate the angle between the projected vector and north:

```
dotE = dot(e,[0.0, -1.0, 0.0])
dotN = dot(n,[0.0, -1.0, 0.0])
headingRad = math.atan2(dotE, dotN)
headingDeg = headingRad / (2*math.pi) * 360
```

Finally, the user gets compass heading as degrees.

Experiment: Hall Switch

A Hall switch (Figure 10-9) detects if a magnet is nearby. They are often used in measuring how fast a wheel spins. Before GPS, many bike speedometers used Hall switches.

As you bring a magnet near the Hall switch, the code in this experiment prints "switch triggered."

There are many Hall switches available from different manufactures. We used an affordable and robust Hall Magnetic Sensor (part number 141363) from *http://dx.com*. That sensor has a nice built-in LED that lights up when a magnet is near. It has weird wire colors: black is data (yes, black goes to D2 rather than the usual GND), blue is ground, and brown is +5 V.

Figure 10-9. *Hall switch*

Hall Switch Code and Connection for Arduino

Figure 10-10 shows the wiring diagram for Arduino. Wire it up as shown, and then run the code shown in Example 10-7.

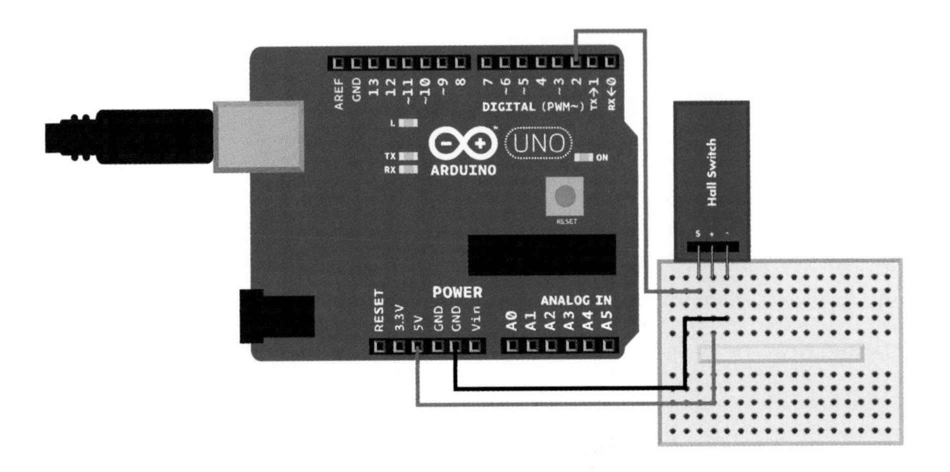

Figure 10-10. *Arduino circuit for Hall switch*

Example 10-7. hall_switch.ino

```
// hall_switch.ino - write to serial if magnet triggers the switch
// (c) BotBook.com - Karvinen, Karvinen, Valtokari

int switchPin=2;
void setup() {
  Serial.begin(115200);
  pinMode(switchPin, INPUT);
  digitalWrite(switchPin, HIGH);
}

void loop() {
  int switchState=digitalRead(switchPin);        // ❶
  if (switchState == LOW) {
    Serial.println("YES, magnet is near");
  } else {
     Serial.println("no");
  }
  delay(50);
}
```

❶ It's a simple digital switch, like a button.

Hall Switch Code and Connection for Raspberry Pi

Figure 10-11 shows the circuit for Raspberry Pi. Hook everything up as shown, and then run the code shown in Example 10-8.

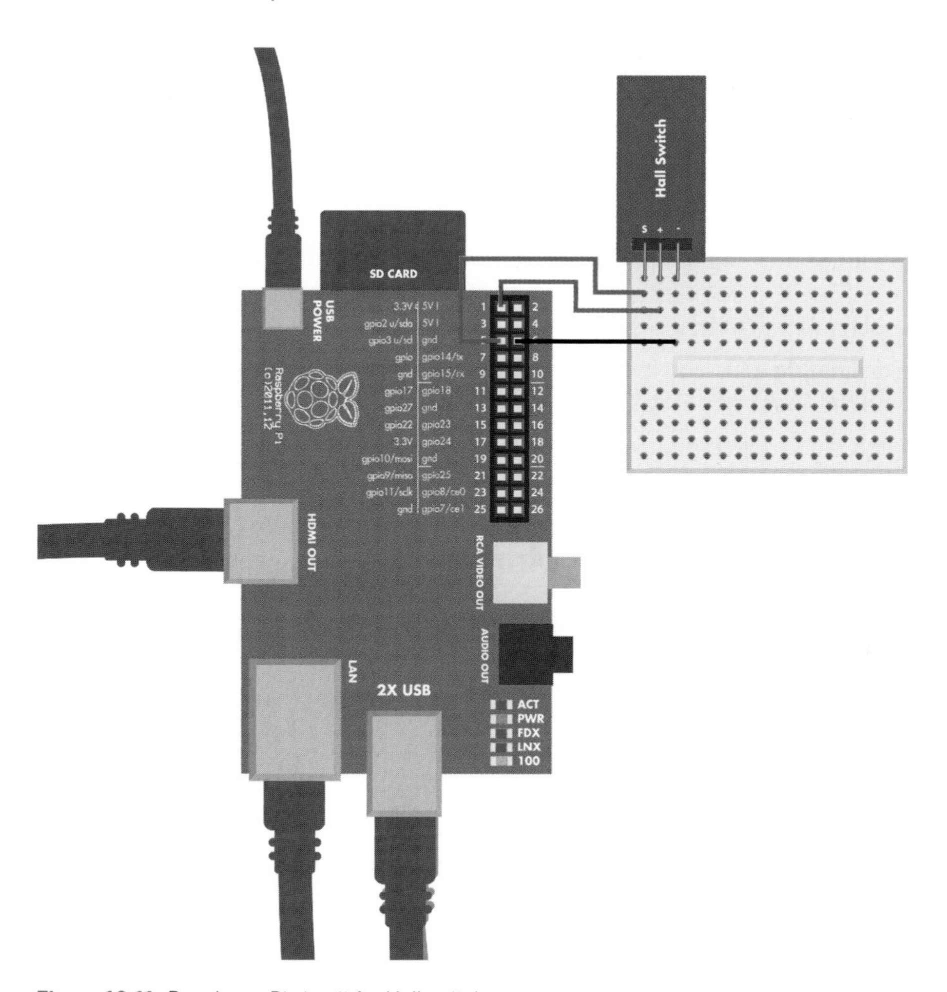

Figure 10-11. *Raspberry Pi circuit for Hall switch*

Example 10-8. *hall_switch.py*

```
# hall_switch.py - write to screen if magnet triggers the switch
# (c) BotBook.com - Karvinen, Karvinen, Valtokari
import time
import botbook_gpio as gpio     # ❶

def main():
        switchPin = 3   # has internal pull-up  # ❷
        gpio.mode(switchPin, "in")
```

```
        while (True):
                switchState = gpio.read(switchPin)        # ❸
                if(switchState == gpio.LOW):
                        print "switch triggered"

                time.sleep(0.3)

if __name__ == "__main__":
        main()
```

❶ Make sure there's a copy of the *botbook_gpio.py* library in the same directory as this program. You can download this library along with all the example code from *http://botbook.com*. See "GPIO Without Root" on page 19 for information on configuring your Raspberry Pi for GPIO access.

❷ To avoid a floating pin, you need a pull-up resistor. Fortunately, pull-ups are included on the Raspberry Pi's GPIO pins 2 and 3.

❸ A Hall switch is a simple digital switch sensor, so it works like a button.

Test Project: Solar Cell Web Monitor

Turn your Raspberry Pi into a web server and monitor the voltage of your solar cells remotely (Figure 10-12).

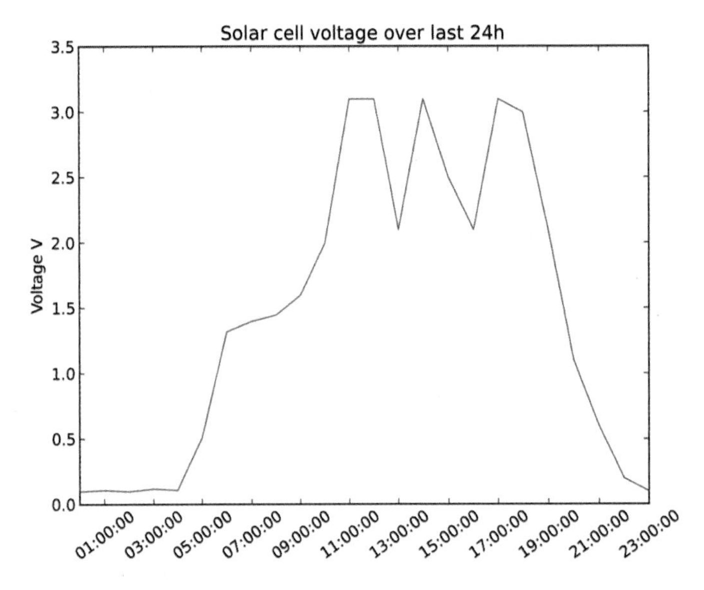

Figure 10-12. *Solar cell power graph*

What You'll Learn

In the *Solar Cell Web Monitor* project, you'll learn how to:

- Measure voltage of your solar cells, and then report it on your own web server.
- Turn the Raspberry Pi into a web server—using the most popular web server in the world!
- Create timed tasks using the *cron* scheduler that keep running even if Raspberry is rebooted.
- Draw graphs with Python matplotlib.

Big, public web servers use many of the same techniques you learn here. We teach Apache and cron to Linux students who work on servers, so these techniques aren't specific to embedded systems or robots.

Connecting Solar Cells

Do you still remember IKEA's Solvinden lamp, which we used to build Chameleon Dome? In this project we're going to use the leftover solar panels from it. If you didn't use Solvinden in the first place, just adjust these instructions to suit the solar cells you are using. First, desolder the red wire marked in Figure 10-13.

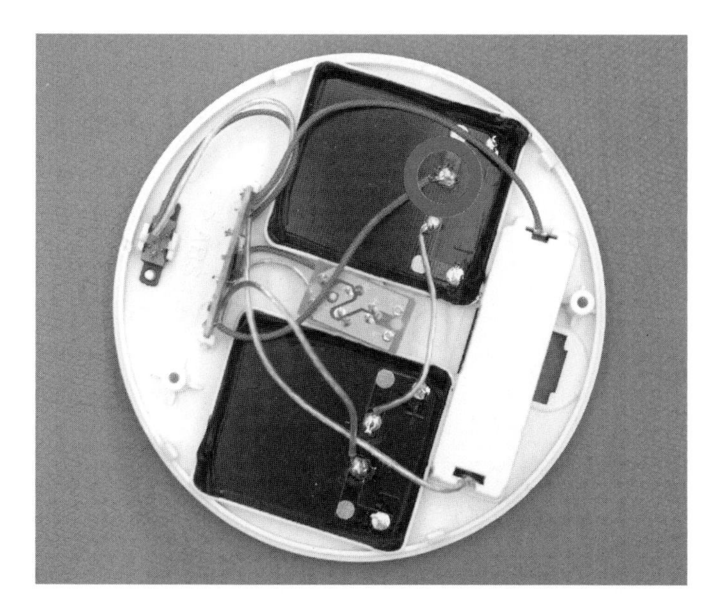

Figure 10-13. *Desolder red wire*

You don't need IKEA's Solvinden lamp to build this. Just connect a solar cell according to the circuit diagram Figure 10-17. Because you're using a current sensor with a maximum measured voltage of 13.6 V, make sure the solar cell or panel doesn't put out more than 13.6 V. If you're using a different solar cell, it probably has power leads, so you won't need to follow the Solvinden disassembly steps.

Cut the jumper wires as shown in Figure 10-14 and solder them to the current sensor and to the solar cells as shown in Figure 10-15. The final product is shown in Figure 10-16.

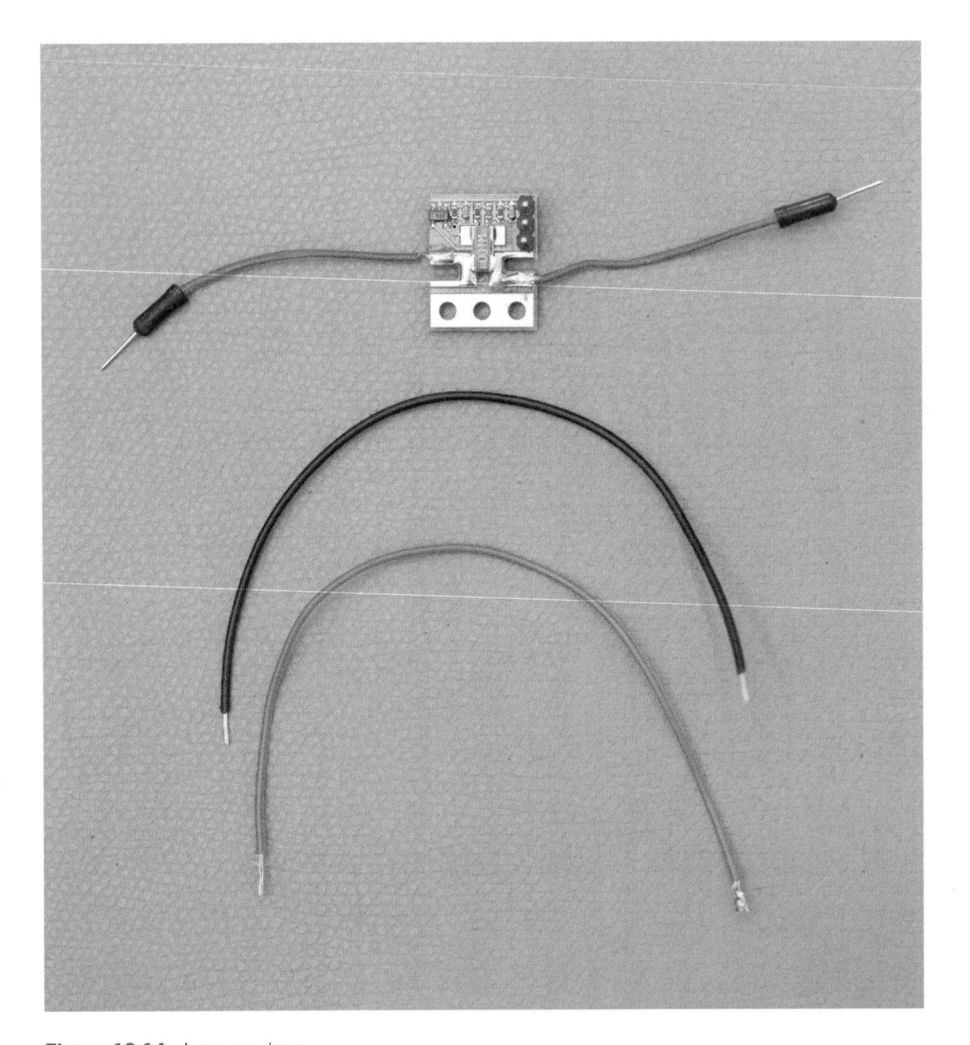

Figure 10-14. *Jumper wires*

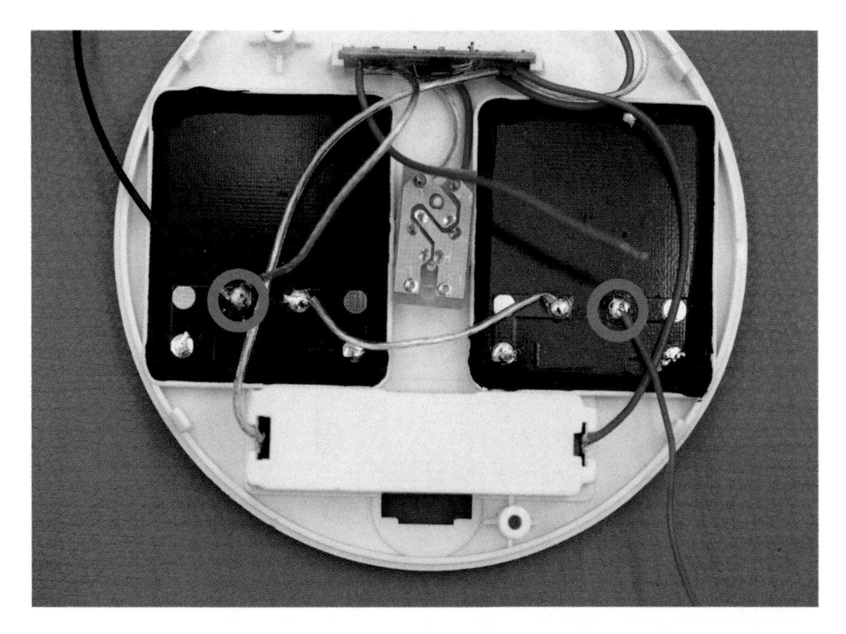

Figure 10-15. *Jumper wires soldered to solar cells*

Figure 10-16. *Everything connected*

Use the AttoPilot Compact DC Voltage and Current Sense code you tried earlier to test that your connection works. Use a flashlight to see how much current your solar panels collect. We got from 1-3 V from ours.

Hook up the Raspberry Pi to the current sensor as shown in Figure 10-17.

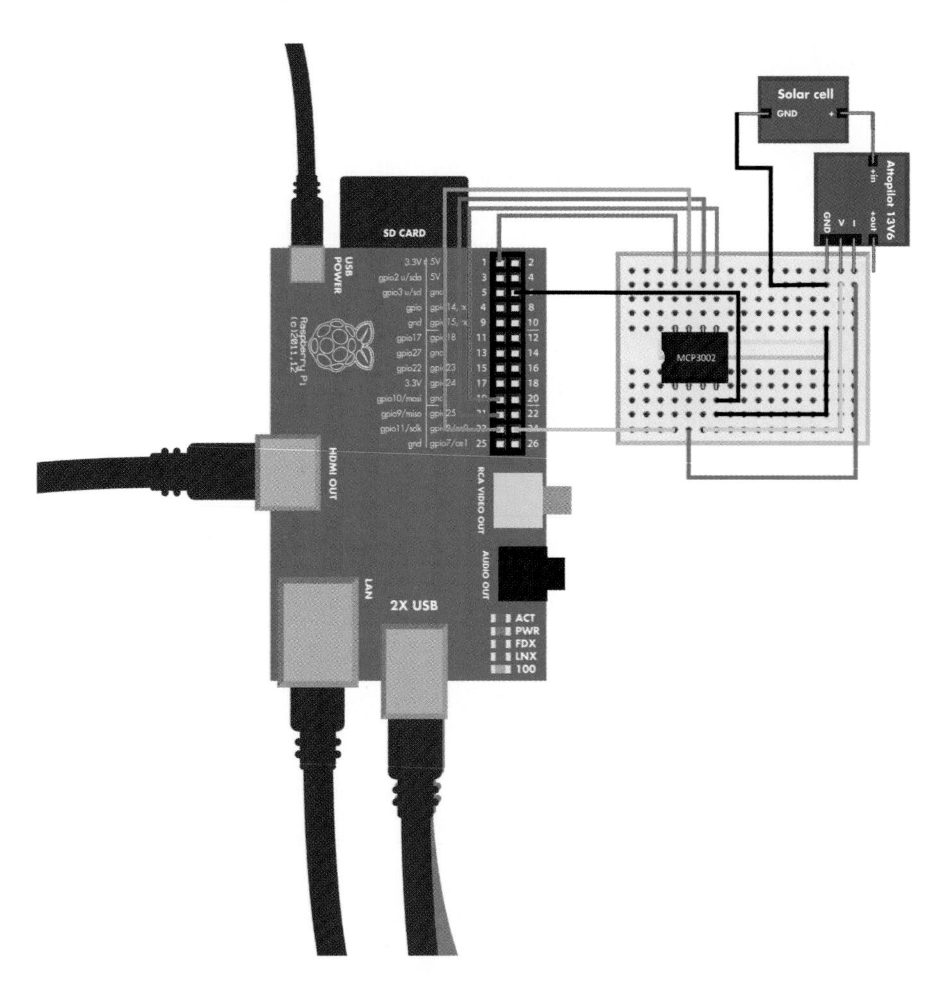

Figure 10-17. *Solar cells connected*

Turn Raspberry Pi into Web Server

Apache is one of the most popular web servers on the planet. According to the Netcraft web server survey (*http://news.netcraft.com*), at many times in its lifetime, it has been more popular than all competing web servers together.

These web server instructions have a lot of explanations. If you are well-versed in Linux, you can just read the commands to get the job done in a couple of minutes.

To turn Raspberry Pi into a web server, you must install Apache. The steps to install Apache on Raspberry Pi are exactly the same as you would use when installing Apache on a physical server, virtual server, or a development laptop. If you are using Debian or Ubuntu on your laptop, you can try the same steps there.

On your Raspbian desktop, open the command-line interface (LXTerminal; you'll find the icon on the left side of the desktop).

Update the list of available packages, and then install Apache:

```
$ sudo apt-get update
$ sudo apt-get -y install apache2
```

The web server is now installed and running. Try it out with a web browser, such as Midori. The Midori icon is on the left side of the desktop. Browse to

```
http://localhost/
```

Do you see a web page? "It works"? Congratulations, you now run a web server:

Finding Your IP Address

To access your web server from other computers, check your IP address:

```
$ ifconfig
```

Look for `inet addr` in the output. It's usually under the eth0 or wlan0 interface.

Your public IP address is not 127.0.0.1. That's localhost address, and every computer refers to itself as localhost.

You can try browsing to this address, too. Just open Midori, and type "http://" and the IP address as the URL. For example, *http://10.0.0.1*. Obviously, you must use your own address.

This address can also work on your local network. You can try browsing to the address with other computers on your network. Can you see your web server from your laptop or desktop?

Making Your Home Page on Raspberry Pi

It's easy to set up and maintain a web page if you create it as a *user home page*. There are two steps: enabling users to make home pages and making the home pages.

Allow users to make home pages. As this changes system-wide settings, you need to use sudo. The following commands enable the Apache user directory module, and then restart Apache:

```
$ sudo a2enmod userdir
$ sudo service apache2 restart
```

Now it's time to create your (pi) home page directory. This doesn't require sudo privileges. The name of your home page directory, *public_html*, consists of the word "public", underscore ("_"), and the word "html". It must be written correctly.

```
$ cd /home/pi/
$ mkdir public_html
```

If you want, you can make a test page:

```
$ echo "botbook">/home/pi/public_html/hello.txt
```

Try it with the Midori web browser:

```
http://localhost/~pi/hello.txt
```

If you feel adventurous, try your IP address instead of localhost. You could even try visiting your page from your desktop computer on the same network.

Well done—you have now turned Raspberry Pi into a web server! And you've already got a home page there. Let's look at the program, and then we'll set it up to run periodically.

Solar Panel Monitor Code and Connection for Raspberry Pi

This code requires *matplotlib*, a great free Python library for mathematical graphing. Install it with these commands:

```
$ sudo apt-get update
$ sudo apt-get -y install python-matplotlib
```

Another useful tool for data visualization is Plotly (https://plot.ly/). You can see an example project at Instructables (http://bit.ly/1f0Goqp).

Running *voltage_record.py* once records a new data point and creates one graph, and then exits. No loop is needed, because you'll see how to use cron to run the program every five minutes.

The measurement history is kept in */home/pi/record.csv*.

The generated plot file is put into */home/pi/public_html/history.png*. Because you've already installed Apache, this file is published at the URL *http://localhost/~pi/history.png*. To visit that page from another device, you need to replace localhost with your Raspberry Pi's IP address.

Example 10-9. voltage_record.py

```
# voltage_record.py - record voltage from solar cell and print history to png
# (c) BotBook.com - Karvinen, Karvinen, Valtokari

import time
import os
import matplotlib         # 1
matplotlib.use("AGG")    # 2
import matplotlib.pyplot as plt
import numpy
from datetime import datetime
from datetime import date
from datetime import timedelta
import attopilot_voltage         # 3
import shelve

historyFile = "/home/pi/record" # 4
plotFile = "/home/pi/public_html/history.png"    # 5

def measureNewDatapoint():
        return attopilot_voltage.readVoltage()  # 6

def main():
        history = shelve.open(historyFile)        # 7

        if not history.has_key("x"):     # 8
                history["x"] = []        # 9
                history["y"] = []

        history["x"] += [datetime.now()]          # 10
        history["y"] += [measureNewDatapoint()] # 11

        now = datetime.now()     # 12
        sampleCount = 24 * 60 / 5
        history['x'] = history["x"][-sampleCount:]
        history['y'] = history["y"][-sampleCount:]

        plot = plt.figure()     # 13
        plt.title('Solar cell voltage over last 24h')
        plt.ylabel('Voltage V')
        plt.xlabel('Time')
        plt.setp(plt.gca().get_xticklabels(), rotation=35)
        plt.plot(history["x"], history["y"], color='#4884ff')    # 14
        plt.savefig(plotFile)    # 15
        history.close() # 16

if __name__ == '__main__':
        main()
```

❶ This library must be installed as directed earlier in this section.

❷ AGG (Anti-Grain Geometry) is a good matplotlib backend for saving pixel graphics (PNG).

❸ The example you created earlier must be in an *attopilot_voltage/* directory that's stored in the same directory as this program (*voltage_record.py*). See "Experiment: Voltage and Current" on page 279.

❹ This is the "shelve" file, where numerical data is stored. Python's `shelve` function (which you'll see in a moment) makes it very easy to store values.

❺ The plot (a PNG image) to be created. To publish it as a web page, we put it in a folder that's served up by the Apache web server.

❻ This is the actual measurement command. To measure anything else, just change this command.

❼ Open the history file. The file is created automatically if needed.

❽ If our variables aren't in the shelve yet…

❾ …create them. The shelve history is a *dictionary*. The keys x and y store one list each.

❿ Append timestamp to x. Notice how we don't have to care about date formatting or parsing.

⓫ Append measured value to y. The x and y dictionaries are completely different, but values with the same index are related. For example, the data point stored in y[12] is from the time stored in x[12].

⓬ This stanza drops records older than 24 hours. A simpler, alternative way would be to just store 100 data points and cut after that.

⓭ Create a new matplotlib plot to draw on.

⓮ Draw the actual graph.

⓯ Save the whole plot as a PNG image.

⓰ Close the shelve to write our values to disk.

Timed Tasks with Cron

Cron is our favorite way of performing timed tasks. Even if your Raspberry Pi shuts down, cron tasks are resumed after reboot.

First make sure that you can run the Python program on the command line, with full path. Use the path to wherever you have downloaded or saved *voltage_record.py* (see Example 10-9).

```
$ python /home/pi/voltage_record/voltage_record.py
```

Don't you remember where you put voltage_record.py? Run this: `cd /home/pi; find -iname voltage_record.py`.

If it executes normally, it's time to add it as a timed task to cron. Edit your user cron file with this:

```
$ crontab -e
```

Add the following as the last lines of the file:

```
*/5 * * * * /home/pi/voltage_record/voltage_record.py
*/1 * * * * touch /tmp/botbook-cron
```

Save the file. If you edited it with nano, press Control-X, y, then Enter to save.

The first line tells cron to run the program whenever minutes are divisible by 5 (:00, :05...), ignoring hour, day of month, month, and day of week.

The second line just creates an empty file */tmp/botbook-cron* every minute. It's for checking up on cron, and you can delete this line later if you want. Wait for a minute, and then check if the file is there:

```
$ ls /tmp/botbook*
/tmp/botbook-cron
```

Did `ls` show your file? Well done! You successfully added a timed task to cron.

Whenever you want Raspberry Pi to automatically do something, use cron. Even if the power is cut, the timed tasks are run again when you boot up the Raspberry Pi.

If this cron job writes too much output to your logs, you can append >/dev/null to your command in crontab. This hides (deletes) everything the command prints to standard output.

What's Next?

You have now played with electricity. Your gadgets can measure voltage and current, even at levels that would break your microcontroller board if connected directly. Magnetic fields can be detected simply, as a Hall switch detects a magnet, or in an advanced way as you did with the three-dimensional, tilt-corrected compass.

The solar cell web monitor you built can be adapted to any sensor project. Simply by changing the measurement function, you can store data from any sensor. In your project, you published the graph to the Web. Try adding password protection or even use ssh to view the graph for projects requiring more secrecy.

In the next chapter, you move from electromagnetism to sound waves. It's time to give ears to your gadgets!

Sound 11

Sound waves in the air are a compression and decompression of the medium. This movement can be detected and analyzed.

Raspberry Pi can record and analyze sounds. In one project, you'll use a fast Fourier transform (FFT) to split sound into frequencies. This calculation can extract frequencies from any wave. With Arduino, you'll use a microphone to measure volume.

Experiment: Hearing Voices/Volume Level

A microphone can detect sound level. This first experiment simply reads and prints values from a microphone.

For many projects, you probably want to manipulate the numbers read from the microphone. You could set a threshold, apply a moving average, or detect minimum and maximum values. Later examples in this chapter let you try setting a threshold for sound. The component we used is the Breakout Board for Electret Microphone (BOB-09964) from Sparkfun. See Figure 11-1.

Figure 11-1. *Microphone on a breakout board*

Microphone Breakout Code and Connection for Arduino

Figure 11-2 shows the Arduino circuit for the microphone. Wire it up as shown, and then run the code shown in Example 11-1.

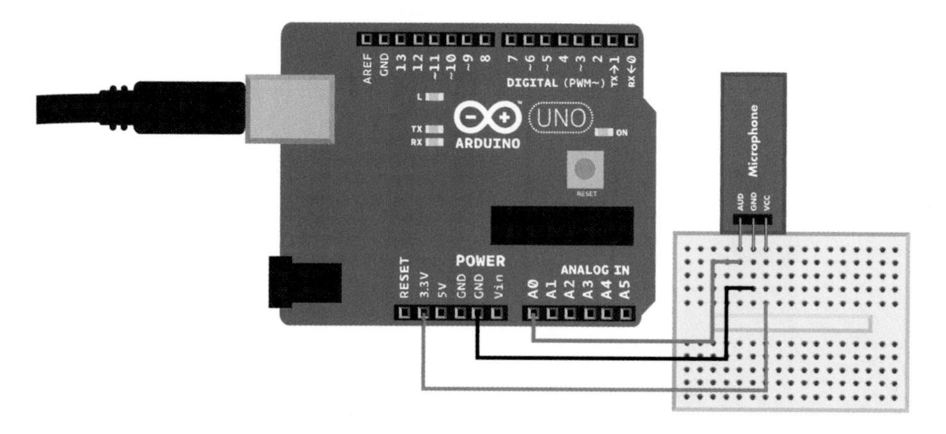

Figure 11-2. *Arduino circuit for microphone*

Example 11-1. microphone.ino

```
// microphone.ino - print audio volume level to serial. Print "Sound" on loud sound.
// (c) BotBook.com - Karvinen, Karvinen, Valtokari


const int audioPin = A0;
const int sensitivity = 850;

void setup() {
  Serial.begin(115200);
}

void loop() {
  int soundWave = analogRead(audioPin); // ❶
  Serial.println(soundWave);
  if (soundWave>sensitivity) {  // ❷
    Serial.println("Sound!");
    delay(500);
  }
  delay(10);
}
```

❶ You can read the microphone breakout like any analog resistance sensor. `analogRead()` returns raw values between 0 and 1023, where a bigger number means a louder sound. The connection and the program are similar to a potentiometer connection.

❷ Take action if it's loud enough.

Microphone Breakout Code and Connection for Raspberry Pi

Figure 11-3 shows the wiring diagram for the Raspberry Pi. Hook everything up, and then run the code shown in Example 11-2.

Example 11-2. microphone.py

```
# microphone.py - read sound from analog and print it
# (c) BotBook.com - Karvinen, Karvinen, Valtokari

import time
import botbook_mcp3002 as mcp   # ❶

def readSound(samples):
        buff = []       # ❷
        for i in range(samples):        # ❸
                buff.append(mcp.readAnalog())   # ❹
        return buff

def main():
        while True:
                sound = readSound(1024) # ❺
```

```
        print(sound)
        time.sleep(1)   # s

if __name__ == "__main__":
    main()
```

❶ The library (*botbook_mcp3002.py*) must be in the same directory as this program. You must also install the *spidev* library, which is imported by *botbook_mcp3002*. See the comments in the beginning of *botbook_mcp3002/botbook_mcp3002.py* or "Installing SpiDev" on page 56.

❷ Declare a new empty list.

❸ Repeat the block below, so that `i` gets assigned each of the values. In this program, this becomes: `for i in [0, 1, 2 ... 1023]:`

❹ Read the value from the microphone. Append a new item to the list (`buff`).

❺ Read 1024 samples, and get a list of values returned.

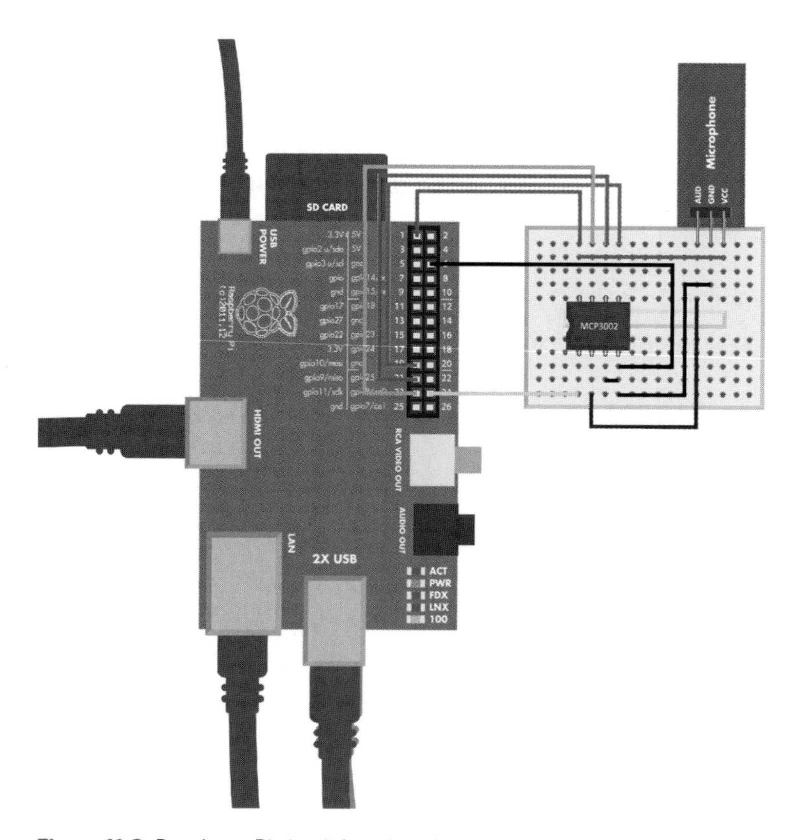

Figure 11-3. *Raspberry Pi circuit for microphone*

Environment Experiment: Can You Hear a Pin Drop?

Can you hear a pin drop? Let's solve this question once and for all with a sound sensor. Connect the sensor to Arduino as you did in "Microphone Breakout Code and Connection for Arduino" on page 316 and upload the code. Place the sensor on a solid plane such as a wooden table or a floor so that the microphone part points in the direction where you are going to drop the pin. Do all you can to minimize any background noise. Change the sensitivity value in the code so that the "sound" message is not triggered when you don't make any sound. Carefully find a value that is just on the edge of reacting to sound without responding to silence.

Figure 11-4. *Pin dropping on floor*

Open the Serial Monitor in the Arduino IDE and drop a pin on the plane. Did you get the "sound" message in the serial monitor? If yes, then you really can hear a pin drop. If you didn't get any reaction from the sensor, make sure that the room is as quiet as possible, drop the pin closer to the sensor, and make sure that you have adjusted the sensitivity properly.

Sensitivity to quiet sounds depends on the background noise level. Try putting some loud music on and readjust the sensitivity before dropping the pin. Can the sensor still hear it?

Test Project: Visualize Sound over HDMI

Have you always wanted a 50" graphical equalizer? In this project, you'll analyze sound with Raspberry Pi and show the result on your television. Sound frequencies are shown as a colorful, animated graph (Figure 11-5).

Figure 11-5. *Animated graph on a big screen*

What You'll Learn

In the *Visualize sound on HDMI* project, you'll learn how to:

- Analyze sound numerically.
- Do very fast calculations in Python, using SciPy.
- Extract frequencies with fast Fourier transform (FFT).

You'll also refresh your skills on pyGame and drawing Full HD graphics to HDMI output, like your television or a video projector (see "Test Project: Pong" on page 147).

Enabling the Serial Port in Raspberry Pi

To use the serial port in Raspberry Pi, you must release it first. Otherwise, it's used by a login shell that you can connect to over a serial cable:

```
$ sudoedit /etc/inittab
```

Comment out the last line that grabs the serial port. You can comment out a line by putting a hash in front of it, which causes the line to be ignored.

```
# T0:23:respawn:/sbin/getty -L ttyAMA0 115200 vt100
```

Reboot the Raspberry Pi.

Visualizer Code and Connection for Raspberry Pi

Install prerequisites. On your Raspberry Pi, open the terminal and run these commands:

```
$ sudo apt-get update
$ sudo apt-get -y install python-pygame python-numpy
```

Figure 11-6 shows the circuit diagram for Raspberry Pi. Hook it up, and then run the code shown in Example 11-3.

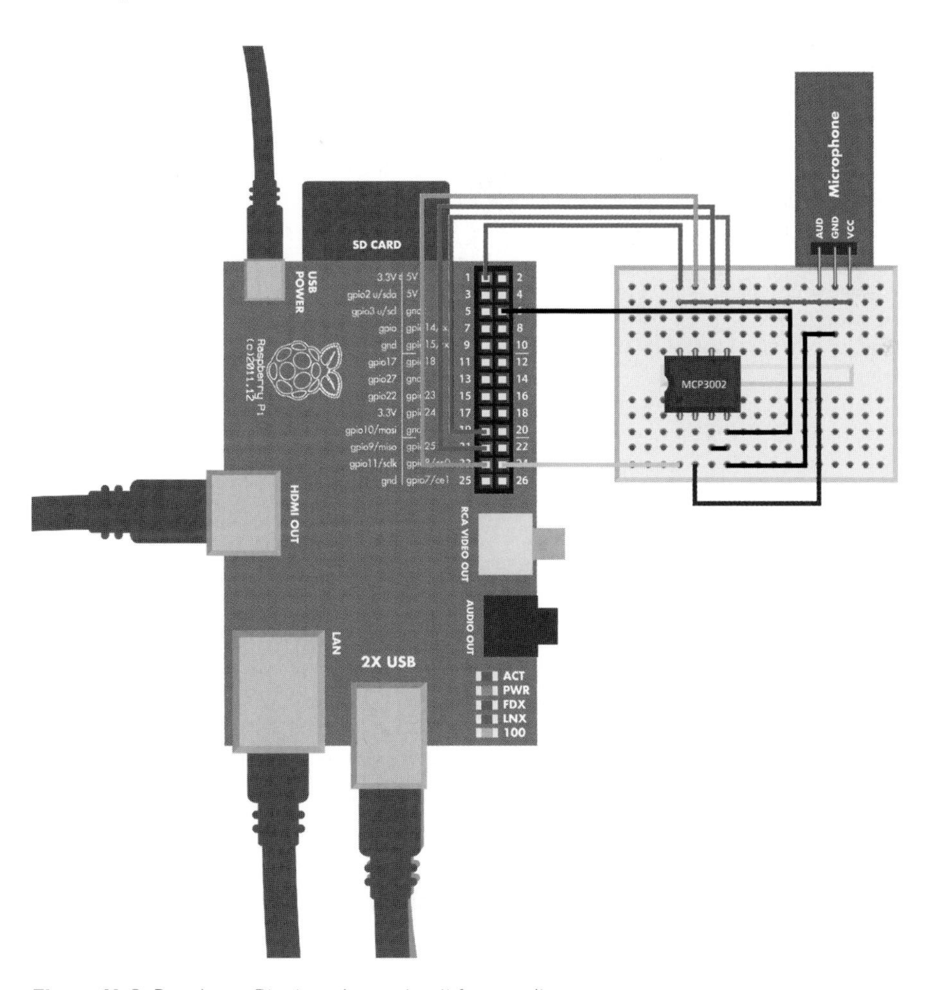

Figure 11-6. *Raspberry Pi microphone circuit for equalizer*

Example 11-3. equalizer.py

```
# equalizer.py - show equalizer based on microphone input
# (c) BotBook.com - Karvinen, Karvinen, Valtokari


import pygame  # sudo apt-get -y install python-pygame
import math
import numpy  # sudo apt-get -y install python-numpy


import microphone    # ❶
from pygame.locals import *

pygame.init()

width = 800
height = 640

size = width, height
background = 0, 0, 0

screen = pygame.display.set_mode(size, pygame.FULLSCREEN)
fullBar = pygame.image.load("equalizer-full-small4.jpg")    # ❷
emptyBar = pygame.image.load("equalizer-empty-small4.jpg")
clock = pygame.time.Clock()
pygame.mouse.set_visible(False)
mainloop = True

barHeight = 36
barWidth = 80
barGraphHeight = 327
barPos = [55, 130]

sampleLength = 16


def fftCalculations(data):    # ❸
    data2 = numpy.array(data) / 4    # ❹
    fourier = numpy.fft.rfft(data2)    # ❺
    ffty = numpy.abs(fourier)    # ❻
    ffty = ffty / 256.0        # ❼
    return ffty

while mainloop:
    buff = microphone.readSound(sampleLength)    # ❽

    barData = fftCalculations(buff)    # ❾

    for event in pygame.event.get():
        if event.type == pygame.QUIT:
            mainloop = False
```

```
        if (event.type == KEYUP) or (event.type == KEYDOWN):
            if event.key == K_ESCAPE:
                mainloop = False
    screen.fill(background)
    # Draw data to pillars
    for i in range(8):    # ❿
        bars = barData[i+1]    # ⓫
        bars = int(math.floor(bars*10))    # ⓬
        if bars > 10:
            bars = 10
        bars -= 1
        screen.blit(emptyBar, (barPos[0]+i*(barWidth+10), barPos[1]),
                    (0, 0, barWidth, barHeight*(10-bars)))
        if bars >= 0:
            barStartPos = (barPos[0] + i * (barWidth + 10),
                          barPos[1] + barGraphHeight - barHeight * bars + 6)
            barSourceBlit = (0, barGraphHeight - barHeight * bars+6,
                            barWidth, barHeight*bars)
            screen.blit(fullBar, barStartPos, barSourceBlit)    # ⓭
    pygame.display.update()
```

❶ The *microphone.py* program from "Microphone Breakout Code and Connection for Raspberry Pi" on page 317 must be in the same directory as this program (*equalizer.py*).

❷ PyGame can load normal JPG images. You can draw them in any program, such as Inkscape, Gimp, Photoshop, or Illustrator.

❸ When this function is called from the main program, `data` will contain 16 samples, each in the range of 0..1024.

❹ Divide the samples to put them into the range 0..256.

❺ Perform an FFT on the recorded sound samples. This gives the frequency ranges of the samples. The `rfft()` function returns an array that has `16/2+1` (9) cells, each representing a frequency.

❻ Get rid of the imaginary part of the numbers, so that they can be plotted later. For example, `abs(1+1j)` is approximately 1.4.

❼ Convert the Fourier-transformed values to percentages (in the range 0.0 to 1.0).

❽ Read 16 samples from the microphone, using the botbook.com microphone library.

❾ Save the FFT of the sound sample to `barData`.

❿ For each of the eight vertical bars (0 to 7)…

⓫ …get the frequency percentage. Because of the `i+1` here, the first cell number (0) is ignored. The first, ignored cell contains the DC component, the average value between AC wave's positive and negative peaks.

⓬ Count the number of bars needed (e.g., 1.0 (100 %) gets the maximum, nine bars).

⓭ Draw the full bars on screen.

Fast Fourier Transformation

An FFT extracts individual frequencies from a wave. It's used for creating frequency diagrams in graphic equalizers, spectrum analyzers, and oscilloscopes.

Imagine playing two notes, one lower (5 Hz) and one higher (40 Hz). In real life, audible sounds are from 20 Hz to 20,000 Hz, but we picked smaller numbers for this example.

In Figure 11-7, the horizontal time axis grows to the right. Vertical axis is amplitude, which is compression and decompression of air.

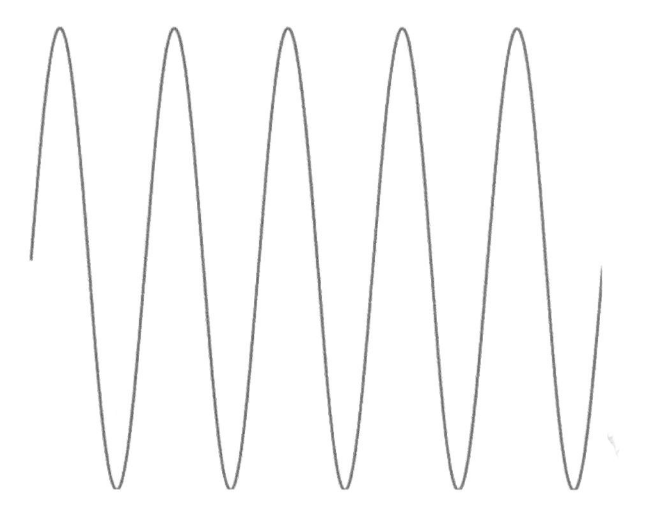

Figure 11-7. *A low note, 5 Hz sine wave*

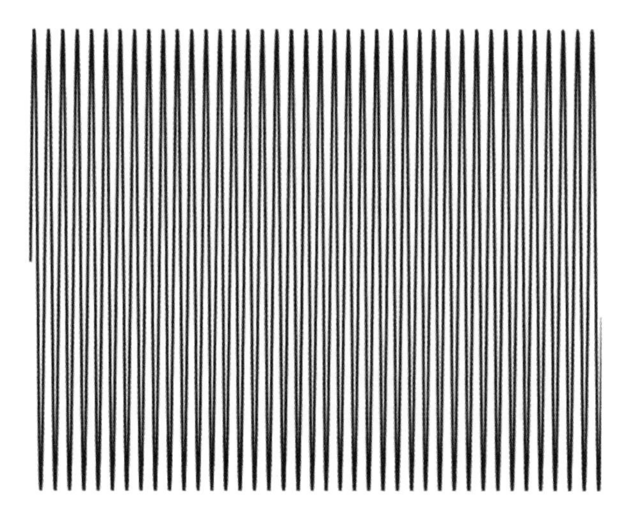

Figure 11-8. *A higher note, 40 Hz sine wave*

When they play at the same time, they form one wave. This combined wave is like a little ripple on top of a big wave. The lower note, 5 Hz, forms the big wave. The small ripple is the higher 40 Hz frequency. The combined wave is just a sum of the two waves, combined = low + high.

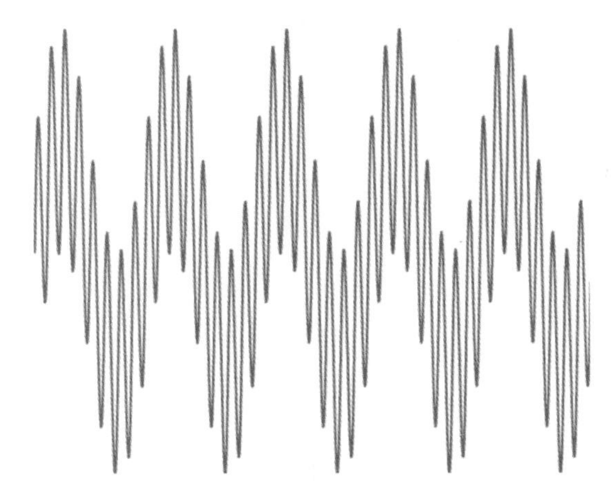

Figure 11-9. *Both notes combined*

When you record sound, you get this kind of combined wave. So this is similar to the typical starting material with sound samples, such as mp3, wav, or ogg. Given a combined wave like this, how do you get the original two sine waves back? Fourier to the rescue!

Perform a fast Fourier transform of the combined wave. As you've seen in the code of this project, you can simply call an FFT function, with the wave data as the parameter. See Figure 11-10.

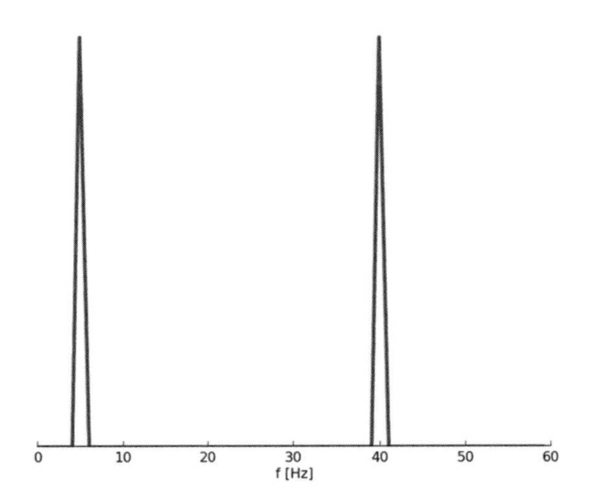

Figure 11-10. *FFT reveals the frequencies of the original notes*

The FFT gave the frequencies of the original sine waves, 5 Hz and 40 Hz. You have now broken a complex wave into frequencies. Figure 11-11 shows a sample equalizer display.

Figure 11-11. *Equalizer*

What Next?

You have now played with sound. With Arduino, you detected volume level and noticed a volume over a certain threshold. With Raspberry Pi, you recorded sound and analyzed it in real time. As you built the project, you improved your pyGame skills and used fast Fourier transform to break a wave into components.

Next, it's time to look at air on a bigger scale. You'll measure weather and climate, from local room temperature and humidity to predicting weather in your part of town.

Weather and Climate

12

Weather is one of the few things we humans can't control. Yet.

The immediately obvious components of weather include temperature and humidity. To forecast weather, you can measure the atmospheric pressure.

Experiment: Is It Hot in Here?

An LM35 sensor reports temperature with varying voltage. Because it's an analog resistance sensor, it's easy to use with Arduino. Using it with Raspberry Pi requires an analog-to-digital converter chip.

The LM35 has three leads (Figure 12-1), similar to a potentiometer. The output lead voltage U is converted to temperature T with a simple formula:

```
T = 100*U
```

Some example values are in Table 12-1. From the table, you can see that the LM35 can measure temperatures between 2 C and 150 C. The output voltage varies from about 0 V to 1.5 V.

Do you need to measure subzero temperatures? Use LM36 or check for alternative connection options on the LM35 data sheet.

Table 12-1. LM35 reports temperature with voltage

Temperature C	Voltage V	Comment
2 C	0.02 V	Minimum measured temperature
20 C	0.2 V	Room temperature
100 C	1.0 V	Boiling water
150 C	1.5 V	Maximum measured temperature

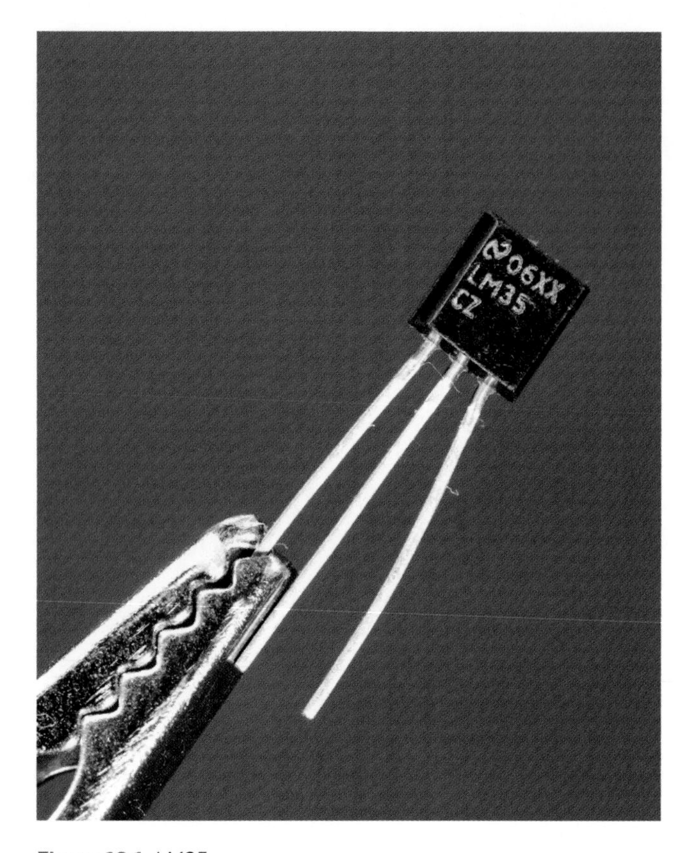

Figure 12-1. *LM35*

LM35 Code and Connection for Arduino

Figure 12-2 shows the wiring diagram for Arduino. Hook it up as shown, and then run the code from Example 12-1.

Example 12-1. temperature_lm35.ino

```
// temperature_lm35.ino - measure temperature (Celsius) with LM35 and print it
// (c) BotBook.com - Karvinen, Karvinen, Valtokari


int lmPin = A0;

void setup()
{
  Serial.begin(115200);
  pinMode(lmPin, INPUT);
}

float tempC()
```

```
{
  float raw = analogRead(lmPin);  // ❶
  float percent = raw/1023.0; // ❷
  float volts = percent*5.0; // ❸
  return 100.0*volts;  // ❹
}

void loop()
{
  Serial.println(tempC());
  delay(200); // ms
}
```

❶ LM35 is just an analog resistance sensor, so you can measure its voltage with `analogRead()`. It returns a raw value between 0 and 1023.

❷ Convert the raw value (0 to 1023) to a percentage of the HIGH voltage (5 V). The `percent` variable will be a float between 0.0 and 1.0.

❸ Convert percentage to volts (between 0 and 5 V).

❹ Return the temperature in Celsius. The formula we used comes from the LM35's data sheet.

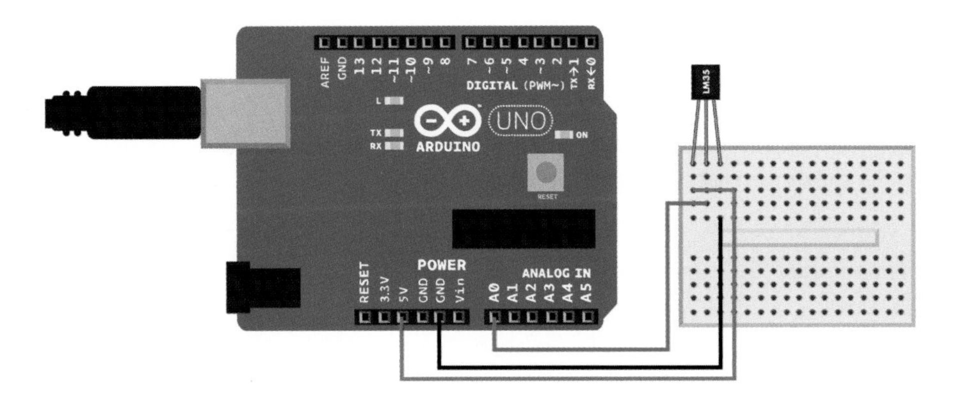

Figure 12-2. *Arduino/LM35 connections*

LM35 Code and Connection for Raspberry Pi

Figure 12-3 shows the connection diagram for the Raspberry Pi. Hook it up as shown, and then run the code in Example 12-2.

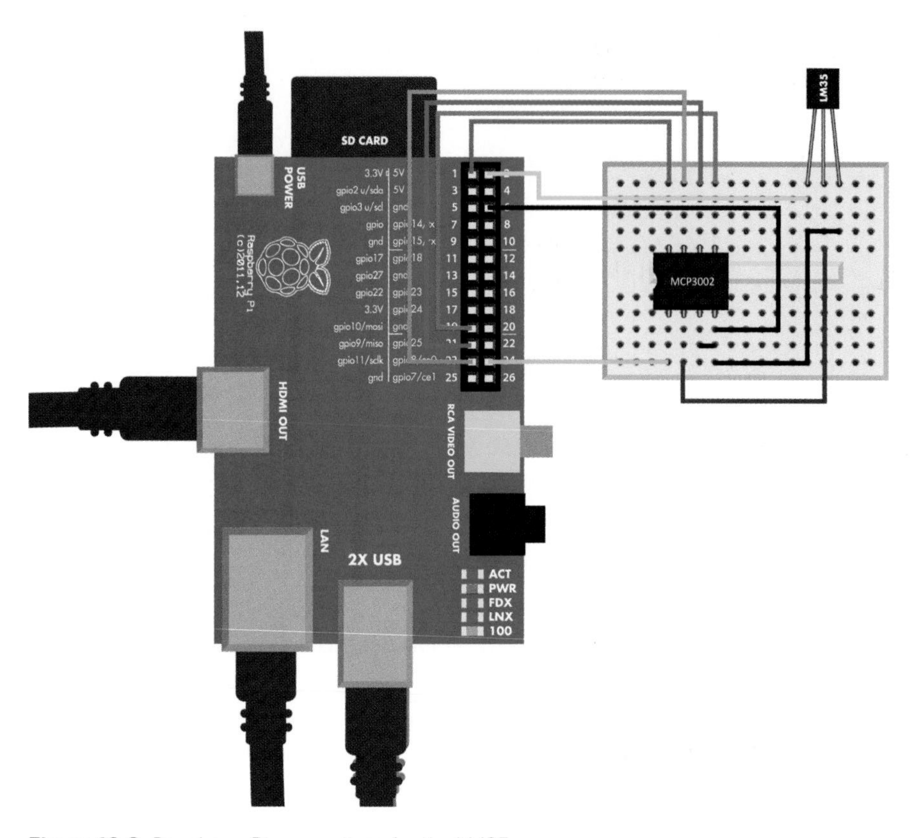

Figure 12-3. *Raspberry Pi connections for the LM35*

Because the maximum output voltage of LM35 is 1.5 V at 150 C, there is no danger of going over Raspberry Pi's maximum safe input voltage level of 3.3 V.

Example 12-2. ***temperature_lm35.py***

```
# temperature_lm35.py - print temperature
# (c) BotBook.com - Karvinen, Karvinen, Valtokari
import time
import botbook_mcp3002 as mcp   # ❶
#LM35 data pin voltage keeps under 2V so no
#level conversion needed
def main():
        while True:
                rawData = mcp.readAnalog()      # ❷
                percent = rawData / 1023.0      # ❸
                volts = percent * 3.3   # ❹
                temperature = 100.0 * volts     # ❺
                print("Current temperature is %.1f C " % temperature)
                time.sleep(0.5)
```

```
if __name__ == "__main__":
        main()
```

❶ The *botbook_mcp3002.py* library file must be in the same directory as this program (*temperature_lm35.py*). You must also install the *spidev* library, which is imported by *botbook_mcp3002*. See the comments at the top of *botbook_mcp3002/botbook_mcp3002.py* or "Installing SpiDev" on page 56.

❷ Read voltage using MCP3002 analog-to-digital converter and *botbook_mcp3002* library. The raw value is an integer between 0 and 1023.

❸ Convert the raw value to a percentage of 5 V.

❹ Convert `percent` to a voltage (between 0 and 3.3 V).

❺ Convert to Celsius using the formula we obtained from the component's data sheet. See examples in Table 12-1.

Environment Experiment: Changing Temperature

Air is a very good insulator. You'll find that your patience will be tested if you take your LM35 temperature sensor (Figure 12-4) to the sauna, fridge, and balcony—it seems to change so slowly.

Figure 12-4. *Measuring cold temperatures*

To get quick changes, press an ice cube directly against the LM35. Avoid wetting your wires, microcontrollers, or breadboard. Now the heat is conducted away from LM35, and the change of temperature is rapid.

Experiment: Is It Humid in Here?

DHT11 measures humidity and temperature (Figure 12-5).

Figure 12-5. *DHT11 humidity sensor*

The protocol that the DHT11 uses is weird. It sends bits as very short pulses very rapidly. Arduino doesn't have an operating system, so it's more real-time than Raspberry Pi and can easily read these pulses. But even Arduino needs to be coded in a special way, as even the built-in `pulseIn()` function is not fast enough.

Raspberry Pi is less real-time and has a difficult time reading the pulses reliably. That's why you'll connect Arduino to Raspberry Pi and use Arduino to do the actual reading of the DHT11. The connection is made using serial over USB. If you want, you can easily apply this technique to any other sensor.

The data pin is normally HIGH. To start reading, the microcontroller sends a LOW pulse of 18 milliseconds.

The DHT11 sends 5-byte packets. As each byte is 8 bits, the packet is 5 B * 8 bit/B = 40 bits.

How Humid Is Your Breath?

How can we change the measured humidity? Just buy an ultrasonic humidifier that uses a piezoelectric transducer to create mist from water, or book a flight to Thailand. No, wait, there is an easier way. You have a built-in humidifier in your lungs. Bring the sensor close to your mouth and breathe (Figure 12-6). Keep checking to see how values change by watching the Arduino Serial Monitor.

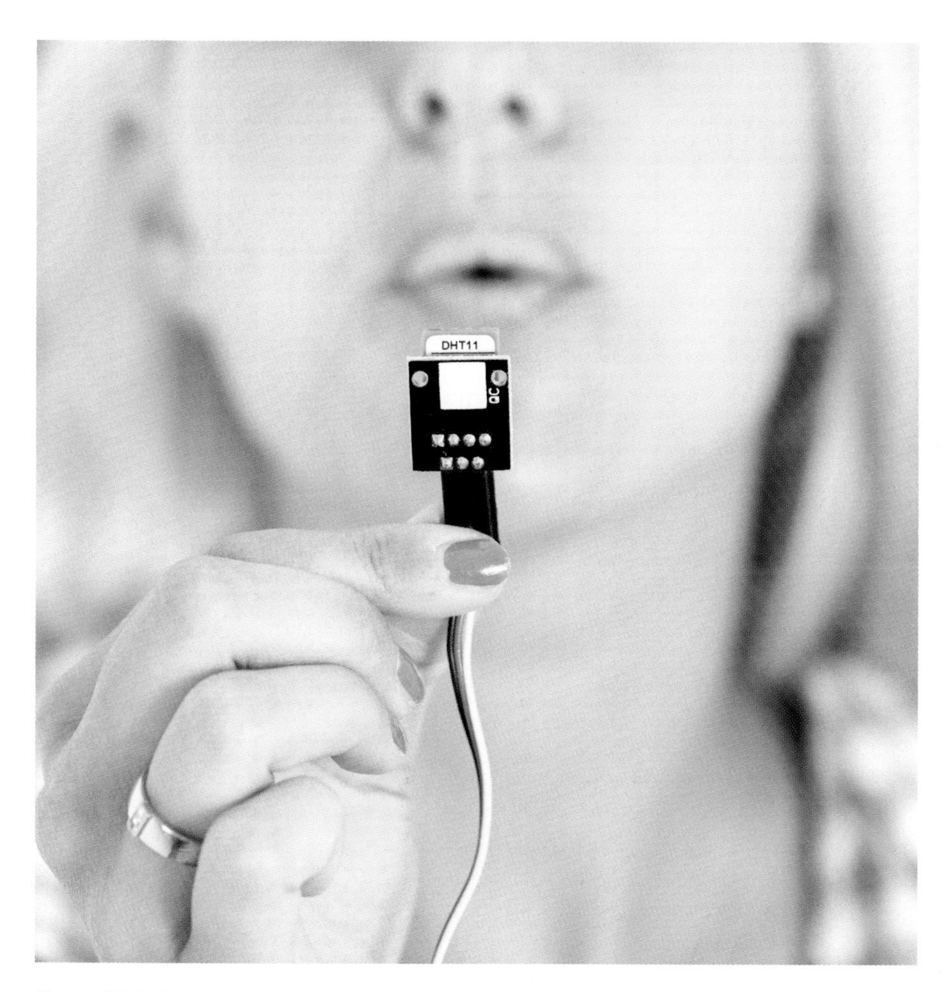

Figure 12-6. *Breath to humidity sensor*

DHT11 Code and Connection for Arduino

Figure 12-7 shows the connections for Arduino. Wire them up as shown, and then run the code shown in Example 12-3.

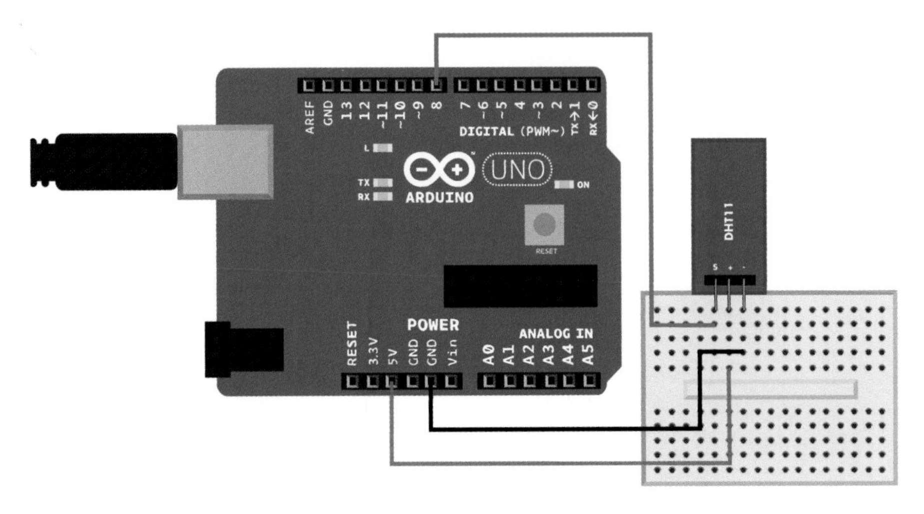

Figure 12-7. *DHT11 humidity sensor connected to Arduino*

Example 12-3. dht11.ino

```
// dht11.ino - print humidity and temperature using DHT11 sensor
// (c) BotBook.com - Karvinen, Karvinen, Valtokari

const int dataPin = 8;

int temperature = -1;
int humidity = -1;

void setup() {
  Serial.begin(115200);
}

int readDHT11() {
  uint8_t recvBuffer[5];          // ❶
  uint8_t cnt = 7;
  uint8_t idx = 0;
  for(int i = 0; i < 5; i++) recvBuffer[i] = 0; // ❷

  // Start communications with LOW pulse
  pinMode(dataPin, OUTPUT);
  digitalWrite(dataPin, LOW);
  delay(18);
  digitalWrite(dataPin, HIGH);

  delayMicroseconds(40);          // ❸
  pinMode(dataPin, INPUT);
```

```
  pulseIn(dataPin, HIGH);    // ❹
  // Read data packet
  unsigned int timeout = 10000; // loop iterations
  for (int i=0; i<40; i++)      // ❺
  {
        timeout = 10000;
        while(digitalRead(dataPin) == LOW) {
                if (timeout == 0) return -1;
                timeout--;
        }

        unsigned long t = micros();    // ❻

        timeout = 10000;
        while(digitalRead(dataPin) == HIGH) { // ❼
                if (timeout == 0) return -1;
                timeout--;
        }

        if ((micros() - t) > 40) recvBuffer[idx] |= (1 << cnt);       // ❽
        if (cnt == 0)   // next byte?
        {
                cnt = 7;    // restart at MSB
                idx++;      // next byte!
        }
        else cnt--;
  }

  humidity = recvBuffer[0];     // %    // ❾
  temperature = recvBuffer[2];  // C
  uint8_t sum = recvBuffer[0] + recvBuffer[2];
  if(recvBuffer[4] != sum) return -2;    // ❿
  return 0;
}

void loop() {
  int ret = readDHT11();
  if(ret != 0) Serial.println(ret);
  Serial.print("Humidity: "); Serial.print(humidity); Serial.println(" %");
  Serial.print("Temperature: "); Serial.print(temperature); Serial.println(" C");
  delay(2000); // ms
}
```

❶ The packet from DHT11 is 5 bytes (40 bits) in length.

❷ Initialize the buffer with zeroes.

❸ Wait a very short time for the DHT11 to start sending data.

❹ Receive the DHT11's response signal.

❺ For each of the 40 bits, perform the operations inside the curly braces.

❻ Uptime in microseconds (μs); this is stored to measure the duration of the pulse.

7. Measure the pulse with a loop.
8. If the pulse was longer than 40 µs, it was a 1. Otherwise, that bit is left as its initialized value, 0. To toggle the corresponding bit in the corresponding byte, a new number is created with bit shifting (e.g., `1<<2 == 0b100`). This bit is then combined with the previous byte value using a *bitwise inplace OR*. So if the value was already a 1, it stays set at 1. If it was a zero, but the new number (0b100) was 1, then this sets it to 1. See "Bitwise Operations" on page 221 for more details.
9. Humidity is the first byte of the received packet.
10. The checksum (byte 4) is the sum of humidity and temperature.

DHT11 Code and Connection for Raspberry Pi

Would you like to combine Arduino and Raspberry Pi to get the best of both worlds? We hope so, because DHT11 doesn't like the Raspberry Pi. This code shows you how you can read data from Arduino, in a way you can easily adapt to other projects.

Even if it's possible to use DHT11 from Raspberry Pi with complicated code, the result is not satisfactory. Luckily, Arduino can be used for fast low-level tasks, letting Raspberry Pi concentrate on the big picture.

After you have tried the code, you should read "Talking to Arduino from Raspberry Pi" on page 337.

To use the serial port in Raspberry Pi, you must first release it from its default use as a login terminal. See "Enabling the Serial Port in Raspberry Pi" on page 320 for instructions. Wire them up as shown in Figure 12-8, and then run the code shown in Example 12-4.

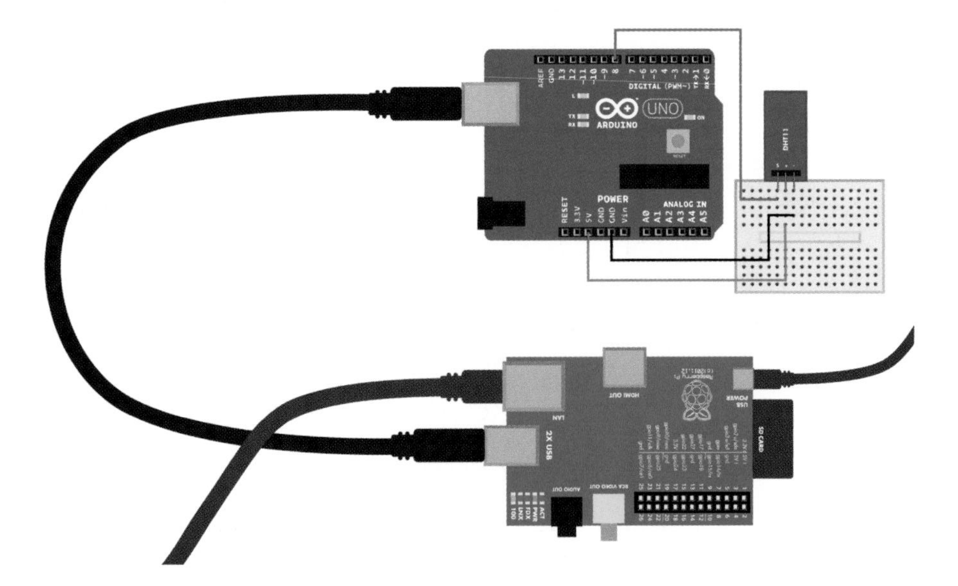

Figure 12-8. *Arduino connected to Raspberry Pi*

Example 12-4. dht11_serial.py

```
# dht11_serial.py - print humidity and temperature using DHT11 sensor
# (c) BotBook.com - Karvinen, Karvinen, Valtokari

import time
import serial   # ❶

def main():
        port = serial.Serial("/dev/ttyACM0", baudrate=115200, timeout=None)     # ❷
        while True:
                line = port.readline()  # ❸
                arr = line.split()      # ❹
                if len(arr) < 3:        # ❺
                        continue        # ❻
                dataType = arr[2]
                data = float(arr[1])    # ❼
                if dataType == '%':
                        print("Humidity: %.1f %%" % data)
                else:
                        print("Temperature: %.1f C" % data)


                time.sleep(0.01)

if __name__ == "__main__":
        main()
```

❶ You might need to run these commands to install this library: `sudo apt-get update && sudo apt-get -y install python-serial`.

❷ Open the USB serial for reading. The Arduino program must use the same speed, 115,200 bit/s to communicate.

❸ Read one line, until reaching a newline. The `readline()` function is a *blocking function*, so it will wait here until it reads something in.

❹ Split the text string into a list. This makes it easy to parse the human-readable input from the Arduino. For example, `"Humidity: 25 %".split()` creates a convenient list: `['Humidity:', '25', '%']`.

❺ If the line length is incorrect…

❻ …ignore this line and jump to the next iteration of the `while` loop.

❼ Typecast (convert) the string ("25") to a number (25.0).

Talking to Arduino from Raspberry Pi

Arduino is real-time, robust, and simple. Also, Arduino has a built-in ADC (analog-to-digital converter). Raspberry Pi is high level, and runs a whole Linux operating system. Why not combine the benefits?

You can use Arduino to directly interact with sensors and outputs, and control Arduino from Raspberry Pi.

First, make the sensors work with Arduino, and print the readings to serial. Use your normal desktop computer and Arduino IDE for this. You can write to serial with `Serial.println()`. Use the Serial Monitor (Tools→Serial Monitor) to see what your program prints. Only proceed to the Raspberry Pi part once you are happy with how the Arduino part of your project works.

The connection between Arduino and Raspberry Pi is simple: just connect them with a USB cable.

For a more intimate combination of the Arduino platform and Raspberry Pi, check out the AlaMode for Raspberry Pi (http://bit.ly/1icPd0z), which incorporates an Arduino-compatible microcontroller board into a Raspberry Pi expansion that snaps right on top of your Pi. There's no need to connect them by USB, since they use the Raspberry Pi's expansion header.

Arduino communicates with serial over USB. To read serial from Raspberry Pi, use Python and PySerial. For a complete example of using PySerial, see Figure 12-8 or Chapter 7 of *Make: Arduino Bots and Gadgets*.

You'll have Arduino print normal text, then extract the numbers with Python's string handling. In many prototyping applications, you don't have to design a new low-level protocol.

For example, consider an Arduino program printing the following:

```
Humidity: 83 %
```

If you have read this into a Python string, use `split()` to separate the words into a list:

```
>>> s="Humidity: 83 %"
>>> s.split()
['Humidity:', '83', '%']
```

You can refer to items of the list with square brackets, `[]`. The first item is number zero. This prints the second item:

```
>>> x = s.split()[1]
>>> x
'83'
```

Finally, convert the string "83" into integer 83 so you can process it as a number. You can perform sensible mathematical operations and comparisons with a number, but not with a string.

```
>>> int(x)
83
```

You can use any Arduino-compatible sensor from Raspberry Pi this way.

Atmospheric Pressure GY65

Atmospheric pressure can help you forecast weather. High pressure means sunny, clear weather. Low pressure means rainy, snowy, or cloudy weather.

Figure 12-9. *The GY65 barometric pressure sensor*

Normal air pressure is about 1,000 hPa (hectopascals). The exact value depends on your altitude.

```
1,000 hPa = 100,000 Pa = 1,000 mbar = 1 bar
```

The pressure doesn't vary much, even in extreme weather. The world records are a high of 1,084 hPa and a low of 870 hPa (inside a typhoon). Typical changes in normal weather are a few hectopascals.

Pressure changes in the lower atmosphere are tiny compared with other everyday pressure changes. A hand-pumped bike tire has pressure of 4,000 hPa. Outside a passenger airplane over 30,000 feet, the pressure is less than 300 hPa.

So what level of atmospheric pressure is considered high, then? Meteorologists say that high pressure is just relatively high, compared with the pressure of surrounding areas.

You can still predict weather with pressure over time. If pressure goes up, better weather is coming. The faster it changes, the sunnier it will be.

For a simpler forecast, you can compare current pressure to expected pressure for your altitude. You can find out your altitude with a GPS or Google Maps. According to SparkFun, the distributor of the GY65 module we used, a difference of +2.5 hPa means great, sunny weather. Conversely, low pressure of 2.5 hPa under the normal pressure means bad, rainy weather.

GY65 is a breakout board for the BMP085 sensor. You can find the data sheet by searching for "BMP085 digital pressure sensor data sheet."

GY65 Code and Connection for Arduino

Figure 12-10 shows the connection diagram for Arduino. Hook it up as shown, and then run the code shown in Example 12-5.

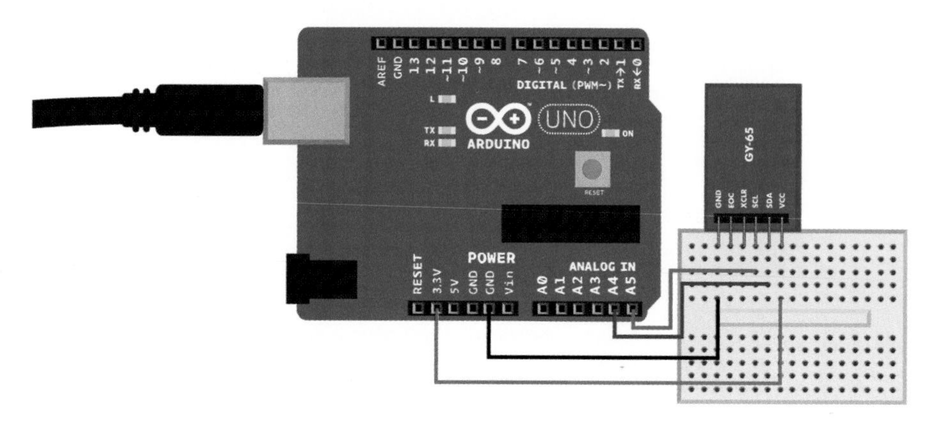

Figure 12-10. *GY65 atmospheric pressure sensor connected to Arduino*

Arduino code uses the *gy_65* library from *http://botbook.com*. If you want an in-depth, technical view how the I2C/SMBus protocol works, have a look at "GY65 Code and Connection for Raspberry Pi" on page 346.

Example 12-5. *gy_65.ino*

```
// gy_65.ino - print altitude, pressure and temperature with GY-65 BMP085
// (c) BotBook.com - Karvinen, Karvinen, Valtokari

#include <Wire.h>
#include <gy_65.h>       // 1

void setup() {
  Serial.begin(115200);
  readCalibrationData(); // 2
}

void loop() {
  float temp = readTemperature();
```

```
    float pressure = readPressure();        // ❸
    float altitude = calculateAltitude(pressure); // ❹

    Serial.print("Altitude: ");
    Serial.print(altitude,2);
    Serial.println(" m");
    Serial.print("Pressure: ");
    Serial.print(pressure,2);
    Serial.println(" Pa");
    Serial.print("Temperature: ");
    Serial.print(temp,2);
    Serial.println("C");
    delay(1000);
}
```

❶ The *gy_65* library makes this sensor very easy to use. It hides the complicated I2C protocol that's behind the sensor. The library folder *gy_65* must be in the same directory as the main program, *gy_65.ino*.

❷ Update global variables.

❸ With the library, it's trivial to retrieve the pressure in Pascals.

❹ Air is thinner when you are higher, so pressure can also be used for altitude estimation.

Using Arduino Libraries

If you have a lot of code, it's best to split it into multiple files. The Arduino code for GY65 is such code. Also, you'll use the same code again in "Test Project: E-paper Weather Forecast" on page 353.

The library is in a folder, within the same directory as the main program (you will probably also encounter libraries that need to be installed into your Arduino sketch folder's *libraries* subdirectory). Here is the folder structure:

```
gy_65.ino          # the main program
gy_65/             # folder that contains the library
gy_65/gy_65.cpp    # code for the library
gy_65/gy_65.h      # prototypes of each library function
```

The location of the main program, *gy_65.ino*, should seem normal to you. Every Arduino project has one main program.

The library is in its own folder. The code for the library is in *gy_65.cpp*. This code looks similar to other Arduino code you have seen.

The header file, *gy_65.h*, contains the *prototypes* of the functions in the *cpp* file. Prototypes are just copies of the first line of each function in *gy_65.cpp*. For example, the header file has the line:

```
float readTemperature();
```

GY65 Arduino Library Explained

You can use the GY65 without going through the implementation details of the library. But if you want a detailed understanding of how communication with the GY65 works, read on.

You can learn the communication protocol by studying the library itself, the Raspberry Pi program (Figure 12-11), or both.

This section also introduces you to creating your own C++ libraries for Arduino. You should also see Arduino's documentation on Writing a Library for Arduino (*http://bit.ly/1iHgFCr*).

The header file *gy_65.h* (see Example 12-6) has prototypes for each function. It's simply a list of functions in the *cpp* file, which is where you'll find the actual implementation of those functions.

Example 12-6. gy_65.h

```
// gy_65.h - library for altitude, pressure and temperature with GY-65 BMP085
// (c) BotBook.com - Karvinen, Karvinen, Valtokari

void readCalibrationData();
float readTemperature();
float readPressure();
float calculateAltitude(float pressure);
```

The implementation of the definitions from the *h* file are in a *cpp* file, *gy_65.cpp* (Example 12-7). If some parts of this code seem demanding, you might want to review the I2C code explanation in Figure 8-5, as well as "Hexadecimal, Binary, and Other Numbering Systems" on page 219 and "Bitwise Operations" on page 221.

Example 12-7. gy_65.py

```
// gy_65.cpp - library for altitude, pressure and temperature with GY-65 BMP085
// (c) BotBook.com - Karvinen, Karvinen, Valtokari
#include <Arduino.h>
#include <Wire.h>
#include "gy_65.h"
const char i2c_address = 0x77;
int OSS = 0; // Oversampling
const long atmosphereSeaLevel = 101325; // Pa

struct calibration_data // ❶
{
  int16_t ac1;
  int16_t ac2;
  int16_t ac3;
  int16_t ac4;
  int16_t ac5;
  int16_t ac6;
```

```
  int16_t b1;
  int16_t b2;
  int16_t mb;
  int16_t mc;
  int16_t md;
};

calibration_data caldata;

long b5;

int16_t swap_int16_t(int16_t value)     // ❷
{
  int16_t left = value << 8;
  int16_t right = value >> 8;
  right = right & 0xFF;
  return left | right ;
}

unsigned char read_i2c_unsigned_char(unsigned char address)     // ❸
{
  unsigned char data;
  Wire.beginTransmission(i2c_address);
  Wire.write(address);
  Wire.endTransmission();
  Wire.requestFrom(i2c_address,1);
  while(!Wire.available());
    return Wire.read();
}
void read_i2c(unsigned char point, uint8_t *buffer, int size)
{
  Wire.beginTransmission(i2c_address);
  Wire.write(point);
  Wire.endTransmission();

  Wire.requestFrom(i2c_address,size);

  int i = 0;

  while(Wire.available() && i < size) {
    buffer[i] = Wire.read();
    i++;
  }

  if(i != size) {
    Serial.println("Error reading from i2c");
  }

}

int read_i2c_int(unsigned char address) {
    int16_t data;
    read_i2c(address,(uint8_t *)&data,sizeof(int16_t));
    data = swap_int16_t(data);
```

```
    return data;
}

void readCalibrationData()      // ❹
{
  Wire.begin();
  read_i2c(0xAA,(uint8_t *)&caldata,sizeof(calibration_data)); // ❺

  uint16_t *p = (uint16_t*)&caldata;    // ❻
  for(int i = 0; i < 11; i++) { // ❼
    p[i] = swap_int16_t(p[i]);
  }
}

float readTemperature() {       // ❽
  // Read raw temperature
  Wire.beginTransmission(i2c_address);
  Wire.write(0xF4); // Register
  Wire.write(0x2E); // Value
  Wire.endTransmission();
  delay(5); // ❾
  unsigned int rawTemp = read_i2c_int(0xF6);

  // Calculate true temperature
  long x1,x2;
  float t;
  x1 = (((long)rawTemp - (long)caldata.ac6) * (long)caldata.ac5) / pow(2,15);
  long mc = caldata.mc;
  int md = caldata.md;
  x2 = (mc * pow(2,11)) / (x1 + md);
  b5 = x1 + x2;
  t = (b5 + 8) / pow(2,4);
  t = t / 10;
  return t;     // Celsius
}

long getRealPressure(unsigned long up){ // ❿
  long x1, x2, x3, b3, b6, p;
  unsigned long b4, b7;
  int b1 = caldata.b1;
  int b2 = caldata.b2;
  long ac1 = caldata.ac1;
  int ac2 = caldata.ac2;
  int ac3 = caldata.ac3;
  unsigned int ac4 = caldata.ac4;

  b6 = b5 - 4000;
  x1 = (b2 * (b6 * b6) / pow(2,12)) / pow(2,11);
  x2 = (ac2 * b6) / pow(2,11);
  x3 = x1 + x2;

  b3 = (((ac1*4 + x3) << OSS) + 2) / 4;
  x1 = (ac3 * b6) / pow(2,13);
  x2 = (b1 * ((b6 * b6) / pow(2,12)) ) / pow(2,16);
```

```
  x3 = ((x1 + x2) + 2) / 4;
  b4 = (ac4 * (unsigned long)(x3 + 32768) ) / pow(2,15);

  b7 = ((unsigned long)up - b3) * (50000 >> OSS);
  if (b7 < 0x80000000)  p = ( b7 * 2 ) / b4;
  else p = (b7 / b4) * 2;

  x1 = (p / pow(2,8)) * (p / pow(2,8));
  x1 = (x1 * 3038) / pow(2,16);
  x2 = (-7357 * p) / pow(2,16);
  p += (x1 + x2 + 3791) / pow(2,4);

  long temp = p;
  return temp;
}

float readPressure() {  // ⓫
  // Read uncompensated pressure
  Wire.beginTransmission(i2c_address);
  Wire.write(0xF4); // Register
  Wire.write(0x34 + (OSS << 6)); // Value with oversampling setting.
  Wire.endTransmission();

  delay(2 + (3 << OSS));

  unsigned char msb,lsb,xlsb;
  unsigned long rawPressure = 0;
  msb = read_i2c_unsigned_char(0xF6);
  lsb = read_i2c_unsigned_char(0xF7);
  xlsb = read_i2c_unsigned_char(0xF8);

  rawPressure = (((unsigned long) msb << 16) |
        ((unsigned long) lsb << 8) |
        (unsigned long) xlsb) >> (8-OSS);

  return getRealPressure(rawPressure);
}

float calculateAltitude(float pressure) {       // ⓬
  float pressurePart = pressure / atmosphereSeaLevel;
  float power = 1 / 5.255;
  float result = 1 - pow(pressurePart, power);
  float altitude = 44330*result;
  return altitude; // m
}
```

❶ Global structs and variables. The calibration data will be read from the sensor's non-volatile (EEPROM) memory.

❷ Take a two-byte integer, swap the left byte and the right byte. This is needed because the sensor represents numbers differently than the Arduino does.

❸ These are convenience functions: `read_i2c_unsigned_char()`, `read_i2c()`, and `read_i2c_int()`. They are wrappers for functionality in *Wire.h* to make it easier to use in this program.

❹ Read the calibration data from the non-volatile EEPROM memory of the sensor. The format is described on the BMP085 data sheet.

❺ The trick is to use your own `calibration_data` struct to decode the EEPROM data. The bytes from the device are overlaid on the struct residing in Arduino's memory. The struct is created so that each variable has a length that corresponds to each piece of data read from the EEPROM. After this line, `caldata` is filled with bytes from the sensor, and each piece of data is accessible through the struct.

❻ As the sensor has different endianness than Arduino, you must swap the bytes of each two-byte integer. First, get a two-byte pointer to the start of `caldata`...

❼ ...and then walk through `caldata` and swap the bytes. Notice how the two-byte (16 bit) pointer correctly points to a new two-byte integer on each iteration, instead of naively pointing to single bytes.

❽ `readTemperature()` first reads the raw temperature using I2C. Then it calculates the actual temperature, using a formula from the data sheet.

❾ The datasheet specifies that a delay of at least 4.5 ms is needed.

❿ `getRealPressure()` takes the raw temperature read by `readPressure()`, then returns the result in Pascals (Pa). It uses the calibration data read by `readCalibrationData()`, and applies the formula from page 13 of the BMP085 data sheet.

⓫ Read the raw pressure using I2C. The register values and data format are taken from the data sheet.

⓬ Estimate altitude from pressure. The formula is from the international barometric formula, available on the BMP085 data sheet, page 14.

GY65 Code and Connection for Raspberry Pi

This Raspberry Pi code is easy to use, and you should try running it before reading it in detail. If you go through it line by line, you'll see that the code can be quite demanding to understand. With Raspberry Pi, the I2C communication with the GY65 sensor is not in a separate file, so the code is longer than the Arduino example. The code uses some techniques you have already seen, such as the ones described in "Hexadecimal, Binary, and Other Numbering Systems" on page 219 and "Bitwise Operations" on page 221.

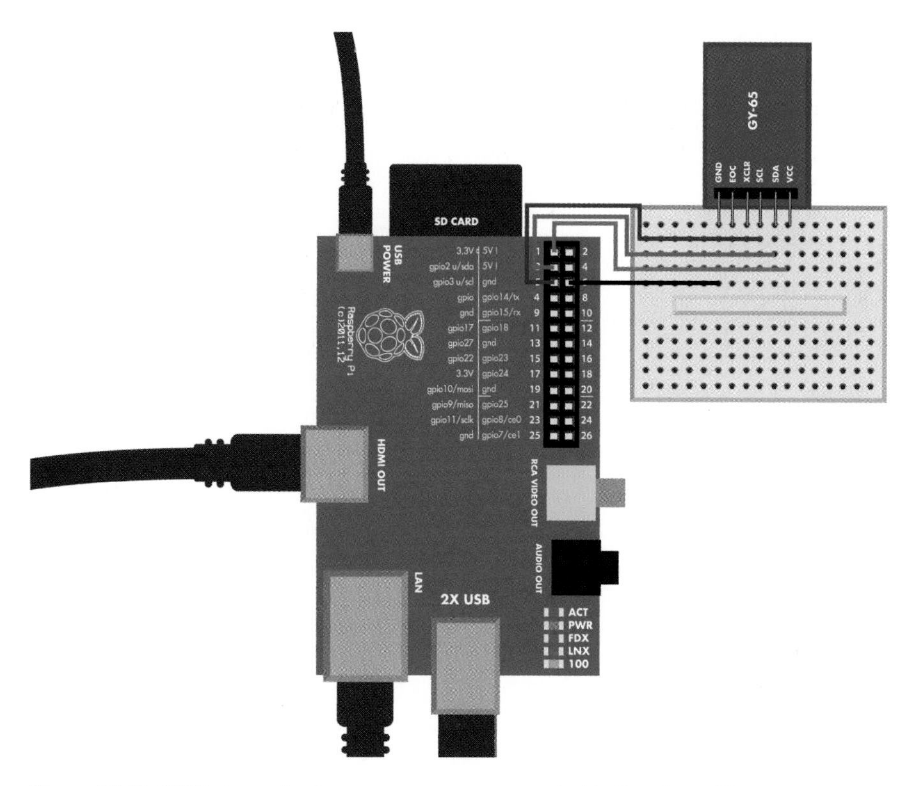

Figure 12-11. *GY65 atmospheric pressure sensor connected to Raspberry Pi*

Example 12-8. gy_65.py

```
# gy_65.py - print altitude,pressure and temperature to serial
# (c) BotBook.com - Karvinen, Karvinen, Valtokari

import smbus # sudo apt-get -y install python-smbus     # ❶
import time
import struct

bus = None
address = 0x77
caldata = None

atmosphereSeaLevel = 101325.0
OSS = 0
b5 = 0

def readCalibrationData():        # ❷
  global bus, caldata
  bus = smbus.SMBus(1)
  rawData = ""

  for i in range(22):
    rawData += chr(bus.read_byte_data(address, 0xAA+i)) # ❸
```

```
  caldata = struct.unpack('>hhhhhhhhhhh', rawData) # ❹

def readTemperature():
  global b5
  bus.write_byte_data(address, 0xF4, 0x2E)      # ❺
  time.sleep(0.005)     # ❻
  rawTemp = bus.read_byte_data(address, 0xF6) << 8      # ❼
  rawTemp = rawTemp | bus.read_byte_data(address, 0xF7)
  x1 = ((rawTemp - caldata[5]) * caldata[4]) / 2**15
  x2 = (caldata[9] * 2**11) / (x1 + caldata[10])
  b5 = x1 + x2
  temp = (b5 + 8) / 2**4
  temp = temp / 10.0
  return temp


def readPressure():     # ❽
  bus.write_byte_data(address, 0xF4, 0x34 + (OSS << 6))
  time.sleep(0.005)
  rawPressure = bus.read_byte_data(address, 0xF6) << 16
  rawPressure = rawPressure | bus.read_byte_data(address, 0xF7) << 8
  rawPressure = rawPressure | bus.read_byte_data(address, 0xF8)
  rawPressure = rawPressure >> (8 - OSS)

  #Calculate real pressure
  b6 = b5 - 4000

  x1 = (caldata[7] * ((b6 * b6) / 2**12 )) / 2**11
  x2 = caldata[1] * b6 / 2**11
  x3 = x1 + x2
  b3 = (((caldata[0] * 4 + x3) << OSS) + 2) / 4
  x1 = caldata[2] * b6 / 2**13
  x2 = (caldata[6] * ((b6 * b6) / 2**12 )) / 2**16
  x3 = ((x1 + x2) + 2) / 2*2
  b4 = (caldata[3] * (x3 + 32768)) / 2**15
  b4 = b4 + 2**16 # convert from signed to unsigned
  b7 = (rawPressure - b3) * (50000 >> OSS)
  if b7 < 0x80000000:
    p = (b7 * 2) / b4
  else:
    p = (b7 / b4) * 2
  x1 = (p / 2**8) * (p / 2**8)
  x1 = (x1 * 3038) / 2**16
  x2 = (-7357 * p) / 2**16
  p = p + (x1 + x2 + 3791) / 2**4
  return p

def calculateAltitude(pressure):        # ❾
  pressurePart = pressure / atmosphereSeaLevel;
  power = 1 / 5.255;
  result = 1 - pressurePart**power;
  altitude = 44330*result;
  return altitude
```

```
def main():
  readCalibrationData()
  while True:
    temperature = readTemperature()
    pressure = readPressure()
    altitude = calculateAltitude(pressure)
    print("Altitude %.2f m" % altitude)
    print("Pressure %.2f Pa" % pressure)          # ⑩
    print("Temperature %.2f C" % temperature)
    time.sleep(10)


if __name__ == "__main__":
  main()
```

❶ The SMBus standard is a subset of I2C. The `smbus` library makes it easy to use. The *python-smbus* package must be installed on Raspberry Pi for this to work (see "SMBus and I2C Without Root" on page 218).

❷ The sensor ships with 176 bits of calibration data stored into its EEPROM.

❸ Read data from address 0xAA, 0xAB, on up through 0xBF. Append these character values to the string. The string ends up having 22 bytes (176 bits). You didn't have to think about how many bits the struct had in Arduino when you were doing this, because the struct's length is defined by the length of the variables that are inside it.

❹ Unpack 11 big endian (>) short two-byte integers (`h`). This consumes all the bytes from the string.

❺ Write the command for "read temperature" (0x2E) to the sensor's control register (0xF4).

❻ Wait for the measurement to finish.

❼ Read the answer, and start manipulating the raw number into a temperature in Celsius. The formulas are from the data sheet.

❽ The `readPressure()` function works just like `readTemperature()`.

❾ The higher you go, the lower the pressure. Based on this, the international barometric formula can give you an estimate of your altitude.

❿ 100,000 Pa = 1 bar

Experiment: Does Your Plant Need Watering? (Build a Soil Humidity Sensor)

A soil humidity sensor is a simple analog resistance sensor. Stick it in the soil to see if your plant needs watering.

Normal tap water and groundwater contain diluted salts and other material. This makes water conductive. The soil humidity sensor (Figure 12-12) simply measures that conductivity.

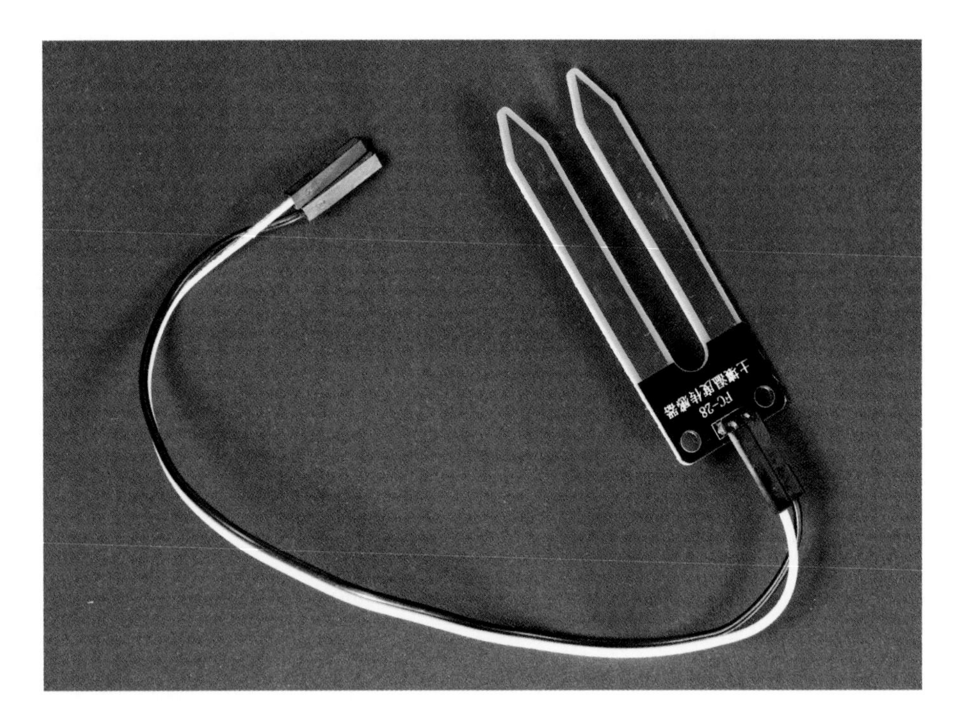

Figure 12-12. *Soil humidity sensor*

Some soil humidity sensors have a built-in circuit. The sensor used here doesn't have built-in electronics, so you could make your own from two pieces of metal (for the sensor probes) and use Arduino or Raspberry Pi to measure the resistance. The circuit uses a 1 megohm resistor (brown-black-green).

Soil Sensor Code and Connection for Arduino

Figure 12-13 shows the wiring diagram for Arduino. Hook it up as shown, and then run the code from Example 12-9.

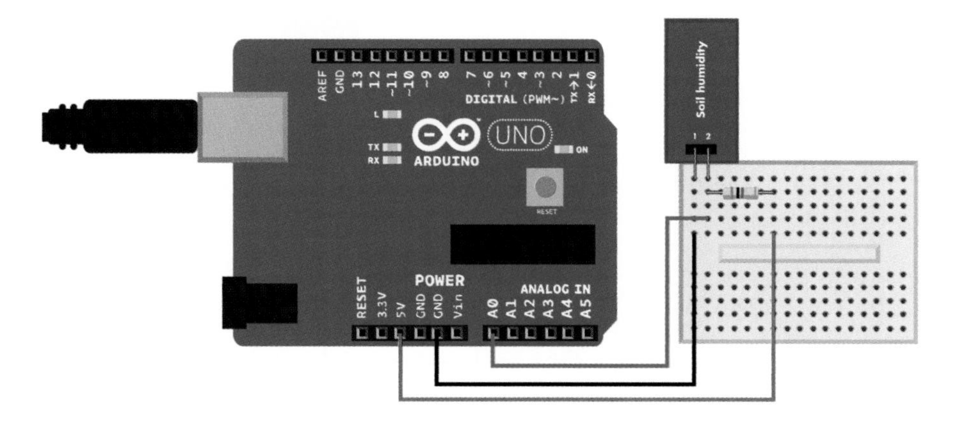

Figure 12-13. *Soil humidity sensor connected to Arduino*

Example 12-9. soil_humidity_sensor.ino

```
// soil_humidity_sensor.ino - read soil humidity by measuring its resistance.
// (c) BotBook.com - Karvinen, Karvinen, Valtokari


const int sensorPin = A0;
int soilHumidity = -1;

void setup() {
  Serial.begin(115200);
}

void loop() {
  soilHumidity = analogRead(sensorPin); // ❶
  Serial.println(soilHumidity);
  delay(100);
}
```

❶ It's a simple analog resistance sensor.

Soil Sensor Code and Connection for Raspberry Pi

Figure 12-14 shows the connection diagram for Raspberry Pi. Hook it up as shown, and then run the code from Example 12-10.

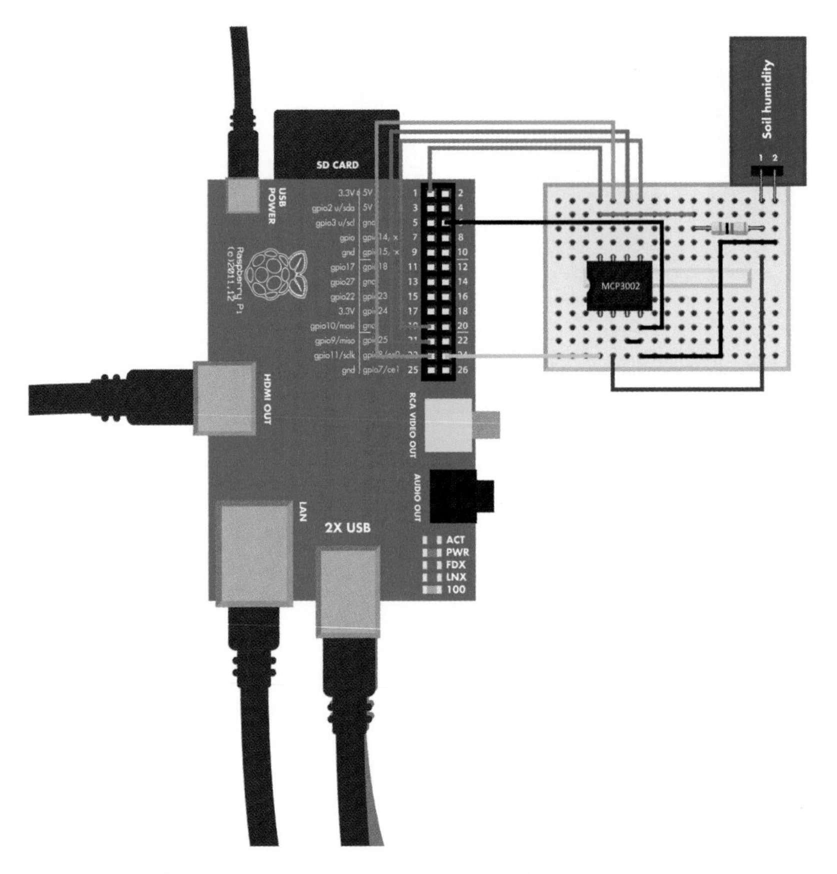

Figure 12-14. *Soil humidity sensor connected to Raspberry Pi*

Example 12-10. ***soil_humidity_sensor.py***

```
# soil_humidity_sensor.py - read soil humidity by measuring its resistance.
# (c) BotBook.com - Karvinen, Karvinen, Valtokari

import time
import botbook_mcp3002 as mcp

def main():
        while True:
                h = mcp.readAnalog()    # ❶
                h = h / 1024 * 100      # ❷
                print("Current humidity is %d %%" % h)
                time.sleep(5)

if __name__ == "__main__":
        main()
```

❶ It's a simple analog resistance sensor. As with other analog resistance sensors in this book, the *botbook_mcp3002.py* library must be in the same directory as this program. You must also install the *spidev* library, which is imported by *botbook_mcp3002*. See the comments at the top of *botbook_mcp3002/botbook_mcp3002.py* or "Installing SpiDev" on page 56.

❷ The raw value is converted to percentage of the maximum measurement. For display purposes, h is actually 100 times percentage, e.g., .53 becomes 53 so it can be displayed as 53%.

Test Project: E-paper Weather Forecast

Create your own weather forecast on e-paper. The weather forecast is based on changes in atmospheric pressure. The e-paper display is quite special: you can see it well in bright light, it looks a bit like paper, and the picture stays on without consuming electricity.

You're reading the hardest project in the book. If you haven't practiced with the easier experiments and projects already, you might want to go back and complete some of those first.

Figure 12-15. *E-paper Weather Forecast*

What You'll Learn

In the *E-paper Weather Forecast* project, you'll learn how to:

- Build a box that shows a graphical weather forecast.
- Predict weather using atmospheric pressure.
- Display graphics on e-paper with zero energy consumption.
- Make Arduino sleep to conserve power.

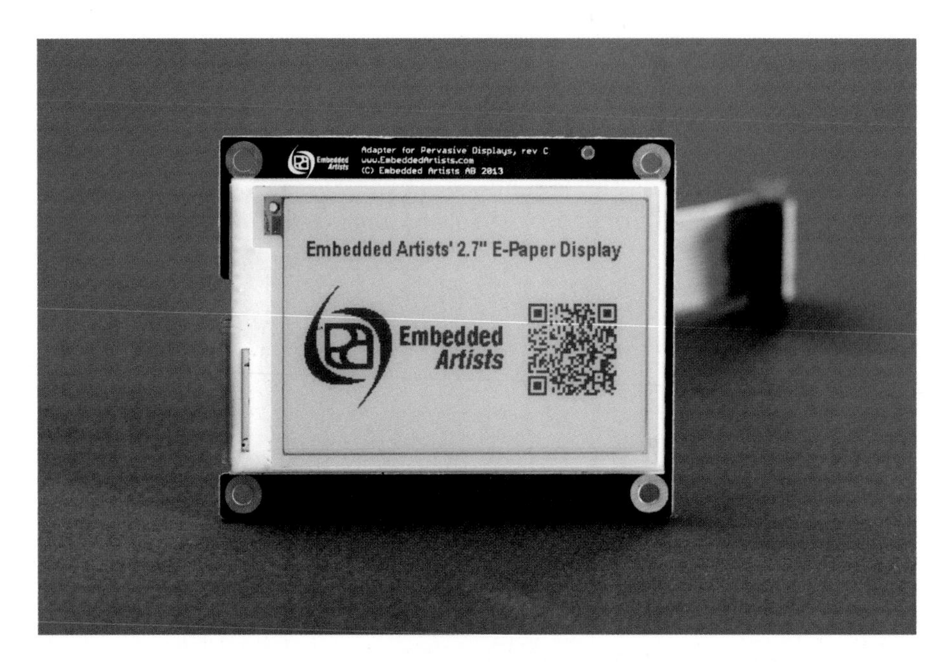

Figure 12-16. *E-paper display*

Weather Forecast Code and Connection for Arduino

The code combines many techniques. You can just build it first, and then learn about the implementation details once it works.

To create your own version, it's not required that you understand all the code. After you have your weather station running, have a look at `drawScreen()`. It's the main function, and quite high level. For example, you could start by changing `pos`, the location where the plus sign is drawn:

```
int pos = 10;
drawCharacter(pos, 70, font,'+');
```

Techniques used in this code:

- Reading data from the GY65 atmospheric pressure sensor ("Atmospheric Pressure GY65" on page 339).
- Working with hexadecimal numbers, binary numbers, and bitwise operations. See ("Hexadecimal, Binary, and Other Numbering Systems" on page 219 and "Bitwise Operations" on page 221).
- Making Arduino sleep to save power. This uses low-level commands to write to ATmega (the chip that powers Arduino) registers.
- Drawing on e-paper display, using the *EPD* library.
- Storing images as header files (as in ***imagename**.h*). This is explained in detail in "Storing Images in Header Files" on page 362.

This code uses the Arduino Mega. It would need modification to work on another board.

Figure 12-17 shows the connections for the Arduino Mega. Wire it up as shown, and then run the code from Example 12-11.

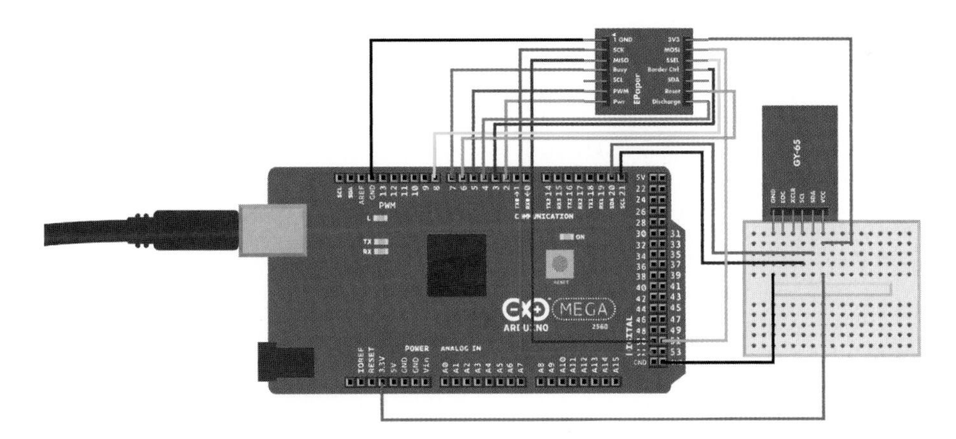

Figure 12-17. *Connections on Arduino Mega*

Example 12-11. ***weather_station.ino***

```
// weather_station.ino - print weather data to epaper
// (c) BotBook.com - Karvinen, Karvinen, Valtokari

#include <inttypes.h>
#include <ctype.h>

#include <SPI.h>
```

```
#include <Wire.h>
#include <EPD.h>          // ❶
#include <gy_65.h>        // ❷
#include <avr/sleep.h>
#include <avr/power.h>

#include "rain.h"         // ❸
#include "sun.h"
#include "suncloud.h"
#include "fonts.h"

uint8_t imageBuffer[5808]; // 264 * 176 / 8

const int pinPanelOn = 2;
const int pinBorder = 3;
const int pinDischarge = 4;
const int pinPWM = 5;
const int pinReset = 6;
const int pinBusy = 7;
const int pinEPDcs = 8;

EPD_Class EPD(EPD_2_7,
              pinPanelOn,
              pinBorder,
              pinDischarge,
              pinPWM,
              pinReset,
              pinBusy,
              pinEPDcs,
              SPI);

float weatherDiff;
float temperature;

const int sleepMaxCount = 10;   // min
volatile int arduinoSleepingCount = sleepMaxCount;

void setup() {
  Serial.begin(115200);
  pinMode(pinPanelOn, OUTPUT);
  pinMode(pinBorder, OUTPUT);
  pinMode(pinDischarge, INPUT);
  pinMode(pinPWM, OUTPUT);
  pinMode(pinReset, OUTPUT);
  pinMode(pinBusy, OUTPUT);
  pinMode(pinEPDcs, OUTPUT);


  digitalWrite(pinPWM, LOW);
  digitalWrite(pinReset, LOW);
  digitalWrite(pinPanelOn, LOW);
  digitalWrite(pinDischarge, LOW);
  digitalWrite(pinBorder, LOW);
  digitalWrite(pinEPDcs, LOW);
```

```
  SPI.begin();  // (4)
  SPI.setBitOrder(MSBFIRST);     // (5)
  SPI.setDataMode(SPI_MODE0);   // (6)
  SPI.setClockDivider(SPI_CLOCK_DIV4); // (7)

  WDTCSR |= (1<<WDCE) | (1<<WDE);        // (8)
  WDTCSR = 1<<WDP0 | 1<<WDP3;   // (9)
  WDTCSR |= _BV(WDIE);  // (10)
  MCUSR &= ~( 1 << WDRF);        // (11)

  for(int i = 0; i < 5808; i++) // (12)
    imageBuffer[i] = 0;

  readCalibrationData();         // (13)

}

char characterMap[14] = {'+', '-', 'C', 'd',
                         '0', '1', '2', '3',
                         '4', '5', '6', '7',
                         '8', '9'};     // (14)

void drawCharacter(int16_t x, int16_t y, const uint8_t *bitmap, char character) {
  int charIndex = -1;
  for(int i = 0; i < 14; i++) { // (15)
    if(character == characterMap[i]) {
      charIndex = i;
      break;
    }
  }
  if(charIndex == -1) return;
  drawBitmap(x,y,bitmap,charIndex*25,0,25,27,350);       // (16)
}

void drawBitmap(int16_t x, int16_t y, const uint8_t *bitmap, int16_t x2,
    int16_t y2, int16_t w, int16_t h, int16_t source_width) {   // (17)

  int16_t i, j, byteWidth = source_width / 8;

  for(j=y2; j<y2+h; j++) {      // (18)
    for(i=x2; i<x2+w; i++ ) {
      byte b= pgm_read_byte(bitmap+j * byteWidth + i / 8);
      if(b & (128 >> (i & 7))) {
            drawPixel(x+i-x2, y+j-y2, true);
      }
    }
  }
}

void drawPixel(int x, int y, bool black) { // (19)
  int bit = x & 0x07;
  int byte = x / 8 + y * (264 / 8);
  int mask = 0x01 << bit;
```

```
  if(black == true) {
    imageBuffer[byte] |= mask;
  } else {
    imageBuffer[byte] &= ~mask;
  }
}

void drawBufferToScreen() {    // ⓴
  for (uint8_t line = 0; line < 176 ; ++line) {
        EPD.line(line, &imageBuffer[line * (264 / 8)], 0, false, EPD_inverse);
  }
  for (uint8_t line = 0; line < 176 ; ++line) {
        EPD.line(line, &imageBuffer[line * (264 / 8)], 0, false, EPD_normal);
  }
}

void drawScreen() {    // ㉑
  EPD.begin();
  EPD.setFactor(temperature);
  EPD.clear();
  if(weatherDiff > 250) {      // Pa
    // Sunny
    drawBitmap(140,30,sun,0,0,117,106,117);    // ㉒
  } else if ((weatherDiff <= 250) || (weatherDiff >= -250)) {
    // Partly cloudy
    drawBitmap(140,30,suncloud,0,0,117,106,117);
  } else if (weatherDiff < -250) {
    // Rain
    drawBitmap(140,30,rain,0,0,117,106,117);
  }
  //Draw temperature
  String temp = String((int)temperature);

  int pos = 10;
  drawCharacter(pos,70,font,'+');      // ㉓
  pos += 25;
  drawCharacter(pos,70,font,temp.charAt(0));
  pos += 25;
  if(abs(temperature) >= 10) {
    drawCharacter(pos,70,font,temp.charAt(1));
    pos += 25;
  }
  drawCharacter(pos,70,font,'d');
  pos += 25;
  drawCharacter(pos,70,font,'C');


  drawBufferToScreen(); // ㉔

  EPD.end();

  for(int i = 0; i < 5808; i++) // ㉕
    imageBuffer[i] = 0;
}
```

```
void loop() {
  Serial.println(temperature);  // ㉖
  if(arduinoSleepingCount >= sleepMaxCount) {   // ㉗
    readWeatherData();  // ㉘
    drawScreen();
    arduinoSleepingCount = 0; // ㉙
    arduinoSleep(); // ㉚
  } else {
    arduinoSleep();
  }
}

const float currentAltitude = 40.00; // Installation altitude in meters
const long atmosphereSeaLevel = 101325; // Pa
const float expectedPressure = atmosphereSeaLevel * pow((1-currentAltitude / 44330), 5.255);

void readWeatherData(){
  temperature = readTemperature();
  float pressure = readPressure();
  weatherDiff = pressure - expectedPressure;
}

ISR(WDT_vect)   // ㉛
{
  arduinoSleepingCount++;
}

void arduinoSleep() { // ㉜
  set_sleep_mode(SLEEP_MODE_PWR_DOWN);
  sleep_enable();
  sleep_mode(); // ㉝
  sleep_disable();       // ㉞
  power_all_enable();
} // ㉟
```

❶ Library for e-paper display. You can download the library from *https://github.com/repaper/gratis/tree/master/Sketches/libraries*. Follow the instructions at *http://arduino.cc/en/Guide/Libraries* to install the library.

❷ This is the *botbook.com* GY65 library. You used it earlier in "Atmospheric Pressure GY65" on page 339. Copy the library directory into your sketch folder (if you downloaded the sample code for this book, it will already be in there).

❸ Images are saved as header files.

❹ Prepare SPI before we talk to the e-paper display.

❺ Configure SPI to have the most significant bit first (see "Bitwise Operations" on page 221).

❻ Configure SPI for SPI_MODE0: this sets the clock polarity to mode CPOL 0, and the clock phase to mode CPHA 0. This means that data is captured on the clock's rising edge (when the clock signal goes from LOW to HIGH). Data is propagated on the falling edge (when the clock signal goes from HIGH to LOW). Modes (SPI_MODE0) are listed in Arduino documentation (Help→Reference→Libraries→SPI). The SPI settings for the sensor are on its data sheet.

❼ SPI_CLOCK_DIV4 sets the SPI clock to 1/4 of the Arduino CPU frequency. This project uses Arduino Mega, so the SPI clock frequency is 16 MHz / 4 = 4 MHz.

❽ Enable the watchdog timer. This is what will allow Arduino to go into a deep sleep. Set WDCE (watch dog change enable) and WDE (watch dog enable) in WDTCSR (Watchdog Timer Control Register). These registers are so low level that they are from ATmega documentation (instead of being part of the core Arduino library). Arduino Mega uses the ATmega 1280 chip, so you can find the documentation by searching for "ATmega 1280 data sheet." The symbol |= is an inplace bitwise XOR (performs an XOR against the variable's current value and replaces the variable's value with the result). The symbol << represents the bit shift operation. See "Bitwise Operations" on page 221.

❾ Set the watchdog to wake up the Arduino every 8 seconds.

❿ Enable the watchdog interrupt. WDIE is "watch dog interrupt enable."

⓫ Clear the watchdog system reset flag (WDRF) from the microcontroller unit status register (MCUSR).

⓬ Initialize the image buffer (where graphics are drawn before being displayed) with zeroes.

⓭ For details about using the GY65 sensor, see the code explanations in "Atmospheric Pressure GY65" on page 339.

⓮ List of the 14 characters that are in *font.h*, a large image that contains these characters.

⓯ Find the index of the character in the *font.h* image.

⓰ Draw a character to screen. In practice, the bitmap will contain a font, a big picture with characters side by side. The calculation just selects one of these characters.

⓱ `drawBitmap()` takes a source bitmap, position, and dimensions. It then draws each pixel of the picture to an intermediate image buffer (but does not yet display it on the e-paper display).

⓲ Traverse the area two dimensionally to draw the bitmap one pixel at a time.

⓳ Draw one pixel to image buffer (but not yet to the e-paper display). The image is stored one bit per pixel, so every byte (8 bits) contains 8 pixels. The width of the display is 264 pixels, so one line is stored in 264/8 = 33 bytes. The code then traverses over each bit, changing it to 1 or 0 as needed.

⑳ *Blit* the already drawn graphics from `imageBuffer` to screen. This uses the EPD library. The display has 176 lines, 264 dots per line. As one bit represents one pixel, the display has 264/8 (33) bytes per line. Blitting the image is a simple matter of drawing each byte of `imageBuffer`, using `EPD.line`.

㉑ `drawScreen()` uses the GY65 library to measure environment, and the EPD library to draw on the e-paper display. The larger the pressure difference, the worse the weather. If you want to create your own version of the program, this `drawScreen()` function is where you should begin your customizations.

㉒ Draw the image from *sun.h* to `imageBuffer`. To put your own images on the screen, use this function. The parameters are `drawBitmap(imageBufferX, imageBufferY, source Image, sourceX, sourceY, sourceWidth, sourceHeight, totalSourceWidth)`. The first two are the target position (`imageBufferX`, `imageBufferY`). The rest of the parameters concern the source bitmap: the source bitmap (`sourceImage`), what to take from the source bitmap (`sourceX`, `sourceY`, `sourceWidth`, `sourceHeight`). Finally, the total width of the source bitmap is listed (`totalSourceWidth`).

㉓ Write a character to image buffer.

㉔ Blit the `imageBuffer` from memory to the e-paper display. Without this, none of the images would be shown.

㉕ Clear the image buffer.

㉖ With an unfamiliar component such as a fancy e-paper display, it's a good idea to confirm sensor data separately by writing it to the serial console.

㉗ The watchdog wakes Arduino periodically. If enough time has passed…

㉘ …it's time to run the meat of the program.

㉙ Then reset the sleep counter…

㉚ …and fall back to sleep to save power.

㉛ The ISR() Interrupt Service Routine is called automatically. ISR() runs every time Arduino wakes up, just before anything else runs. This code wakes up Arduino every 8 seconds, so ISR() runs every 8 seconds.

㉜ Make Arduino fall asleep to conserve power. As you saw earlier, this required quite some preparation.

㉝ This command makes Arduino sleep and stop executing code. Nothing happens after this—until the watchdog wakes up Arduino.

㉞ When the watchdog wakes Arduino, code flow continues on this line and wakeup starts.

㉟ You worked through all this difficult code? Pat yourself on the back—you're on your way to becoming a guru!

Environment Experiment: Look Ma, No Power Supply

E-paper displays use power only to change the picture on the display. They don't need any energy to leave a picture on the screen.

You can try it yourself. Use Arduino to display something on your e-paper display. Then power down Arduino, and even disconnect the e-paper display. The picture stays unchanged. In fact, there is no visible difference at all between keeping e-paper connected or disconnecting it (Figure 12-18).

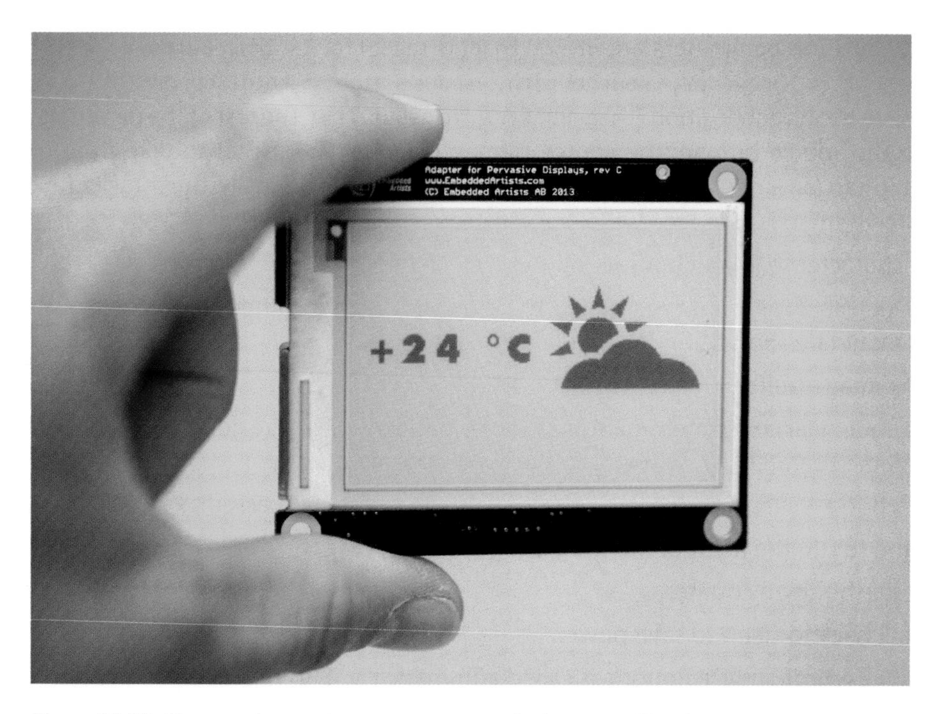

Figure 12-18. *The same image stays on an e-paper display even without power*

Storing Images in Header Files

Your very own e-paper weather station can already show you the sun. What if you want to draw your own graphics? You can draw your images in an image editing program like GIMP or Photoshop, and then convert the saved BMP images to the C code headers we use in the sketch.

First, use your favorite drawing program to draw an image. GIMP is a good, free choice for this. Save your image in the BMP format. Put the image into the same folder with the other images in this project (*images/*).

Next, you'll need to convert the BMP image (*sun.bmp*) to a C header file (*sun.h*). You can use the included *image2c-botbook.py* script for this.

First, install the requirements: Python and Python Imaging Library. Windows users can download a free installer for Python from *http://python.org*. Mac users will already have Python installed, but Mac and Windows users will need to download the Python Imaging Library from *http://www.pythonware.com/products/pil/*. Linux users have it easier. For example, here's how to install the needed libraries:

```
$ sudo apt-get update
$ sudo apt-get -y install python python-imaging
```

If your source image is not called *foo.bmp*, change the filenames in the script. Then you're ready to convert. These commands work in all of Linux, Windows, and Mac. The dollar sign ($) represents the shell prompt, so don't type that. Windows users probably have a different-looking prompt. The first command assumes you're in the subdirectory containing the example code for this book. You can download the sample code from *http://botbook.com*.

```
$ cd arduino/weather_station/
$ python image2c-botbook.py
BMP (117, 106) 1
```

The file is now converted. With the `ls` or `dir` command, you can see that *foo.h* was created in the current working directory. You have now converted your BMP image to a C header file.

In *foo.h*, your image is now C code:

```
// File generated with image2c-botbook.py
const unsigned char sun [] PROGMEM= {
0x00, 0x00, 0x00, 0x00, 0x00, 0x00, 0x00, 0x00, 0x00, // ..
0x7F, 0xF0, 0x00, 0xFF, 0xC0, 0x00, 0x1F, 0xF8, 0x00, // ..
```

This format is convenient for programming. Each byte represents eight pixels, one pixel per bit. For example, hex 0xFC is number 252. In bits, it's 0b11111100. This means eight bits side by side: black black black black, black black white white.

BMP to C Conversion Program

Image2c-botbook.py converts a BMP image to C header file.

If you need some "code golf" (see *http://codegolf.com*), try making it work with PNG source files.

This code uses hexadecimal numbers and bitwise operations. To review them, see "Hexadecimal, Binary, and Other Numbering Systems" on page 219 and "Bitwise Operations" on page 221.

Example 12-12. Image to C header code

```
# image2c-botbook - convert a BMP image to C for use in eInk and LCD displays
# (c) BotBook.com - Karvinen, Karvinen, Valtokari

import Image    # ❶
import math
imageName = "images/foo.bmp"    # ❷
```

```
outputName = "foo.h"    # ❸

im = Image.open(imageName)      # ❹
print im.format, im.size, im.mode

width, height = im.size
pixels = list(im.getdata())     # ❺
length = int(math.floor(width * height / 8.0))  # ❻
carray = length * [0x00]        # ❼
for y in xrange(0, height):     # ❽
        for x in xrange(0, width):      # ❾
                pixel = pixels[y * width + x]   # ❿

                bit = 7 - x & 0x07      # ⓫
                byte = x / 8 + y * (width / 8)  # ⓬

                mask = 0x01 << bit      # ⓭
                if pixel == 0:
                        carray[byte] |= mask    # ⓮
                else:
                        carray[byte] &= ~mask


fileHandle = open(outputName, "w")      # ⓯
index = 0
fileHandle.write("// File generated with image2c-botbook.py\n")
fileHandle.write("const unsigned char sun [] PROGMEM= {\n")
for b in carray:
        fileHandle.write("0x%02X, " % b)        # ⓰
        index += 1
        if index > 15:
                fileHandle.write("\n")
                index = 0
fileHandle.write("};\n")
```

❶ PIL, the Python Imaging Library, must be installed as described earlier.

❷ The source image, saved in BMP format. Change this to match the name and location of your file if needed.

❸ Output filename. Change this to match the file you want to create.

❹ PIL can open many formats.

❺ Break the image into individual pixels. This is half of the work.

❻ Length of the whole image in bytes (1 B == 8 bit)

❼ Create the target C array, and initialize it with zeroes.

❽ For each line…

❾ …and for each pixel in the current line, perform the indented operations.

⑩ The current pixel.

⑪ The index of bit in the current byte.

⑫ The current byte index.

⑬ Create the bit mask for current bit in current byte. For example, the mask 0b00000001 would be created to change the last bit of a byte to zero.

⑭ Change the bit.

⑮ C headers are just text files.

⑯ Write each of the bytes as a hex code.

Enclosure Tips

We used a plastic box made by Hammond for our weather forecaster. But how do you make an odd-shaped hole like the one we need here? First draw the shape you want to cut out on the box lid. Drill holes in each corner of the shape with a large drill bit (10-20 mm). Then start going from corner to corner with a jigsaw blade. Finish the hole with file and sandpaper (see Figure 12-19).

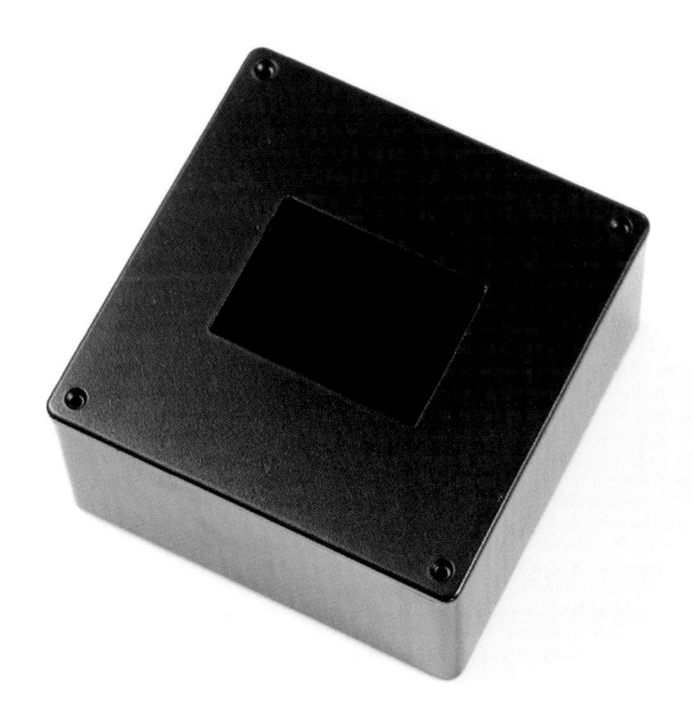

Figure 12-19. *Hole for the screen*

Use hot glue to attach the screen to the box as shown in Figure 12-20.

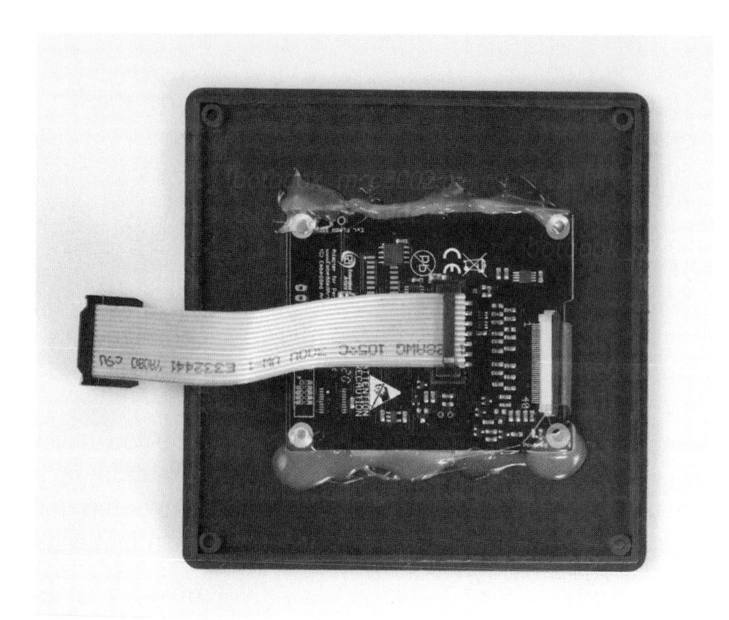

Figure 12-20. *Screen hot glued*

On the back of the box, we made a pattern of small holes (see Figure 12-21). Without those, air would not be able to get to the sensor. The finished gadget is shown in Figure 12-22.

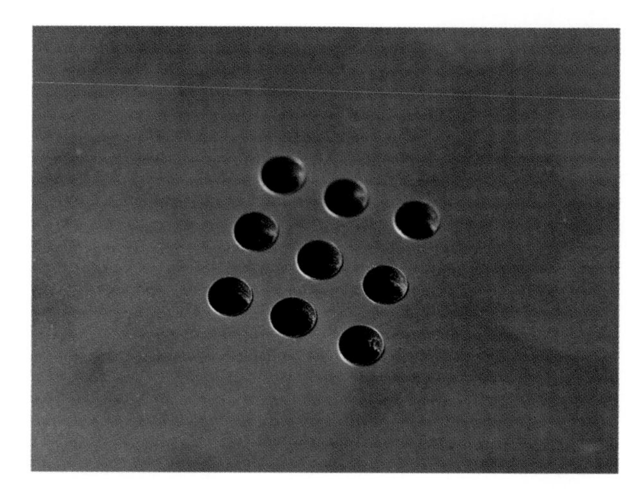

Figure 12-21. *Holes to let the air in*

Figure 12-22. *The finished gadget*

Congratulations, you now have your own weather prophet with e-paper display!

This might be the end of this book, but it's just the beginning for your projects. Now that you can work with so many sensors and many outputs too, what are you going to build?

Good luck with your projects!

Raspberry Pi Linux Quick Reference

Table A-1 shows some useful and common Linux commands. Table A-2 shows the directories you'll be working in much of the time.

Table A-1. Some Linux commands

Command	Meaning
`pwd`	Prints your working directory
`ls`	Lists files in working directory
`cd Desktop`	Changes directory to the *Desktop* directory immediately under your current directory
`cd ~/Desktop`	Changes directory to the *Desktop* directory immediately under your *home* directory
`cd ~`	Changes to your home directory
`nano foo.txt`	Edits the text file *foo.txt*. Use Control-X, then type `y` followed by Enter or Return to save the file.
`passwd`	Change your password (asks for the old one first, but doesn't echo what you type to the screen)
`startx`	Launch the graphical desktop from a command-line-only Linux session
`sudo apt-get update`	Updates the list of software you can install (requires a network connection)
`sudo apt-get install ipython`	Installs the program *ipython*
`sudo shutdown -P now`	Prepares the Raspberry Pi to be powered off in a safe way
`sudoedit /etc/motd`	Edits a file as root, with more safety checks than `nano sudo /etc/motd`
`ifconfig`	Shows the Pi's IP addresses (127.0.0.1 is localhost, which is used for connections between programs running on the Pi; use the other one, which is your Ethernet or WiFi adapter)
`sudo raspi-config`	Calls up the configuration menu of most common Raspberry Pi settings (you'll usually need to reboot after you run it)
`mkdir bar/`	Create a directory called *bar*
`rm -r bar/`	Remove the directory *bar/* and its contents (there is no Undo)
`ssh login@example.com`	Connects to remote computer example.com with user "login"

Command	Meaning
`exit`	Close the shell or ssh connection
`less /var/log/syslog`	View a text file (press space for the next page, press q to quit)
`tail -F /var/log/syslog`	Follows the specified text file as lines are added to it (use Control-C to kill the *tail* command and return to the shell)
`man ssh`	View the manual page (built-in documentation) of the command "ssh" (press space for the next page, press q to quit)

Table A-2. The most important directories

Directory	Purpose
/home/pi/	Your (you are the user named `pi`) home directory; contains all of your files on the Pi
/var/log/	Contains all the system-wide log files, such as */var/log/syslog* and */var/log/auth.log*
/etc/	Contains all the system-wide configuration files.
/sys/	A virtual file system for reading and modifying volatile data (things that change continuously as the system is running, such as input and output pins)
/media/	Removable media, such as */media/cdrom/* or */media/usbdisk/*
/	The root directory; contains every directory and file available on the system

Index

Symbols

A

B

C

D

E

F

G

H

I

J

K

L

M

N

O

P

Q

R

S

T

U

V

W

About the Authors

Tero Karvinen teaches Linux and embedded systems in Haaga-Helia University of Applied Sciences, where his work has also included curriculum development and research in wireless networking. He previously worked as a CEO of a small advertising agency. Tero's education includes a Masters of Science in Economics.

Kimmo Karvinen works as a CEO in a leading company specializing in AV automation in Finland. Before that, he worked as CTO for a hardware manufacturer that specializes in smart building technology, as a marketing communications project leader, and as a creative director and partner in an advertising agency. Kimmo's education includes a Masters of Art, and he's currently working toward his D.Sc. at Helsinki University of Technology.

Ville Valtokari works as the head programmer for an automation hardware manufacturer. Before that he designed and programmed cutting-edge AV systems. Countless personal projects include game design and programming, building robots, and discovering how things work.

Colophon

All photographs and the cover photo are by Kimmo Karvinen. The cover and header font is BentonSans, the body font is Myriad Pro, and the code font is UbuntuMono.